Trenchless Pipeline Projects

Practical Applications

Proceedings of the Conference sponsored by the Pipeline Division, ASCE

Co-sponsored by and in cooperation with:
ASCE Construction Division
Boston Society of Civil Engineers Section
North American Society for Trenchless Technology
Water Environment Federation
Trenchless Technology Magazine
National Utility Contractors Association
Utility Contractors Association of New England
National Association of Sewer Service Companies
American Society for Testing and Materials
Plastics Pipe Institute
Directional Crossing Contractors Association
Northeast Trenchless Technology Society
Gulf Coast Trenchless Association
Trenchless Technology Center

Boston, Massachusetts
June 8-11, 1997

Edited by Lynn E. Osborn

Published by
ASCE American Society of Civil Engineers
345 East 47th Street
New York, New York 10017-2398

Abstract:

The proceedings, *Trenchless Pipeline Projects - Practical Applications*, consists of papers presented at the International Conference sponsored by the Pipeline Division of the American Society of Civil Engineers and held in Boston, Massachusetts on June 8-11, 1997. The papers explore topics from over 50 presentations at the conference covering such areas as case histories; microtunnelling; sliplining; pipe jacking; pipe rehabilitation; service laterals; manholes; inspection/evaluation methods; grouting/joint seals; gas pipelines; materials research; new construction; and the future of trenchless technology. Trenchless pipeline technologies are experiencing rapid and dramatic changes. This book provides a great opportunity to learn from the actual on-site experiences from leading experts in the field who describe recent and current projects, the challenges that had to be overcome, and the solutions used. Also presented are the technologies and methodologies being used today - as well as those that will be emerging in the years ahead.

Library of Congress Cataloging-in-Publication Data

Trenchless pipeline projects : practical applications : proceedings of the conference, Boston, Massachusetts, June 8-11, 1997 / sponsored by the Pipeline Division, ASCE ; co-sponsored by and in cooperation with ASCE Construction Division...[et al.] ; edited by Lynn E. Osborn.
p. cm.
Includes indexes.
ISBN 0-7844-0244-2
1. Pipelines--Design and construction--Congresses. 2. Trenchless construction--Congresses. I. Osborn, Lynn E. II. American Society of Civil Engineers. Pipeline Division.
TA660.P55T74 1997 97-16274
621.8'672--DC21 CIP

Library of Congress Catalog Card No: 97-16274
ISBN 0-7844-0244-2
Manufactured in the United States of America.

FOREWORD

The Conference Steering Committee is pleased to present these proceedings for the "***Trenchless Pipeline Projects - Practical Applications***" on behalf of the Pipeline Division and the Boston Section of the American Society of Civil Engineers.

We take great pride in the continuation of the Pipeline Division specialty conferences. For the first time, all Technical Committees of the Pipeline Division have combined efforts to make this Congress an exceptional opportunity for learning more about trenchless technologies as they relate to pipeline infrastructure - crossings - location and installation - research - planning and design. The Congress topics apply to new pipeline installations and rehabilitation of existing pipelines.

Abstracts received for this Congress were reviewed by members of the Conference Steering Committee. Each paper contained in these proceedings was peer reviewed by members of the Steering Committee and Subcommittee task members from each Technical Committee, as well as from the Publications Committee. We thank everyone who contributed in this peer review process. All papers are eligible for discussion in the ASCE Journal of Transportation Engineering and are also eligible for ASCE awards.

Appreciation is extended to all those who worked for over the past two years to bring this conference together. These proceedings and the conference are the culmination of the volunteer efforts of many people, authors, committee and subcommittee members, and ASCE staff. The conference compliments the trenchless industry, and has been broadly supported by numerous co-sponsoring/cooperating organizations - ASCE Construction Division - Boston Society of Civil Engineers Section - North American Society for Trenchless Technology - Water Environment Federation - Trenchless Technology Magazine - National Utility Contractors Association - Utility Contractors Association of New England - National Association of Sewer Service Companies - Plastics Pipe Institute - American Society of Testing and Materials - Directional Crossing Contractors Association - Trenchless Technology Center - Northeast Trenchless Technology Society - Gulf Coast Trenchless Association.

On behalf of the Committee...

John J. Struzziery, P.E., Conference Chairman

<u>Conference Steering Committee:</u>

Lynn E. Osborn, P.E.	-	Technical Program Coordinator and PINS Co-Chair
Bruce A. Bennett, P.E.	-	Exhibits Coordinator
Mark B. Pickell, P.E.	-	Publicity/Marketing Coordinator
William R. Ledbetter, Jr., P.E.	-	Pipeline Crossings Co-Chair
Malcolm N. Stevens, P.E.	-	Pipeline Location and Installation Co-Chair
John G. Bomba, P.E.	-	Pipeline Research Co-Chair
Randall C. Conner	-	Planning and Design Co-chair
Thomas Iseley, Ph.D., P.E.	-	Construction Division Coordinator
Mary H. Wall	-	Local Arrangements Coordinator
Arthur A. Spruch, P.E.	-	Conference Liaison

CONTENTS

* Manuscript not available at time of printing

* Manuscript not available at time of printing

* Manuscript not available at time of printing

MICROTUNNELING: AN OWNER'S PERSPECTIVE

Tim Jonasson, Member [1]
Mark Watkins, Member [1]

Abstract

In April of 1994, the City of San Buenaventura (Ventura), California, completed construction of the Bristol Relief Sewer Project: a $6.2 million, 4,000 meter (13,000 foot) trunk sewer, two-thirds of which was installed by microtunneling. This was one of the first microtunneling projects on the western coast of the United States. This paper will detail some of the problems and solutions the City encountered using this pipe installation technique. These problems included varying site conditions, the suitability of microtunneling to the site conditions, unmarked utility facilities, and the reluctance by local contractors to embrace new construction technologies. The City of Ventura's experience on this project and the technique of microtunneling from the owner's perspective would be of interest to any municipality considering microtunneling installation of underground pipelines.

Design

Design of the Bristol Relief Sewer began in 1986. Several alignments were investigated for the sewerline which was to serve the entire eastern side of the City of Ventura (total population approximately 100,000). After several preliminary engineering reports, an alignment was selected for the project which was broken into phases, the first of which was a 4,000 meter (13,000 foot) section which stretched from agricultural land located in a flood plain through an industrial section of town to a residential area (see map). The alignment crossed beneath a major county road, a state highway, two active railroad lines and ran under and across several city streets.

1

Tim Jonasson, Associate Civil Engineer and Mark Watkins, Maintenance Services Superintendent, City of San Buenaventura, P.O. Box 99, Ventura, CA 93002

Montgomery Watson Engineers (formerly James M. Montgomery Consulting Engineers) was hired for the design of the project. Fugro West., Inc. (formerly Staal, Gardner and Dunne, Inc.) was retained to analyze soil characteristics on the alignment and found groundwater from approximately 1.2 to 15.2 meters (4 to 50 feet) below the ground surface. (Powell, Staal, 1987) This proposed a problem as the section of sewerline outside the flood plain area was to be constructed from approximately 4.6 to 6.1 meters (15 to 20 feet) below ground level. (Bhamrah, Skager, Sprouse, 1988) Because continuous sheating and dewatering was cost prohibitive and would be very disruptive to local businesses, and because the depth of the required trench may have posed future settling problems, the design engineer recommended that trenchless technologies be investigated for the project. Horizontal boring machines by Akkerman Equipment and microtunneling equipment by American Thrustboring, a Division of Iseki Poly-tech Inc., were examined. After further investigation of the high ground water condition, the design engineer recommended that only microtunneling be used for the sewerline installation.

According to the design engineer, microtunneling, which is a remotely operated, laser-guided earth boring machine which is propelled through the ground by jacking pipe behind it, offered the following advantages for our project over conventional horizontal boring.

1. Heaving and subsidence would be reduced on streets above the pipe installation.

2. Dewatering requirements would be less.

3. Protection of existing utilities and structures from settling would be greater.

4. Workers would be safer using microtunneling since the microtunneling machine is operated remotely.

These advantages are largely due to the microtunneling machine having a sealed head and soil pressure balancing system to insure a uniform removal rate of the soil from in front of the cutting head. We realized these advantages would come at an additional cost since microtunneling machines are very expensive and few, if any, contractors in the western United States had any experience at the time with this technology. We believed the cost was justified, however, by the fact that microtunneling would provide the City with a "cleaner" pipe installation over that of conventional boring and certainly over that of open cut. Reduced disruption and protection of underground utilities were paramount in our decision to require microtunneling on this project.

The Bristol Relief Sewer Project was advertised for bids in August of 1992. Due to topography, the project was bid as open cut installation for the section of the project in the flood plain area where the pipe was relatively shallow and microtunneling for the remaining deeper section of the project. Local contractors repeatedly petitioned to be allowed to substitute horizontal boring for the specified microtunneling. The City, following the recommendations of the design engineer, insisted the contractor use a microtunneling machine for the pipe installation which required a considerable up-front expenditure by the contractor to purchase or lease the machine. Because of the soil conditions, we believed that allowing any other pipe installation method on the project would result in either the contractor going bankrupt when he/she was unable to complete the project with the other method or the sewerline would settle uncontrollably either after or during installation.

In October of 1992, Mladen Buntich Construction of Sunland, California, was low bidder and awarded the project for $6.2 million.

Construction

Reinforced concrete pipe production and poor weather at the beginning of the year delayed construction of the project until April 1993 during which time key personnel from both Mladen Buntich Construction and the City attended a partnering session to facilitate communication during construction. The contractor took this opportunity to again solicit the City's engineering staff to allow him to use the Akkerman horizontal boring machine in lieu of the specified microtunneling machine and offered the City a credit of $250,000.

One of the arguments made by the contractor, that the City had not previously considered, was that a microtunneling machine is not retrievable through the pipe should it run into an unknown object or freeze up in the soil. Should this occur, it would become necessary to excavate at the head of the machine to remove the obstruction or to remove the machine in order to bore back in the other direction. In certain areas, such as under the state highway, this would not be possible. For this reason, the City allowed the Akkerman machine to be used beneath the county road and state highway only.

Initial pushes were very slow. The contractor chose a microtunneling machine manufactured by Herrenknect International which was capable of installing pipe at a rate of 0.9 to 1.5 meters (3 to 5 feet) per hour but was producing at less than one half this rate in what appeared to be a much dryer stiffer clay material than anticipated by the soils report. This continued for the first 450 meters (1500 feet) of the microtunneled section of the project. The contractor was also experiencing difficulty settling the silty/clay soil material from the bentonite slurry used to pressurize the cutting head and remove the spoil material.

The contractor, who was microtunneling 16-20 hours per day, tried adjusting the slurry pressure and pushing force to achieve greater productivity. A representative from Herrenknect made some modifications to the cutting teeth of the microtunneling machine and oversaw operation of the machine for the next push to the north. The City meanwhile, at the request of the contractor, had the soils engineer verify that the soil conditions had not changed since the original soils report of 1987.

The adjustments and generally wetter soils encountered to the north increased production in the next section of the project. Unfortunately, a utility vault was marked incorrectly at the beginning of the push which resulted in approximately $100,000 damage to the underground power facility. The microtunneling machine had actually missed the vault, but had passed so close that the increased soil pressure caused by the slurry at the head of the machine collapsed one of the sides of the vault.

The north push on Golf Course Drive passed through an industrial area of Ventura. Although the street was not removed over the pipe, 4.6 by 9.1 meters (15 by 30 foot) jacking and receiving pits were dug approximately every 180 to 300 meters (600 to 1000 feet) due to direction changes to avoid utilities in the pipe alignment. These pits often encroached into driveways for businesses and access was difficult on Golf Course Drive for several weeks. Several business owners filed claims for lost revenues and damage to their property from the construction. In several areas along this section of the project, the slurry pressure from the microtunneling machine elevated curbs, undermined landscaping and percolated through the ground surface wherever it could. Clearly, the microtunneling machine had fallen short of providing a "clean" pipe installation.

The microtunneling machine generally performed better as the moisture content of the soil increased and the soil type went from a stiff clay to a more granular material. This was particularly true in the area near the state highway where the water table was very high and the soil was sandy, allowing free transport of water and better separation of the spoil material. However, for fear of the microtunneling machine becoming trapped beneath the state highway, the horizontal boring machine was substituted. The horizontal boring machine struggled installing the steel casing for the sewerline below the freeway because the open head allowed the soupy soil to flow uncontrollably from in front of the machine. The contractor was forced to slow production in order to avoid subsidence of the highway.

Once north of the freeway, the microtunneling machine was again placed back in service and performed flawlessly in completing the project. The wetter sandy soil and straight alignment greatly enhanced the performance of the machine tunneling over three meters (9.8 feet) per hour. Construction of the project was completed in April 1994 at a total construction cost of $6.6 million including 23 change orders.

Conclusions

The importance of thorough knowledge of soil conditions for microtunneling cannot be overstated. As Bernard P. Krzys stated in Trenchless Technology (see Reference 3):

> "In trenchless new installation work, understanding the geotechnical data is more important than ever. It is a cardinal rule in new installation trenchless work that you need to know the geotechnical information before you launch the project. Plus, you need to have a resolution process as part of the contract documents." (Krzys, 1996)

Daily stand-by time is very expensive for microtunneling machines, therefore, you do not want surprises such as unanticipated varying soils conditions to delay your project. It was our experience that microtunneling worked best with the combination of granular soils and high water content.

When notifying utility companies of your pipeline project, inform them of your intent to bore the pipeline rather than use open cut. Some utility companies prefer to wait until you have excavated to relocate their facilities simultaneously, which is not an option with microtunneling. Your notice to utility companies should emphasize the importance of marking their underground facility accurately since you will not see it before you hit it.

Try to keep your pipe alignment as straight as possible. This allows for longer runs between jacking and receiving pits and greater flexibility as to the locations of the pits. Be prepared for surface disruption and pavement degradation along the pipe alignment, particularly around jacking and receiving pits.

Consider allowing horizontal boring in stiffer soils as well as under major streets and highways where extraction of a microtunneling machine would be particularly difficult. You do not want an expensive microtunneling machine to get stuck where you cannot access it.

Microtunneling, like many new technologies, is beneficial when used in the proper situation. It is a worthwhile technology for an engineer to be aware of and to utilize when its attributes are required. Like other technologies, though, it also has its limitations and weaknesses and since it costs a premium, these limitations must be thoroughly analyzed prior to specification.

Appendix

1) Powell, John R and Staal, Ivar, Geotechnical Feasibility Study, Bristol Relief Sewer, December 1987, p. 11.

2) Bhamrah, Ajit S., Skager, Paul S. And Sprouse, George B., Preliminary Engineering Report for the Bristol Relief Sewer, November 1988, p. 3-2.

3) Krzys, Bernard P., "Are Lawsuits Hurting the Trenchless Market," Trenchless Technology, May 1996, p. 4.

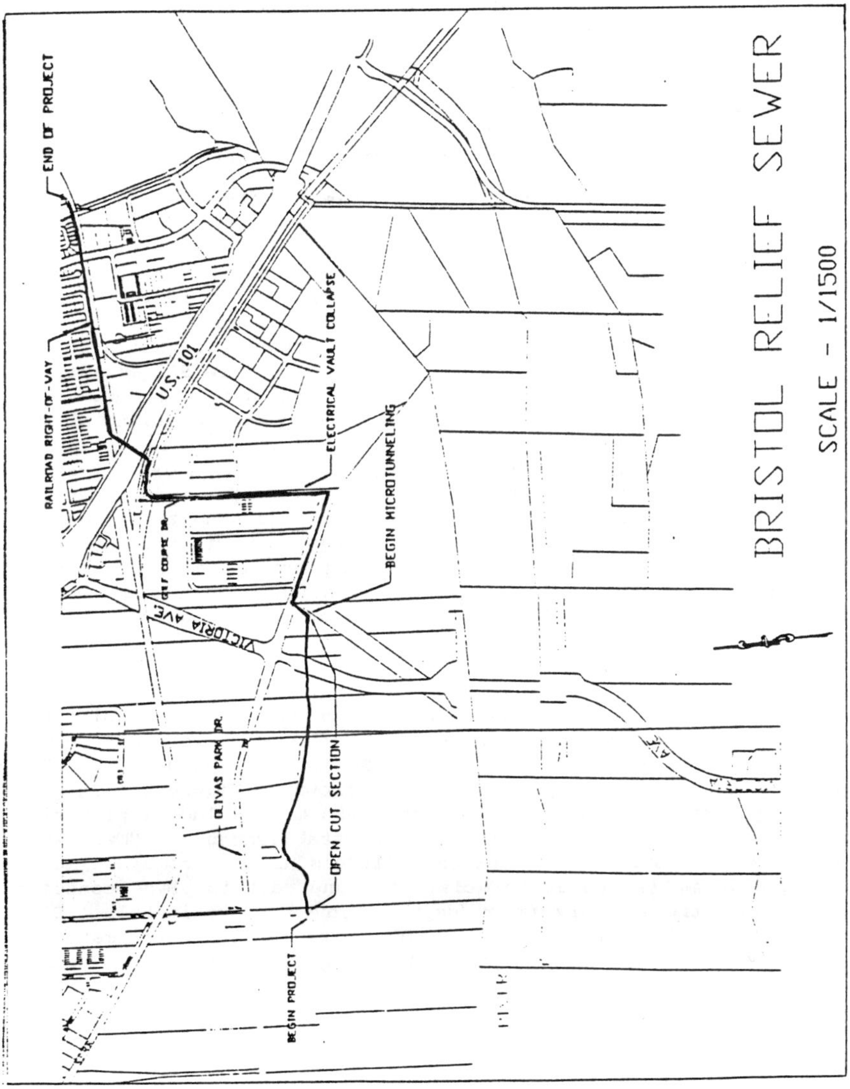
END OF PROJECT
RAILROAD RIGHT-OF-WAY
U.S. 101
ELECTRICAL VAULT COLLAPSE
BEGIN MICROTUNNELING
VICTORIA AVE.
OLIVAS PARK DR.
OPEN CUT SECTION
BEGIN PROJECT
BRISTOL RELIEF SEWER
SCALE - 1/1500

A CASE STUDY IN URBAN MICROTUNNELING
THE NEW ST. JAMES AVENUE INTERCEPTOR PROJECT
BOSTON, MASSACHUSETTS

Daniel J. Dobbels[1] and Michael J. Atwood[1]

ABSTRACT

The New St. James Avenue Interceptor Project included about 4,270 m (14,000 ft) of trenchless applications to replace or rehabilitate sanitary, storm, and combined sewers of various types and sizes. Approximately 1,100 m (3,600 ft) of new 915 mm (36 in) diameter pipeline were installed using microtunneling methods. The project included the construction of 19 jacking and receiving shafts from which 13 microtunneling drives, ranging from 30 to 150 m (100 to 500 ft) in length, were completed. This paper presents general background information relative to the microtunneling planning and design aspects for the project. Following this, jacking and receiving shaft construction methodology is reviewed. Relative to the microtunneling operations, the equipment and methods used by the Contractor are described along with: jacking loads, actual ground conditions, production rates, and surface impacts.

INTRODUCTION

The St. James Avenue Sewer System is a complicated interconnected system of sanitary sewers, storm drains and combined sewers in the Back Bay section of Boston, Massachusetts. The general location of the project area is shown on **Figure 1**. This is a highly congested area with adjacent land usage including high-rise office buildings, hotels, and retail establishments. The sewer system was constructed in the early 1900's using a combination of 300 to 600 mm (12 to 24 in) clay tile and cast-in-place concrete pipes, with the sanitary sewer (depth to invert approximately 4.5 to 6 m [15 to 20 ft]) and storm drain (depth to invert approximately 3 to 4.5 m [10 to 15 ft]) generally being installed in a common trench.

[1]Senior Engineer, Haley & Aldrich, Inc., 58 Charles Street, Cambridge, Massachusetts 02141

In response to longstanding operations and maintenance issues with the existing system, the Boston Water and Sewer Commission, through their prime consultant, undertook a study to identify specific problems with the system and design corrective measures. Based on the results of those studies and evaluations, it was concluded that system improvements using trenchless rehabilitation/ replacement methods were the best alternative. More specifically as regards to this paper, approximately 1,100 m (3,600 ft) of new 915 mm (36 in) diameter sanitary sewers and storm drains were installed by microtunneling. Detailed design work was completed in the fall of 1994 and construction was completed in late 1996.

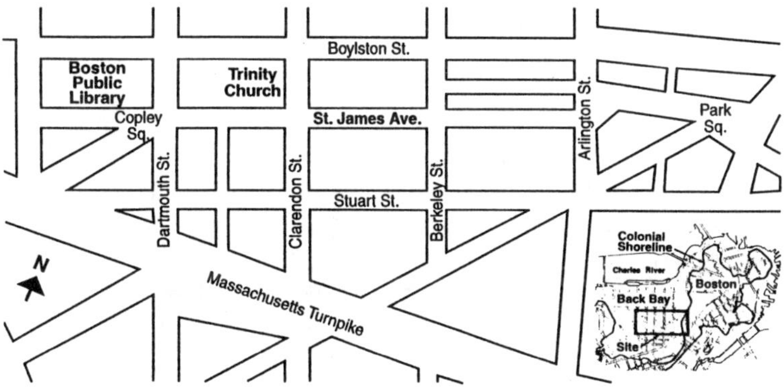

Figure 1. Project Location

PLANNING AND DESIGN

Subsurface Conditions

The project area is a former tidal estuary lying to the west of the original shoreline of the Boston Peninsula. Filling of the area began in the early 1800's and generally progressed from east to west. The area east of Berkeley Street (see **Figure 1**) was originally developed as a railroad terminal. In the early 1900's the structures associated with the rail facilities were demolished and the current street pattern was laid out.

Subsurface investigations undertaken during detailed design indicated that the vast majority of the microtunneling would take place in the Fill soils. The thickness of the Fill ranges between 3 m (10 ft) to greater than 6 m (20 ft) and it can typically be described as loose to medium dense coarse to fine sand, little coarse to fine gravel, with cobbles and varying amounts of silt, clay, glass, cinders, concrete, brick, wood and slag. In addition, the Fill present over the entire project area contains objects such as granite and concrete blocks, building foundations and debris, wood piles and wood sheating from previous below grade construction activities.

Groundwater levels were observed to be 3 to 4.5 m (10 to 15 ft) below ground surface or 0 to 1.5 m (0 to 5 ft) above the invert of the sanitary sewer.

Microtunneling Alignments

The proposed microtunneling alignments are shown schematically on **Figure 2**. To allow for consistency in microtunneling equipment throughout the project, a single pipe diameter of 915 mm (36 in) was planned for all microtunneling work. Because the St. James Avenue Sewer System is a gravity system and the elevation of the downstream end of the system could not be altered, the vertical alignment of new sanitary sewers and drains was designed for approximately the same as that for existing pipelines.

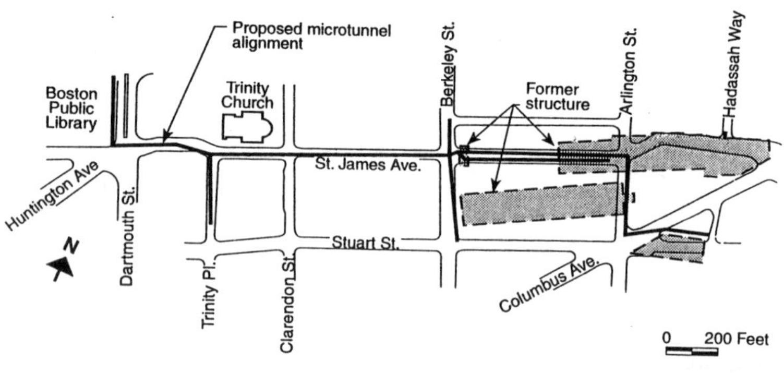

Figure 2. Proposed Microtunneling Alignments

While there was some flexibility relative to establishing the horizontal microtunnel alignments this flexibility was constrained largely by jacking and receiving shaft considerations. Relative to geotechnical issues, the factor that had the most significant impact on horizontal alignment was potential obstructions in the Fill. The horizontal alignments were established to reduce the potential for encountering obstructions, to the extent practicable.

One unique aspect of this project in connection with minimizing the potential for encountering obstructions occurred east of Berkeley Street. In this area, most of the microtunneling was planned to take place along existing clay tile pipeline alignments, referred to as in-line replacement. The reason for this was to reduce the risk for encountering obstructions. This is the area where the remnants of previous structures were anticipated. Due to the uncertainty in the layout of the former structures in the area, it was determined that the alignment with the least risk for obstructions was along the existing pipelines. The logic behind this approach was that any portion of the previous structures within the limits of the trenches excavated for installation of the existing pipelines should have been removed during excavation of those trenches.

Jacking and receiving shafts were located at changes in direction in the alignment, at tie-ins to existing lines, and at the ends of individual reaches. To the extent possible, shafts were located to reduce surface disruption and impacts on existing utilities. Where possible and practical, shafts were located in areas where the alignments crossed known obstructions so they would not impact microtunneling.

Microtunneling Methods

In general, microtunneling for the sanitary sewer was planned to take place in predominantly granular soils 4.5 to 6 m (15 to 20 ft) below ground surface and approximately 1.5 m (5 ft) below groundwater. Microtunneling for the storm drain was planned to take place about 1.5 m (5 ft) above the sanitary sewer or just above groundwater. Selection of specific microtunneling equipment and methods was the responsibility of the Contractor, subject to criteria identified in the Contract Documents. The most important criteria were the need to maintain a stable face to minimize loss of ground and resulting settlement, and control groundwater drawdown.

Obstructions

While the horizontal microtunneling alignments were established to minimize the potential for encountering obstructions, the risk for encountering them could not be totally eliminated. Moreover, it was recognized that the subsurface conditions were such that the risk for encountering obstructions could not be quantified in any practicable manner with a reasonable degree of certainty. It was also recognized that there needed to be some mechanism in the Contract Documents to minimize the cost and schedule impacts to the Contractor and the Owner relative to encountering obstructions.

The approach ultimately implemented in the Contract Documents was essentially a form of risk sharing. The Contract Documents indicated that the costs for removing obstructions consisting of: granite blocks and boulders greater than 1 foot in diameter, metal, reinforced concrete, and timber pile clusters that stopped the forward progress of the microtunnel boring machine (MTBM) would be borne by the Owner and would be negotiated on a case by case basis. The cost for any other objects that stopped the forward progress of the MTBM would be borne by the Contractor. To simplify the obstruction removal and negotiation process, the Contract Documents required that the Contractor's proposed obstruction removal method and crew and equipment rates be included as part of the pre-construction submittal. Regardless of whether the Contractor was entitled to payment, the costs associated with this approach were determined to be significantly less.

CONSTRUCTION

Jacking and Receiving Shafts

The project required construction of 19 jacking and receiving shafts. Detailed design of the shafts was the responsibility of the General Contractor who chose to construct most of the shafts using vertical wood sheeting. This method was selected over other methods due primarily to its flexibility to accommodate the numerous near surface utilities in the area.

Internal bracing for the shafts was provided by either steel or timber wales and struts. Toe penetration of the vertical sheeting and placement of an approximately 150 mm (6 in) thick concrete base slab in each shaft provided a sufficient bottom stability and acceptable working surface for the microtunneling equipment. Typical depths to the base of the shafts were 5.5 to 6.7 m (18 to 22 ft) for sanitary sewer installations, and 4.5 to 5.2 m (15 to 17 ft) for storm drain installations. Groundwater control during shaft excavation was provided by pumping from temporary sumps inside the shafts.

Jacking and receiving shafts were typically rectangular-shaped, although a generally L-shaped configuration was used for some shafts from which multiple drives were conducted. The range in as-built dimensions for both jacking and receiving shafts is shown below in **Table 1**:

Table 1: Jacking and Receiving Shaft As-Built Dimensions

	Minimum	Maximum	Typical
Jacking Shaft Size	2.7 by 4.9 m (9 by 16 ft)	4.9 by 8.2 m (16 by 27 ft)	4.0 by 6.4 m (13 by 21 ft)
Receiving Shaft Size	2.7 by 3.4 m (9 by 11 ft)	5.5 by 5.5 m (18 by 18 ft)	3.7 by 4.6 m (12 by 15 ft)

Reinforced concrete thrust walls and exit walls were designed and constructed by the Microtunneling Subcontractor after shaft construction was completed. Thrust walls were formed with horizontal and steel reinforcing, and typically provided a reaction area of 9.3 m^2 (100 ft^2). The thickness of the thrust walls varied from 0.5 to 0.8 m (1.5 to 2.5 ft); exit wall thickness varied from 0.5 to 1.8 m (1.5 to 6 ft). Exit wall openings were formed using wood knockouts to ease launching of the MTBM. In jacking shafts having multiple drives, an exit wall was often also used as a thrust wall.

Microtunneling Equipment and Procedure

The Microtunneling Subcontractor used a RVS 400AS MTBM manufactured by Soltau Microtunneling Inc., of Lunenberg, Germany. The RVS 400AS features a slurry-type system with a hydraulically driven crusher head. The unit was delivered with a nominal body diameter of 940 mm (37 in) OD that was "skinned" on site for 1145 mm (45 in) OD reinforced concrete jacking pipe. Comprising two sections, the Lead and Tail, the total length of the MTBM was approximately 4.3 m (14 ft).

The articulated Lead section included a center-driven cutting head attached to an inverted cone-shaped crushing chamber. Around the outer crusher cone were four high pressure jetting nozzles for clearing the crushing chamber. The cutting head, capable of 11,000 N-m (66,000 ft-lbs) of torque at 15 rpms, was initially configured with two types of cutting teeth, referred to on the project as "bullet" and "claw" teeth. The first drive of the project used a cutting head fitted with an equal number of bullet and claw teeth **(Photograph No. 1)**. However, after completing the first drive in which several granite obstructions and a significant amount of buried wood was encountered, the Microtunneling Subcontractor chose to redesign the head by replacing most of the "claw" teeth with "bullet" teeth.

Photograph No. 1. MTBM Cutting Head

The jacking frame used with the RVS 400AS consisted of dual 1,700 kN (190 ton) capacity rams, with a maximum cylinder stroke of 760 mm (30 in). The frame was designed for pipe lengths up to 2.4 m (8 ft), with a maximum 1,170 mm (46 in) OD pipe.

A positive seal around the MTBM was created by launching through a steel-flanged rubber ring secured to the exit wall (**Photograph No. 2**). Launching of the MTBM was conducted by simultaneous jacking the Lead Section and rotating the cutting head (in-the-dry) - usually without removing the excavation support sheeting boards between the exit wall and the ground. Intermittent flushing of the cut wood fibers while slowly advancing through the sheeting into the ground was done to prevent plugging of the head and chambers.

Photograph No. 2. MTBM Launch

The MTBM used a slurry-type spoil removal system to transport cuttings to the surface for separation and disposal. The MTBM and each trailing pipe segment contained a charge and discharge line piped to the surface. At the surface, the discharge line transported pumped flow to a single-meshed shale shaker unit for fluid separation. Solids were deposited into a 15 m^3 (20 $yds.^3$) open spoils bin, and fines remaining in the drilling fluid were left to settle in the discharge side of the baffled 37,800 L (10,000 gallon) sedimentation tank. On occasion, a dual four-banked hydrocyclone unit was used to expedite slurry/sediment separation. The overflow of the slurry was diverted into the charge side of the sedimentation tank through which the flow was piped to the charge pump and pumped back to the MTBM to close the circuit. The drilling fluid consisted only of water and re-circulated fines not settled out in the sedimentation tank. During freezing weather conditions, it often became necessary to circulate the slurry continuously when not microtunneling (i.e., between pipe segments and work shifts) to keep the slurry from freezing.

To reduce jacking loads to a minimum, a lubricant consisting of a mixture of bentonite, polymer, and water was pumped through an injection port manifold at the rear of the MTBM Tail section. From the manifold, the lubricant was discharged through three ports: one port at the crown, and two ports offset approximately 0.6 m (2 ft) either side from the crown. In each drive, the leading pipe was cast with three grout ports (at the crown and offset longitudinally and circumferentially either side), and typically every fourth pipe segment was fitted with a single grout port at the crown of the pipe. Lubricant was always pumped from the injection port flange (after the leading pipe segment was buried through the completion of the drive). For several long drives (greater than 61 m [200 ft]), lubricant was also pumped to grout ports cast into selected jacked pipe segments.

Operation of the slurry flow rate and pressure, the hydraulically driven jacking frame and cutting/steering head, the laser guidance system, and the computer-controlled data acquisition system were done at the surface from the unit's main control/power pack.

During jacking operations, the surface area occupied by the microtunneling equipment - including that area occupied by the jacking shaft, but excluding the laydown area required for jacking pipe - was between 275 and 325 m^2 (3,000 and 3,500 ft^2).

Preparation for In-Line Replacement

Before microtunneling for in-line replacement, service connections into the existing pipelines had to be temporarily bypassed and the Contract Documents required that the existing pipeline be filled. Pipe filling was successfully completed using flow fill - a low strength mixture of cement, fly ash and water. The first step in the process was to cap the ends of the pipeline segment exposed in the jacking and receiving shafts. Then flow fill was end-dumped into existing manholes along the pipeline segment. The consistency of the flow fill was such that it flowed freely through the pipeline segment, displacing the water and soft sediments in the pipeline through vent pipes in the ends of the exposed pipeline.

Jacking Loads

Figure 3 shows the actual minimum, average, and maximum jacking loads versus MTBM cutting head location for new line replacement and in-line replacement. Overall, each plot shows that jacking loads increased as drive length increased. The longer in-line drives experienced the highest jacking loads. However, on average, and over the same drive length, new line drives experienced higher jacking loads.

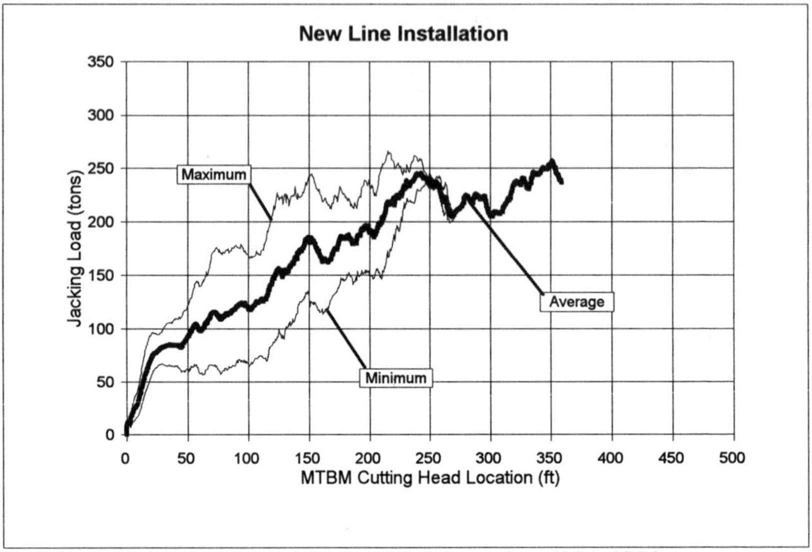

Note: Includes data from nine drives ranging in length from 100 to 350 ft.

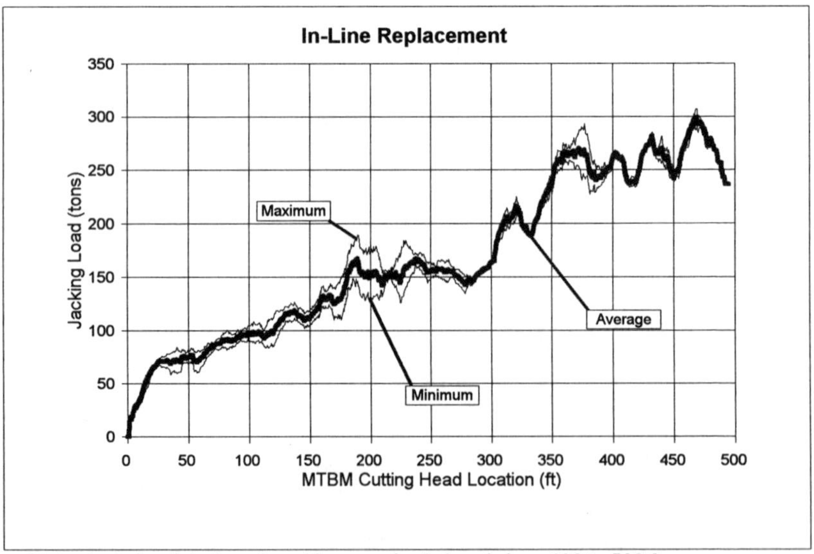

Note: Includes data from four drives ranging in length from 130 to 500 ft.

Figure 3.
Summary of Jacking Loads vs. MTBM Location
(New Line and In-Line Replacement)

The erratic nature of the jacking loads for new line replacement is believed to be simply due to the composition of the mined material. The Fill, as described previously, consists primarily of coarse to fine sand with varying amounts of cobbles and building foundation remnants. New line replacement was more likely to encounter Fill material containing cobbles and building rubble, and material that varied in composition over short lengths.

Obstructions

The general procedure followed when an obstruction was encountered was to excavate an unsupported test pit down to the head of the MTBM and investigate the nature of the obstruction. If the obstruction was small, it was removed from in front of the MTBM, the test pit backfilled and microtunneling operations started again. If the obstruction was larger, such that it could not be removed from the test pit, the test pit was backfilled and a sheeted excavation was made around and in front of the MTBM to remove the obstruction. The majority of the obstructions encountered consisted of granite blocks that appeared to be remnants of former structures in the area. It should be noted that no obstructions were encountered in the in-line replacement work.

On one occasion an abandoned tieback, composed of multiple seven wire high strength steel strands was encountered. The MTBM successfully advanced past the tieback, however the individual pieces of the steel strand, typically 100 to 200 mm (4 to 8 in) long, became lodged inside the slurry chamber and piping system, rendering the MTBM inoperable. Several attempts were made to clear the slurry system lines from inside the machine but were unsuccessful, and ultimately a shaft was constructed to gain access to the front of the machine to clear the pieces of wire strand.

Production Rates

For this project, looking at microtunneling production rates from two perspectives was informative: including downtime and excluding downtime. The term downtime as used in this paper was defined as that non-productive time attributed to obstruction removal and equipment problems. Average microtunneling production rates calculated <u>excluding</u> and <u>including</u> downtime are summarized in **Table 3**.

Table 3: Summary of Microtunneling Production Rates

	Avg. Footage per 10 hr. shift*		
	New Line	In-Line	Combined
Excluding Downtime	10.4 m (34.0 ft)	12.2 m (40.3 ft)	11.0 m (36.0 ft)
Including Downtime	6.7 m (22.0 ft)	11.9 m (39.2 ft)	8.2 m (26.9 ft)

* The best production for a 10 hr shift, was 14.8 m (48.5 ft).

The vast majority of downtime was associated with obstruction removal. When considered in the context of the overall duration of microtunneling work, downtime accounted for approximately 10 percent of the total time that the microtunneling subcontractor was on site.

Surface Impacts

Impacts to vehicular and pedestrian traffic due to microtunneling were not significant. Each equipment setup at the jacking shafts required the Contractor to limit the impact on pedestrian and vehicular traffic, and maintain building access. Often, mobilizing during off-peak hours was required, and operating hours were limited when microtunneling from jacking shafts near hotels and residential structures.

To date, little to no settlement or heave were observed from the many survey points installed before construction, and groundwater levels were typically unaffected.

ACKNOWLEDGMENTS

The design and construction management team for the New St. James Avenue Interceptor Project consisted of the Boston Water and Sewer Commission (Owner), Bryant Associates, Inc., Boston, MA (Prime Consultant); Haley & Aldrich, Inc., Cambridge, MA (Geotechnical and Microtunneling Subconsultant); and Howard Stein Hudson, Boston, MA (Traffic and Community Relations Subconsultant). The General Contractor for the Project was Modern Continental Construction Company, Cambridge, MA, and the Microtunneling Subcontractor was Westcon/Dillingham Microtunneling, Pleasant Grove, UT.

Incorporating Microtunnels Into The Design of Pipeline Networks

Glenn M. Boyce, Ph.D., P.E., Member[1]

Abstract

Pipeline designs incorporating trenchless methods at the outset can provide benefits to a project, including cost effectiveness. However, most sanitary sewer or storm drainage systems are designed following alignments which rely on trenching as the construction method. If the cost of microtunneling or other trenchless methods is compared directly to the cost of trenching along the same alignment, the trenchless method, in most cases, will be more expensive.

When microtunnels are incorporated into the design and the designer is not forced to follow a "trenched" alignment, sufficient benefits are realized. This paper illustrates the ways microtunnels are incorporated into designs and are cost competitive. Two case histories are discussed in the paper. Designing with microtunnels can results in simpler pipe networks, shorter pipe lengths, less environmental impacts, less surface disruptions, minimal permitting efforts, and lower overall project costs.

Introduction

Storm drains and sanitary sewers are designed as gravity systems whenever possible to keep operational costs to a minimum. The system is a network of pipelines, with downstream pipelines becoming larger and placed at lower elevations. The alignment of the sewers are typically placed in the publicly owned right-of-ways (ROW). These ROWs are streets or easements. When the streets were originally developed with houses, the land was open and the sewers installed by digging trenches. As the sanitary sewers were modified or expanded, the trenches were redug and the sewer lines again installed. This method of sewer installation has lead to the typical configuration of manholes at a uniform spacing and laterals connected at points perpendicular to the trunk line. Figure 1 shows this typical configuration.

Occasionally segments of these pipeline networks will encounter special circumstances, forcing consideration of a trenchless construction method. These special situations include:

- Road crossings when more than a few thousand vehicles could be affected daily;
- Difficult ground conditions, like running sands, soft clays, or rock;

[1]Sr. Professional Associate, Parsons Brinckerhoff, 303 Second Street, Suite 700 N, San Francisco, CA 94107

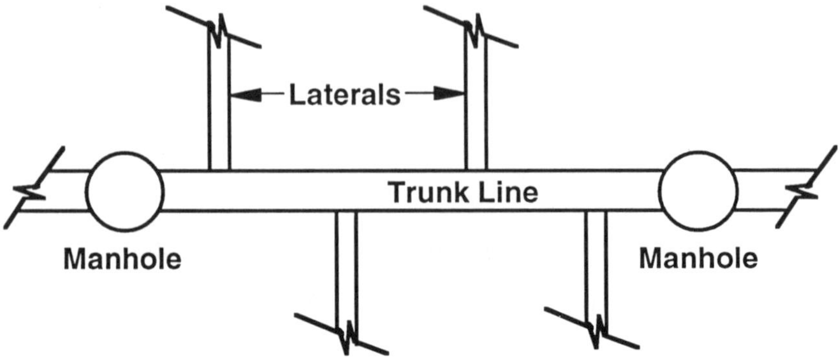

Figure 1 Typical Approach to Sewer System Layout

- Ground conditions where settlements could be hazardous to structures, property, or utilities;
- Below the groundwater table where dewatering may be difficult or expensive;
- At depths greater than 4.6 meters (15 feet); or
- In contaminated soils.

In these situations, the cost of microtunneling or other trenchless methods can be compared directly to the cost of trenching the same alignment. The trenchless method may or may not be less expensive. Sometimes, the social and environmental impacts are too compelling and trenchless methods are required (Boyce and Bried, 1994; Budhu and Iseley, 1994; Coleman, Davidson, and Bandoni, 1995; Currie, Schumacher, and Thornton, 1993). When microtunnels are considered earlier in the design and not forced to follow the trenched alignment, sufficient benefits could be realized.

To make trenchless techniques cost competitive, we can not look at pipelines being installed in a trench only. We need to look at the advantages trenchless techniques provide and incorporate these advantages in the design. These advantages include:

- No penalty for being deeper
- Straighter alignments between two points both vertically and horizontally thus being independent of surface topography
- Using jacking and receiving pits for manholes.

If all new installations were with trenchless techniques, the new approach would be to:

- Convert jacking pits into manholes
- Install trunk lines with no laterals
- Connect laterals to the manholes

Figures 2 and 3 show the configuration for the new approach for layout and design of gravity drainage/sewer systems. Connecting laterals to manholes could decrease the amount of piping and greatly improve operations and maintenance. This design has been used in Berlin, Germany (Stein, Mollers, and Bielecki, 1989). This new approach allows easier rehab of the trunk lines in the future because there are no laterals to reconnect. Disconnecting the laterals from the trunk line should be adopted everywhere, because the benefits can be realized independent of whether trenching or trenchless methods are used.

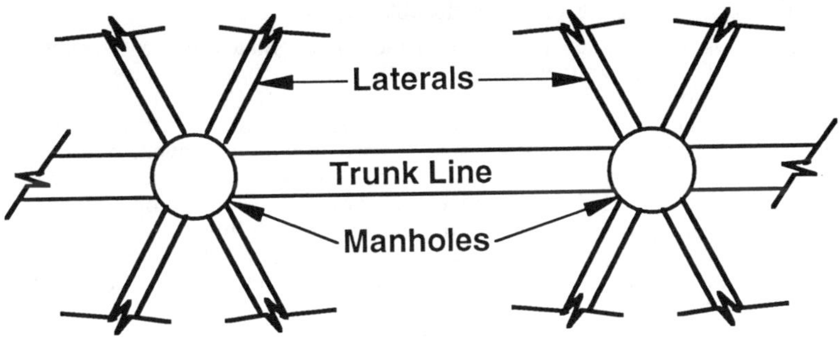

Figure 2 New Approach to Sewer System Layout

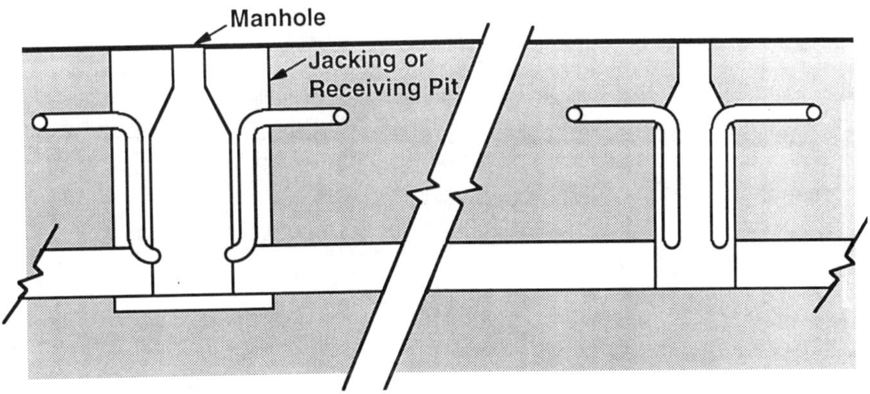

Figure 3 Profile of New Approach to Sewer System Layout

The installation of laterals at various points along the trunk line can lead to other problems. The connection between the trunk line and lateral can become damaged leading to infiltration. In the reverse case, the system could overflow into the laterals. In spite of these problems, the laying out of the systems within streets and laterals has been the standard approach to sewer design.

As streets and urban areas become busier, the use of trenchless technology has become an attractive alternative. To illustrate the benefits of not following the design approach shown in Figure 1, consider the following two case histories.

Nogales, Arizona - Fugitive Wastewater Collection Project

Mexico allows raw sewage to flow into the United States at the Nogales, Arizona border. The dry weather flows enter the United States because of the difference in elevation between the two countries. The US Corps of Engineers studied the problem and concluded inlet structures should be built at the head of the five canyons at the border. The canyons are named: Ephraim Canyon, Nogales Wash, East Street Wash, Trickey Wash, and Cerro Pelon Canyon. The proposed trenched collection system is shown on Figure 4. The proposed solution includes trenched pipelines along the bottoms of the five canyons, following the natural drainage course until the pipelines meet the main/central channel located on the north side of town. The channel would convey to a large diameter outfall which drains to the Nogales treatment plant.

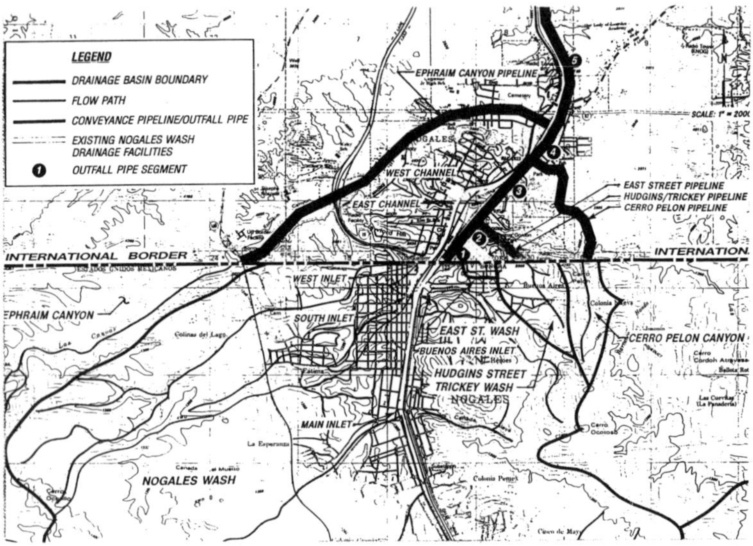

Figure 4 Plan View of Proposed Sewer System for Nogales, Arizona

The proposed plan carried with it many design and construction problems. These problems were:

- Pipeline construction along deep canyons requiring environmental permitting and endangering protected habitats
- Multiple conduits leading into the United States inviting use by illegal aliens
- Right-of-way acquisition

By reopening the design up to using microtunnels and ignoring the concept that the pipeline had to follow the canyon topography, a whole new layout was possible. Instead of running pipelines down canyons perpendicular to the border, microtunneling allowed pipelines to directly connect between the canyons, parallel to the border. The new alignment eliminated multiple long pipelines. The resulting layout is shown in Figure 5.

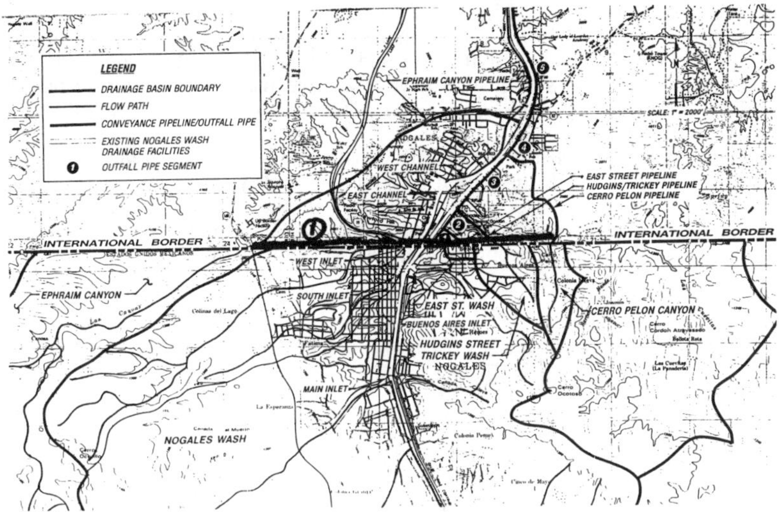

Figure 5 Plan View of Nogales Sewer System Using Microtunnels

The advantages of this new layout are:

- Elimination of 2,565 meters (8,300 feet) of pipeline construction
- No 404 permitting was needed
- No ROW acquisition was required since the Boundary Commission owned an existing 20-meter (60-foot) wide ROW
- Border patrols were not impacted because half of the ROW could be used for construction and the other half for patrols
- No multiple conduits of entry into the United States.

The ultimate advantage was the new layout had a construction estimate of $38 million versus a trenched layout cost of $57 million.

San Francisco Zoo Infrastructure Project

The San Francisco Zoological Society in 1995 developed an overall Master Plan. The plan is known as the Zoo 2000 Master Plan. Part of the plan addresses the planning, design, and construction of new infrastructure facilities to serve the current zoo and future Zoo 2000 exhibits and facilities. The infrastructure consisted of four main systems: water, electrical/communication, natural gas, and sewer/drainage. The first three systems consisted of either conduits or pressure lines and were establish as a network to follow designated utility corridors. These corridors allow for easy access and maintenance to minimize disruption to the zoo animals, patrons, and staff. The fourth system, sewer/drainage, is designed as a gravity system and will follow a layout different than the other three systems.

The zoo's existing sewer/drainage system is inadequate to handle the zoo's existing flows and projected future Zoo 2000 flows. The existing system backs up and causes localized flooding within the zoo during high flow periods. Both existing and future flows consist of toilet flushing, exhibit washdown, pool drainage, and stormwater flows. Zoo 2000 flows anticipate higher than existing flows because of expansion and increased paved areas.

The design objective for the combined sewer/drainage system is to drain all dry weather flows by gravity out of the zoo and into the City's Westside Transport storage tunnel. The new system will have a wet weather pump station to handle large flow conditions during storm events when the Westside Transport tunnel levels rise.

The originally proposed sewer system consisted of a series of trenched branch lines which collect flows from exhibit laterals and convey flows by gravity to the northwest corner of the zoo. The layout of the proposed facilities and improvements are shown in Figure 6. The system had 1000-mm (39-inch) diameter pipe at the northwest corner, becoming progressively smaller upstream.

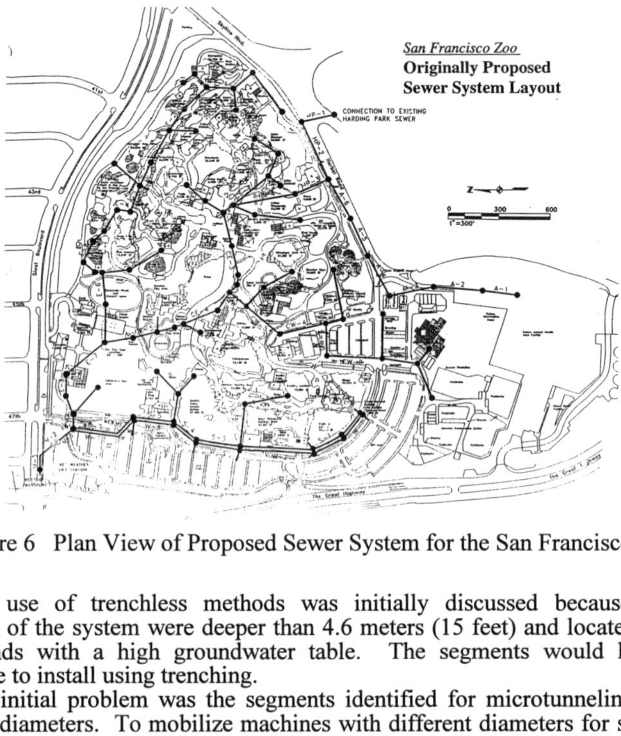

Figure 6 Plan View of Proposed Sewer System for the San Francisco Zoo

The use of trenchless methods was initially discussed because specific segments of the system were deeper than 4.6 meters (15 feet) and located in loose dune sands with a high groundwater table. The segments would have been expensive to install using trenching.

The initial problem was the segments identified for microtunneling were of different diameters. To mobilize machines with different diameters for short drive lengths was cost prohibitive. The ideal use of microtunneling would include:

- One uniform pipe diameter
- Thousands of feet of microtunneling to offset the mobilization costs
- Minimal number of shafts to reduce costs and surface disruptions to zoo operations.

After much discussion, a system consisting of a trunk line system like the layout shown in Figure 7 was proposed and accepted. The system uses uniform trunk lines of 750 mm (30 inches) in diameter. The system involves over 2150

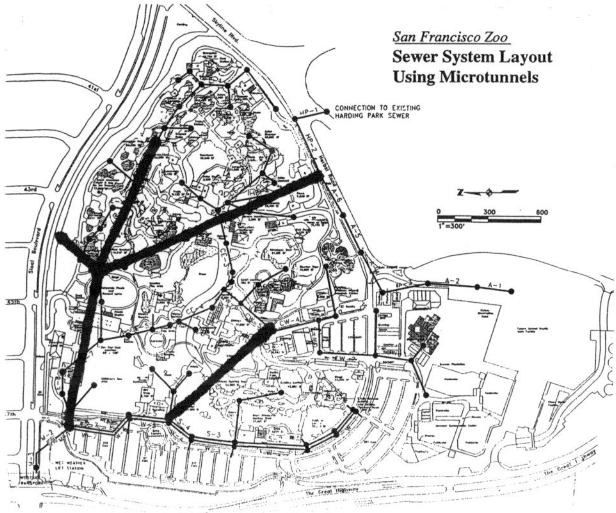

Figure 7 Plan View of the San Francisco Zoo Sewer System Using Microtunnels

meters (6950 feet) of pipelines. Work is laid out of one large starting shaft with drives in four different directions. This layout minimizes the need for large work areas and concentrates the work near the edges of the zoo. Laterals from a new exhibit are connected at the manholes near the surface, see Figure 8a. Laterals can also be connected at intermediate points using the configuration shown in Figure 8b. These configuration will help prevent flooding conditions into the laterals because the laterals are placed at higher elevations. The installation of laterals to the manholes becomes a part of the exhibit costs, helping to defer costs. The new layout has allowed the following advantages:

- Elimination of 830 meters (2,690 feet) of pipeline construction
- Deeper system to reduce overflow conditions
- Simplified layout for maintenance
- Reduced disruptions.

The new layout had a construction estimate of $8.6 million versus a trenched layout cost of $6.3 million. The difference in cost is in part a reflect of a more defined final design estimate versus a planning/conceptual cost estimate.

Conclusions

Microtunneling is an effective means of installing utilities underground. The problem is developing acceptance. For microtunneling to succeed, public agencies and engineers must begin to factor in the environmental and social costs of installing utilities. They must resist the urge to simply "do it the old-fashion way." Designers need to avoid:

- Laying the pipelines to simply follow the topography
- Assuming the construction method is trenching
- Using trenchless methods only at special situations
- Disregarding the need to rehab the pipeline in the future.

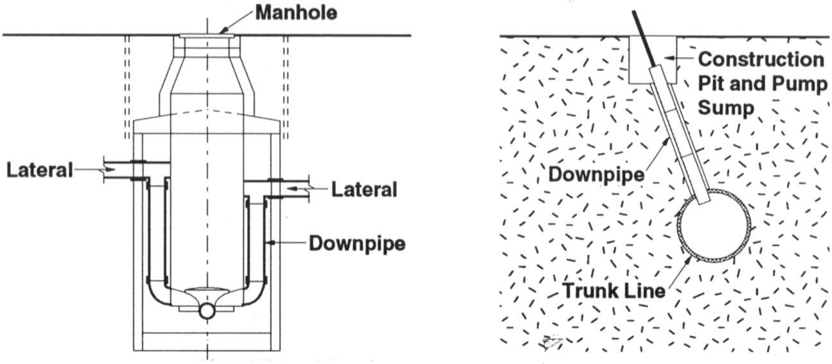

(a) (b)

Figure 8 Manhole Connection Configurations

There are different ways to design pipeline networks, but it requires adopting new concepts. Engineers must begin to fully understand trenchless technology, both the installation by microtunneling and the process of rehabilitation. They must be able to establish the proper layouts of plan, profile, and shaft locations. The design engineer also must make sure the project can be accomplished or, more importantly, recognize when conditions are not right for microtunneling.

This paper illustrates some of the ways microtunnels were incorporated into two designs. The projects were the Nogales Fugitive Wastewater Collection System in Arizona and the San Francisco Zoo Utilities project in California. In both projects, new alignment corridors utilized the advantages of microtunneling. They resulted in simpler pipe networks, shorter pipe lengths, less environmental impacts, less surface disruptions, minimal permitting efforts, and lower overall project costs.

References

Boyce, G.M. and E.M. Bried, 1994. "Benefit-Cost Analysis of Microtunneling in an Urban Area," *Proc. of North American No-Dig '94*, NASTT, Dallas, Paper A1.

Budhu, G. and D.T. Iseley, 1994. "The Economics of Trenchless Technology Versus the Traditional Cut and Fill in High-Activity Urban Corridors - A Research Concept in a Real World Environment," *Proc. of North American No-Dig '94*, NASTT, Dallas, April, Paper A2.

Coleman, D.J., R.H. Davidson, and F. Bandoni, 1995. "Environmentally Appropriate Methods of Sewer Construction," *Proc. of North American No-Dig '95*, NASTT, Toronto, May, Paper 5C-1.

Currie, R., C. Schumacher, and E. Thornton, 1993. "Microtunneling Replaces Open Trench for Suspended Sewer Project," *Proc. of North American No-Dig '93*, NASTT, San Jose, May, Paper N2.

Stein, D., K. Mollers, and R. Bielecki, 1989. *Microtunneling: Installation and Renewal of Nonman-Size Supply and Sewerage Lines by the Trenchless Construction Method*, Ernst & Sohn, Berlin, 352p.

Subsurface Utility Engineering

James F. Noone, PE[1]
M. ASCE

Abstract:
In recent years, tighter budgets have dictated that delays and unforeseen costs be minimized at all stages of a project. One major source of delays and unforeseen costs is the inaccurate or incorrect depiction of existing buried facilities (utilities or structures) on Contract Documents. Information obtained from utility companies is rarely accurate enough for design purposes and may not depict "as-built" conditions. Information for vertical location is almost never available. In addition, work may have been performed which was never documented. This paper shows how Subsurface Utility Engineering (SUE) can be used to deal with design issues relating to existing utilities.

Introduction:

SUE is a professional engineering discipline dedicated to the determination of the exact location of existing underground facilities using Trenchless Technology Methods. The exact information obtained is transferred to the contract documents for the designer to analyze any conflicts between existing and proposed facilities. Too often when this does not happen projects are plagued with contingency bidding, construction claims or change notices. In extreme cases property damage and/or loss of life may occur due to utility breakage. On many projects which have used SUE cost savings in the order of seven dollars for every one dollar spent have been realized.

Use of SUE services, by minimizing inaccurate depiction or omission of buried utilities on final contract drawings, reduces costs, delays, and public inconvenience. In addition, by eliminating the risk of utility breakage, the project will be safer for both construction personnel and the general public, hence reducing liability concerns.

[1]Principal, BSI, 145 Hanover Street, Boston, MA 02108, (617) 722-0040 Fax: (617) 722-3065: Member, ASCE Standards Committee for Collection and Depiction of Existing Utilities on Design and Construction Documents (Subsurface Utility Engineering)

Technical Aspects of SUE:

The SU Engineer identifies, characterizes and maps subsurface utilities. This requires knowledge and experience dealing with historical research, record utility research, electronic location methods, Non-Destructive Excavation Methods, Electronic Field Survey Data Collection, Computer Aided Design and Drafting (CADD), Utility Design, Highway/Structural Design and Graphic Information Systems (GIS). The SU engineer provides designers with the information they require about existing project utilities to do their job effectively and in the most cost effective manner.

The main components of SUE are:

A. Record Research,
B. Electronic Utility Line Location,
C. Non-destructive Air-Vacuum Excavation,
D. Data Collection, and Processing,
E. Interaction with the design team.

A. Record research

Record Research involves performing a diligent effort to obtain all available information on subsurface facilities. This includes, but is not limited to: contacting utility companies, visiting state or local plan rooms, and notifying local one-call centers.

B. **Electronic Utility Line Location**

Electronic Utility Line Location is the process whereby the approximate horizontal location of a utility is determined. Several technologies are used, including, but not limited to, Electromagnetic Line Locators, Radiofrequency Line Locators, Magnetic Locators, Ground Probing Radars and Acoustic Location Methods. The above are placed in order of usefulness for exactly identifying particular Utility Lines. There are cases, however, where due to site conditions, preferred methods cannot be used and alternatives are chosen.

The Electromagnetic and Radiofrequency line locators operate by locating either a background signal or by locating a signal introduced into the utility line using a transmitter. There are three (3) sources of background signals.

1. Utility Line can act as radio antennas. The electrons is an antenna oscillate producing fields which can be picked up with a electromagnetic receiver (Radiofrequency method).
2. Current carrying conductors have a 60HZ signal associated with them which can be picked up by a receiver (Electromagnetic).

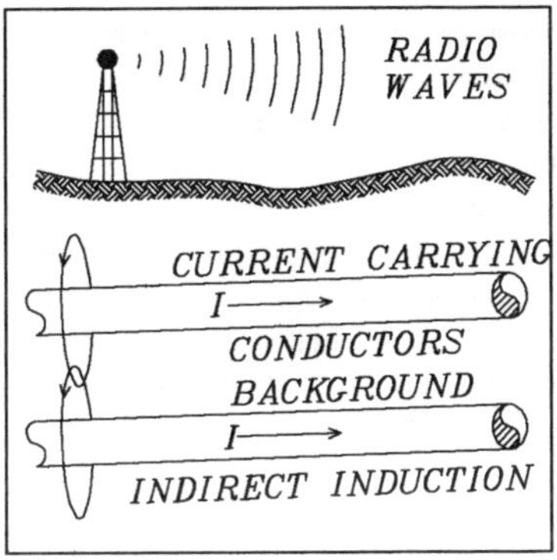

Fig.1.1

3. Utilities in close proximity to power lines or used as grounds may also have a 60HZ signal which can be picked up with a receiver. (Electromagnetic). See Figure 1.1

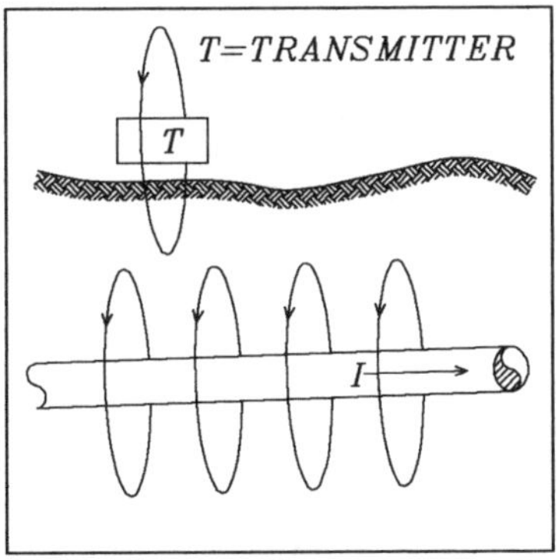

Fig.1.2

A signal can be indirectly induced onto a utility line by placing the transmitter above the line. The transmitted signal must have the same general orientation as the utility line. The optimum position of the transmitter is determined through a trial and error process. See Figure 1.2. A signal can be induced directly using an induction clamp. An induction clamp is a circular clamp which induces only the particular conductor that it is clamped around. This affords the ability to isolate utilities from one another.

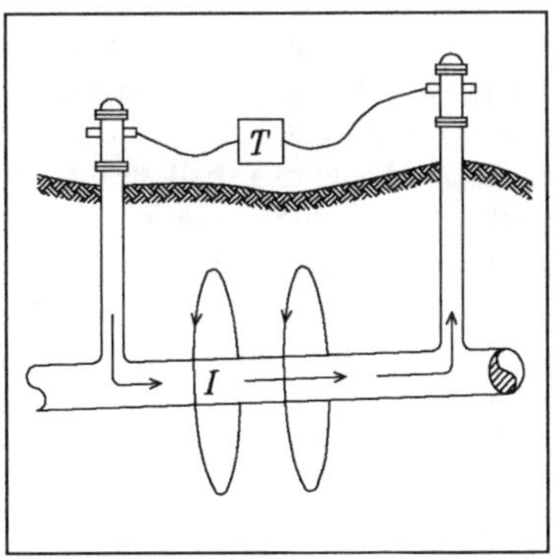

Fig.1.3

The best possible tracing signals are frequently generated using the direct connection method. By virtue of the closed current loop, there is very little chance of the resulting signals being distorted. This is the preferred method of tracing a utility when and where possible. In certain instances, where only one connection point is available on two (2) separate utilities, a connection can be made between them. This allows both utilities to be traced. However, the results should be verified using other tracing methods. See Figure 1.3

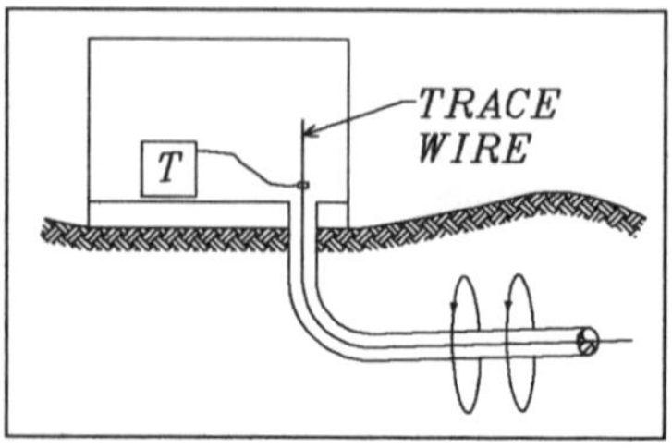

Fig.1.4

When access can be gained to a conduit, a flexible insulated trace wire can be used. The trace wire is fed into the conduit. The transmitter is connected to the trace wire and the signal in the wire can be traced. Alternatively an "in-pipe transmitter" (Sonde) or "Geo-Pig" can be used in the same manner. This method is very useful for non-metallic conduits. See Figure 1.4

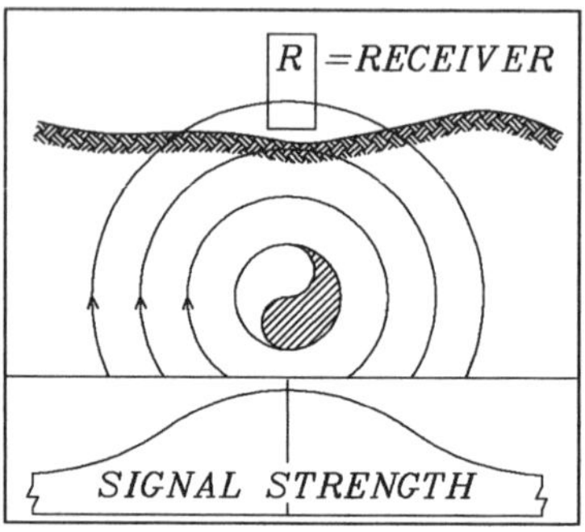

Fig.1.5

The signal carried by the utility line can be located horizontally on the surface using a receiver. The receiver is moved across the estimated location of the utility line. The location of the highest signal strength is marked as the approximate horizontal location of the utility. The receiver is then rotated until the highest signal strength is achieved. This will give the approximate orientation of the utility. See Figure 1.5

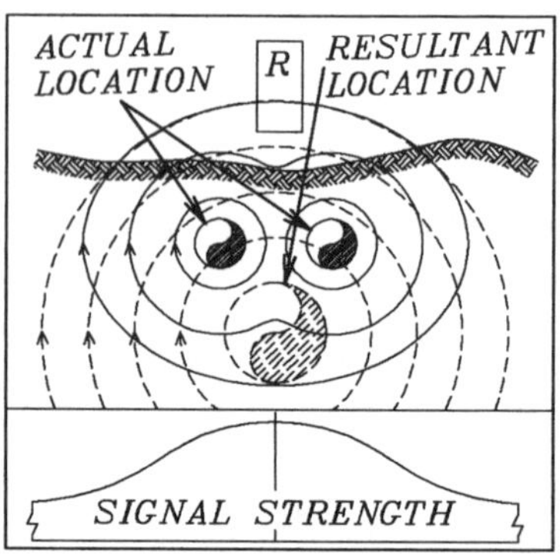

Fig.1.6

Approximate vertical locations can be estimated. They are, however, subject to errors and generally not any better than good judgment would predict. One reason for this is the situations where lines are running side-by-side. What results is a combined field which would indicate a single utility with greater depth than the actual depth. See Figure 1.6 Conversely, the presence of a conductor above the utility being traced could lead to a shallower depth than actual. Many other situations occur which are beyond the scope of this paper.

It should be noted that these technologies can also be used for finding cable faults and determining electrical conductivity of pipe joints.

Ferrous Metal or Magnetic locators operate by indicating relative amounts of buried ferrous metals. They have limited application to locating and identifying utility lines but can be very useful for locating underground storage tanks (UST's).

Ground Probing Radar (GPR), operates by indicating the location of Subsurface Anomalies. A radar signal is sent into the ground via a transmitter. Some portion of the signal will be reflected from subsurface materials and is recorded with a receiver. In general a object which is harder than the surrounding soil will reflect a stronger signal. Utilities, tunnels, underground storage tanks, and footings will reflect a stronger signal than the surrounding soil. This method does not identify specific utilities (i.e., water vs. gas, electrical vs. telephone), hence, verification is necessary using other methods. This method is somewhat limited when used in areas with clay type soils or a high water table. An interesting developing technology is digital signal processing (DSP) of the information obtained from the GPR Receiver. The basic idea is that a utility is a long thin object that should appear on several GPR scans of a particular area. The DSP software suppresses images of discrete objects such as rocks and enhances images of possible utilities, the end product being a map of the subsurface facilities.

Acoustic Location Methods, are generally applied to water lines. A highly sensitive Acoustic Receiver listens for background sounds of water flowing; (at joints, leaks, etc.) or to sounds introduced into the water main using a transducer. This method has good identification capabilities but, can be inaccurate due to background noise.

C. Non-destructive Air-Vacuum Excavation: Non-destructive Air-Vacuum Excavation is used to determine the exact horizontal and vertical location of utilities. The process involves removing the surface material over approximately a 1'x1' area at the electronically determined approximate horizontal location. The air-vacuum process then proceeds with the simultaneous action of compressed air-jets to loosen soil and vacuum extraction of the resulting debris. The process continues until the utility is uncovered. See Figure 2.1 The following information is typically recorded: utility, material, size, depth, condition, location (x,y,z), orientation, roadway section materials and depths, soil type and water table.

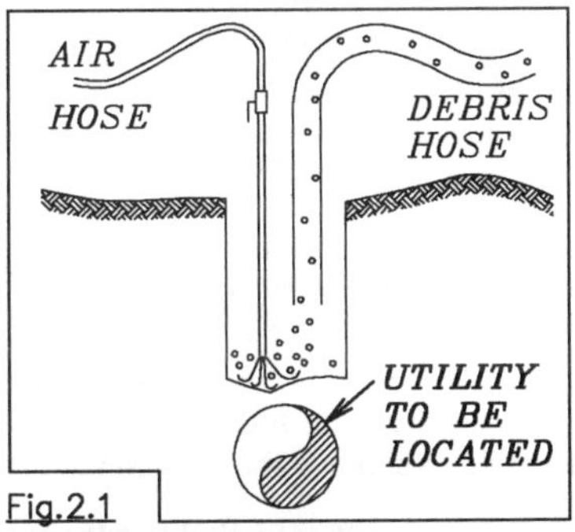

Fig.2.1

Air-Vacuum Excavation can also be used at a proposed boring locations to excavate below the "utility window" which is usually eight feet (8'). This is very useful where the mandated setbacks to utilities can not be maintained or the location of certain utilities is in question. See Figure 2.2

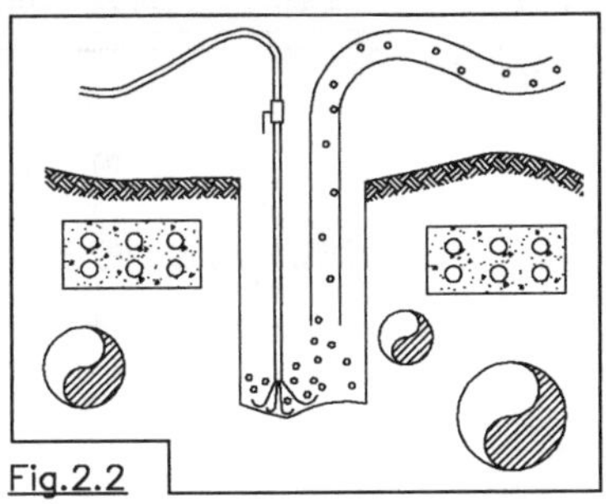
Fig.2.2

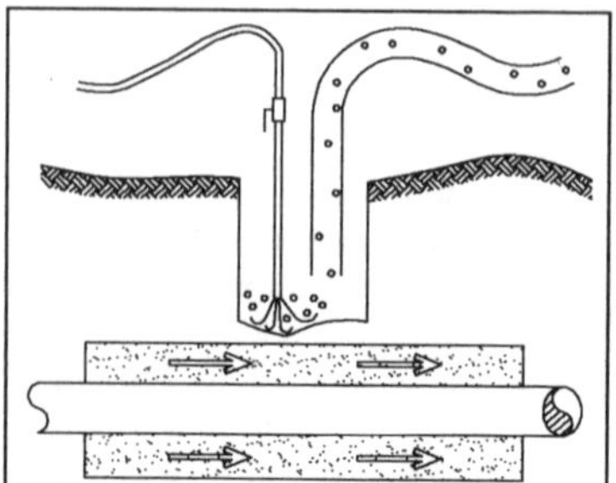

CONTAMINATE FREQUENTLY MOVES ALONG UTILITY LINE TRENCHES

Soil samples can be taken during the air-vacuum excavation process. Frequently, contaminant moves along utility line trenches. Air-vacuum excavation can be used to obtain soil samples from utility line trenches without risking damage to the utility. See Figure 2.3

Fig.2.3

D. Data collection and processing: The collection of data is generally achieved through the use of electronic total stations, Global Positioning Satellite (GPS) receivers and data collectors. Processing involves transferring the data from the data collector into the CADD system from which the contract documents will be generated. The data, particularly the results from the Air-Vacuum Excavation process (utility, material, size, depth, condition, etc.), can also be incorporated into a GIS system for future use by the designer or owner.

E. Interaction with the Design Team: It is the responsibility of the Subsurface Utility Engineer to interact and work with the design team through out all phases of the design. The Subsurface Utility Engineer's purpose is specifically to provide information about existing utilities and how they will affect the design of the project.

Use of SUE in Design

The Subsurface Utility Engineer is responsible for the collection and depiction of existing utilities on contract drawings. The Subsurface Utility Engineer, as a member of a design team, should work with the project engineer through all the stages of a project. There are four (4) distinct stages of a project where different aspects of SUE are used to achieve the desired result.

Sketch Plan [twenty-five percent (25%) design]: During the sketch plan stage the Subsurface Utility Engineer is responsible for collection of all recorded utility information. The Subsurface Utility Engineer depicts the recorded information on a base topographic plan prepared by the project surveyor. This information is used by the project engineer to locate proposed utility corridors or structures. See Figure 3.1

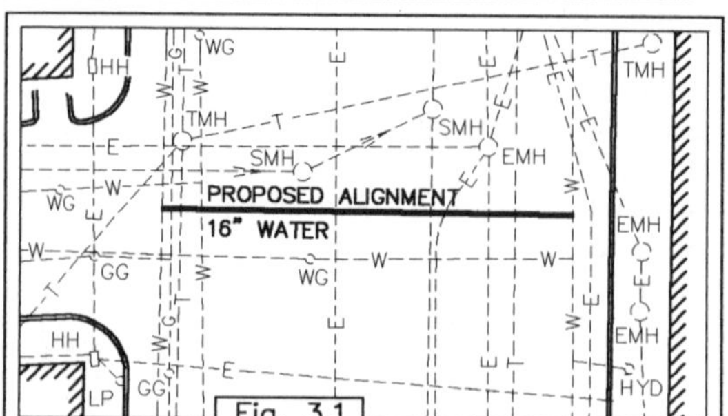

Fig. 3.1

Preliminary Design [seventy-five percent (75%) design]: During the preliminary design stage the Subsurface Utility Engineer locates all existing utilities at proposed areas of work. This is an approximate horizontal location performed using the electronic location methods. The Subsurface Utility Engineer is responsible for determining which methods shall be used for this work. The project engineer can now optimize the location of proposed work with respect to the horizontal location of the existing utilities. In general, the level or accuracy that can be expected for the electronic location methods is in accordance with corresponding ACSM standards. See Figure 3.2

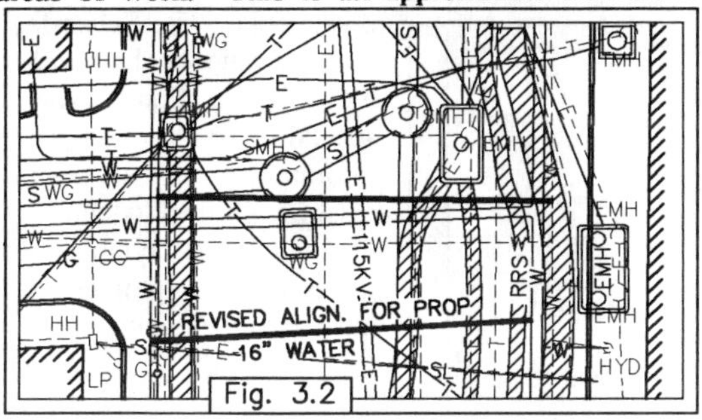

Fig. 3.2

Final Design [one hundred percent (100%) plans]: At this stage the Project Engineer identifies locations where conflicts with existing utilities may occur. At these locations the Subsurface Utility Engineer performs Non-Destructive Air-Vacuum Excavation test holes to accurately locate the existing utilities in question. The accuracy of this location is typically +/- 0.1 (plus or minus one tenth of a foot) or better.

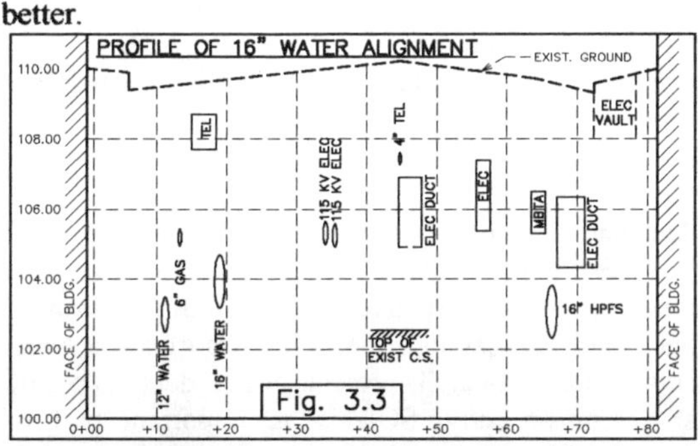

Fig. 3.3

The following information is typically recorded: utility, material, size, depth, condition, soil type, water table, etc. See Figure 3.3

The proposed facility can now be designed by the project engineer so as to minimize the impact to existing utilities. Where utility relocations are required, informed decisions can now be made as to how this should be achieved.
See Figure 3.4

Final Design: Plans, Specifications & Estimate (P, S & E): The Subsurface Utility Engineer shall review the PS & E submission. The Subsurface Utility Engineer shall check that all utility information collected is correctly indicated on the contract documents. The Subsurface Utility Engineer shall add to the specifications, and/or special provisions, any information regarding existing utilities which will facilitate construction of the project and protect the interest of the owner.

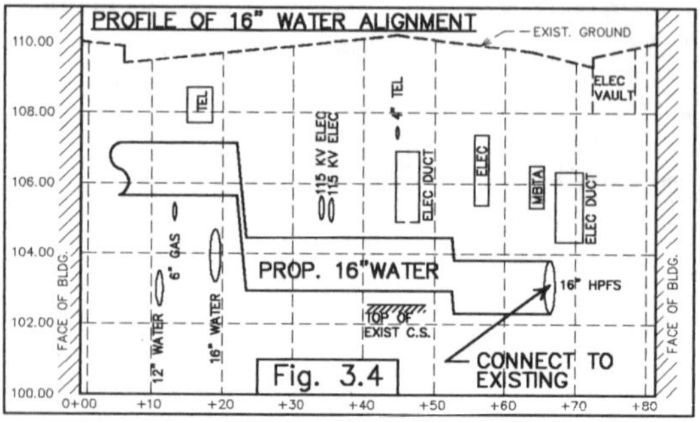

Fig. 3.4

Costs and Savings

SUE, when used as outlined above, costs approximately one percent (1%) of total project cost or ten percent (10%) of design cost. The Federal Highway Administration having reviewed the use of SUE on many projects estimates cost savings to be of the order of seven dollars ($7) for every dollar ($1) spent on SUE.

These savings are based primarily on reduced change notices. The savings realized from reduced public inconvenience, risk of injury or loss of life, and liability risk, are difficult to quantify but are obviously significant.

Conclusion:

SUE provides a method to collect and depict the exact locations of underground utilities on contract documents. Electronic Utility Location techniques provide approximate horizontal location and Non-Destructive Air-Vacuum Excavation techniques provide exact horizontal and vertical location. This information when presented on contract documents reduces overall project costs, delays, public inconvenience and liability risk. Consequently, SUE is an important part of Trenchless Technology.

The Application of Ground Penetrating Radar Technology As Part Of Geotechnical Assessments of Proposed HDD Pipeline River Crossings

Paul M. Tarrant[1]

Abstract

Ground Penetrating Radar (GPR) technology has been used as a geological exploration tool at over 100 planned or proposed Horizontal Directional Drilled (HDD) crossings throughout western Canada, the United States, and abroad. Recent advancements in the design and performance of ground penetrating radar systems have resulted in the ability to profile shallow soil and/or bedrock stratigraphic conditions within the upper 30 to 50 meters from ground surface. PulseEkko ™ GPR instrumentation was used to acquire geophysical information at the crossings. Data was recorded in a digital format and processed on PC computing platforms using seismic based data enhancement routines. GPR data interpretations involved correlation's made with available geotechnical boring data, and geological observations taken in the field. A case study example is presented to illustrate the capability of GPR to detect many subsurface geologic features which, if left undetected, may contribute to higher than anticipated construction costs for HDD pipeline river crossings.

Introduction

Geotechnical investigations at planned or proposed pipeline river crossings play a major role in the success or failure of many horizontal directional drilling (HDD) construction projects. The most common practice in defining geologic conditions at proposed river crossings involves drill core sampling. Drill core results provide standard soil classifications, standard penetration tests (blow counts), sieve analysis of granular material, and unconfined compressive strength of cohesive soil and rock samples (ASCE Manuals and Reports on Engineering Practice No. 89, 1996)[i]. Although this information assists in defining many proposed HDD parameters, it is based on narrow column samples only, and cannot provide reliable assumptions of the lateral continuity of geologic events throughout the length of a proposed drill path. Undetected changes in geologic conditions between drill core sample locations may result in unanticipated construction cost

[1] Vice President, Surface Search GPR Technologies Inc., Suite 700, 700 - 6 Avenue SW, Calgary, AB, T2P 0T8

overruns, or in extreme cases, HDD failure. One solution to this problem is to add more drill core samples along the drill path. However, this may not be practical due to undesirable increases in drilling costs, and/or drill rig access restrictions. Another method then, is desired to supplement the coring data to better define geologic conditions which will be encountered along the proposed HDD drill path. This method must be portable, cost effective, and provide a reasonable amount of stratigraphic detail to detect changing soil or bedrock conditions.

Geophysical techniques have long been used in oil and gas, environmental, mining, and geotechnical engineering fields to provide laterally continuous information between intrusive sampling (e.g. drill core) techniques. The more prevalent, or established geophysical approach to providing continuous profile information between drill core samples is seismic reflection and refraction analysis. Shallow seismic methods are relatively cost effective and portable, yet vertical resolution of the data is often limited to 3 - 4 meters (Jol, 1988[ii]).

Recent advancements in ground penetrating radar (GPR) technology have provided an alternative method to seismic techniques for mapping shallow soil and rock stratigraphic conditions between drill core locations for most river crossing settings. GPR systems are portable, require minimal man power to operate, and provide increased vertical data resolution (decimeter to meter scale) over seismic methods (Jol and Smith, 1991)[iii]. GPR is capable of detecting changes between sands, silts, clays and gravel (Killey and Annan, 1989)[iv], profiling shallow bedrock interfaces (Davis and Annan 1989)[v], identifying the location of voids (Morey, 1974)[vi], and locating buried man made hazards such as utilities, landfills and archaeological grave sites (Unterberger, 1992)[vii].

Method

GPR Concepts and Principles

Ground penetrating radar involves the impulse transmission of radio waves (1 to 1200 MHz) into the subsurface, and the subsequent reception of any reflected energy back to the surface (Figure 1A. & 1B). Much like seismic or sonar systems, GPR measures the elapsed time from when a signal is transmitted into the ground, then reflected back to the surface, and finally, detected by receiver electronics. Radar impulse transmittal and reception is recorded in nano seconds (10^{-9} sec.) and the data can be sampled as fast as every 800 pico seconds (10^{-12} sec.). Transitions within the bulk electrical properties of the soil or rock medium at depth are what gives rise to radar wave reflections. These variations are typically associated with abrupt changes in predominant soil grain size, porosity, compaction, water content, bedrock lithology, as well as the presence of buried man-made objects (e.g. metals, cement, and rubbish).

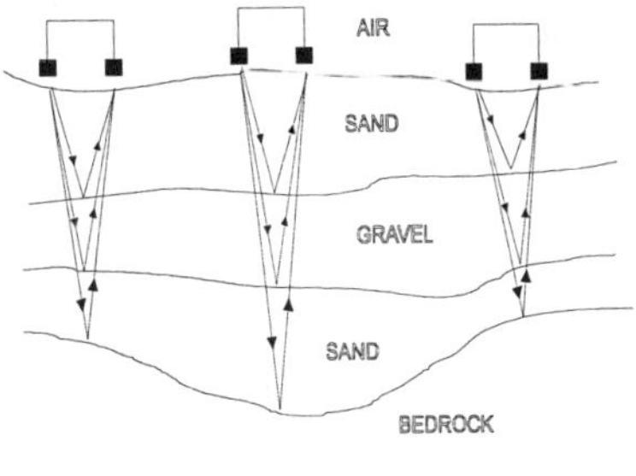

Figure 1A. Conceptual illustration of GPR reflection profiling through heterogeneous soils over bedrock (Modified From Davis and Annan, 1989).

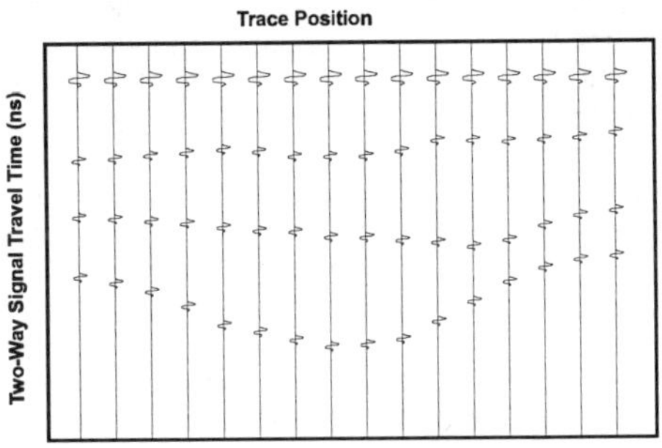

Figure 1B. Resulting radar profile over the idealized stratigraphic sequence depicted in Figure 1A., (Modified From Davis and Annan, 1989).

The degree to which radar signals penetrate the subsurface is controlled primarily by the measure of electrical conductivity associated with the subsurface conditions encountered. Highly conductive mediums such as clay soils, or salt water for example, have the undesired effect of almost completely absorbing GPR signals, whereas relatively resistive mediums (e.g. clean, quartz rich sand and gravel) will allow for detectable signal returns of up to 50 meters in depth (Jol and Smith, 1991)[viii]. Signal depth penetrations are also affected by the antennae frequencies selected for a survey. Higher frequency systems (i.e. 250 - 1200 MHz) will experience higher rates of signal attenuation compared with low frequency systems (i.e. 200 - 12.5 MHz). For GPR applications requiring penetration depths in excess of 5.0 meters, low frequency systems are, as a rule, better suited over high frequency systems (Annan and Cosway, 1994)[ix]. More detailed descriptions of GPR concepts and principles are documented in papers by Davis and Annan (1989)[x], Ulriksen (1982)[xi] and Daniels (1989)[xii].

Proposed HDD River Crossing GPR Surveys

At over 100 proposed HDD pipeline river crossings throughout Canada, the United States, and abroad, PulseEkko™ GPR systems have been used for geologic exploration in conjunction with drill core sampling programs. The primary objectives of the GPR surveys were to provide in-fill soil stratigraphy and bedrock elevation information between drill core sample locations.

PulseEkko™ GPR systems complete with low center frequency antennae arrays ranging between 12.5 and 200 MHz were used to conduct the surveys. The modular configuration of the antennae allowed for increased instrument portability through rough terrain (e.g. hand-cut lines within forested areas) and were operated by 2 to 3 field personnel. The systems were controlled using portable notebook computers which displayed the digitally acquired data in real-time.

For each radar survey, instrument parameter tests were completed prior to data acquisition to ensure that adequate signal resolution and depth penetration was being achieved. Of primary concern was the selection of the specific antennae frequencies for data acquisition. As previously mentioned, lower frequency antennae (i.e. 12.5, 25, and 50 MHz) provide the greatest signal depth penetration capabilities, although the lower the frequency, the more vertical resolution within the data is compromised. Antennae frequencies greater than 50 MHz provided better near surface resolution, and were capable of detecting shallow targets such as buried pipelines for example, but did not provide signal penetration depths desired for most of the proposed HDD river crossings.

Once survey parameters were established, GPR data profiles were then collected over center line of the proposed HDD alignments. Data acquisition consisted of taking readings at pre-determined station intervals (usually every 0.5 to 1.0 meters) over centerline of the proposed drill paths. On land, data was collected by positioning the antennae on the ground by hand prior to the transmission of a stacked array of electromagnetic radio wave impulses into the subsurface. Signal reflections detected by the receiver antenna were transmitted by fiber optic cables back to the operator, and the data was observed in real-time on the computer screen and also stored in a digital format. Since radio waves propagate through the subsurface between 1/2 to 1/3 the speed of light, the time required to collect a GPR reading was virtually instantaneous. Survey lathe positioned at 25.0 meter intervals, or at topographic inflection points prior to the radar survey, were used as position control reference. This process was extended across water bodies that could be waded on foot, and during the winter months, over ice.

Over open water, the GPR data was collected by mounting the antennae off the sides of a variety river craft (i.e. river jet boats, inflatable zodiacs, pontoon boats, etc.), and data was recorded continuously whilst moving across the river.

Survey position control on the water was accomplished by taking distance measurement readings between the boat and a fixed prism location on the shoreline using a laser range finder. This method provided only values along the x-axis of the proposed alignment, so radio communication was used between the boat operator and personnel on shore to ensure that boat positioning was maintained as close as possible to center line of the alignment across the river. For large water bodies, Differential Global Positioning Satellite (DGPS) receiver systems linked with the radar instrumentation were used to allow for more accurate position control during data acquisition.

Data Processing and Interpretation of Results

The GPR data was processed on PC-based computing platforms using geophysical image enhancement software similar to seismic processing programs. The data was typically displayed in a graphical format known as "variable area trace wiggle format" (also similar to seismic data). One advantage to viewing the radar profiles in real-time in the field, as well as having PC-based processing software, was that the data could be reviewed immediately on site, or in a motel room to ensure desired survey objectives were being satisfied.

Once the data was processed, it was then plotted to hard copy and the interpretation process was initiated. If drill core sampling had not been completed at a crossing prior to the GPR survey, then a preliminary interpretation of the results was provided to the client, and used to recommend drill core sample locations. Upon receipt of the drill core logs, a more detailed intepretation of the

GPR data was completed, and observed subsurface reflection events were classified, then mapped across the length of the crossing.

The final presentation of the survey results included elevation values for all interpreted sedimentary soil and bedrock interfaces from the radar data in correlation with the drill core log information. These results were plotted onto cross-section schematic drawings which would later be used to overlay proposed HDD path geometries.

Case Study Example

Toutle River, South Washington State

A GPR survey was conducted at a proposed HDD crossing of the Toutle River approximately 15 kilometers north east of the city of Longview in southern Washington. The head waters of the Toutle River emanate from Mt. St. Helens, and through recent geologic history, the river has experienced dramatic fluctuations in sediment load.

At the time of the radar survey an unsuccessful attempt at drilling an HDD pilot hole across the river had already been made. The survey objectives were to map bedrock elevations throughout the length of the crossing, and to profile the soil stratigraphy overlying bedrock. Six core logs were completed in advance of the HDD attempt at various locations along the alignment and were used to correlate against the GPR results.

A cross section diagram of the proposed crossing is shown under Figure 2. The ground elevation profile consists of a relatively flat floodplain area to the north of the river, and a steep incline up a mountain slope immediately south of the river. Of primary concern to the drilling contractor was the determination of the bedrock elevation profile beneath the river channel extending onto the south bank where drill core information was unavailable. The drilling contractor also wanted to know if the soil stratigraphy identified by the core logs extended across the river and below the south bank. The location of the drill cores and the extent of the GPR data profile are highlighted on Figure 2. along with the soil and bedrock information taken from the drill core logs.

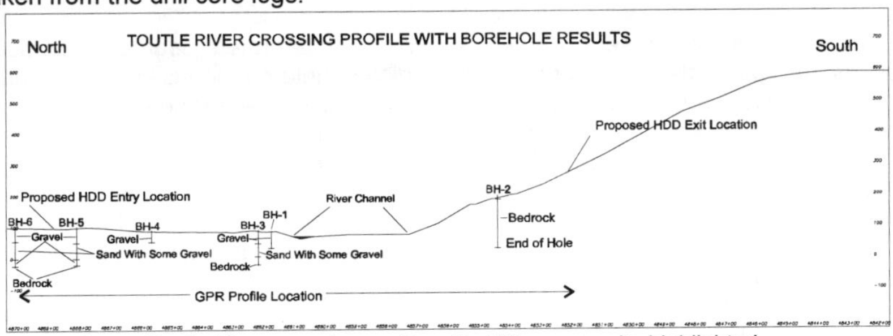

Figure 2. Toutle River crossing profile with drill core results highlighted.

A GPR data profile approximately 500 meters in length was acquired at the crossing using a PulseEkko IV™ GPR system. The profile was collected over center line of the proposed horizontal directional drilling path using 12.5 MHz center frequency antennae. GPR soundings were taken at 1.0 meter intervals, and the antennae were positioned on ground by hand. Data acquired over the river channel was accomplished by wading the radar instrumentation across the water surface.

The processed and interpreted GPR data profile is displayed under Figure 3. The gap in the profile is due to inaccessible terrain conditions on the north side of the river where river bank erosion control structures were situated.

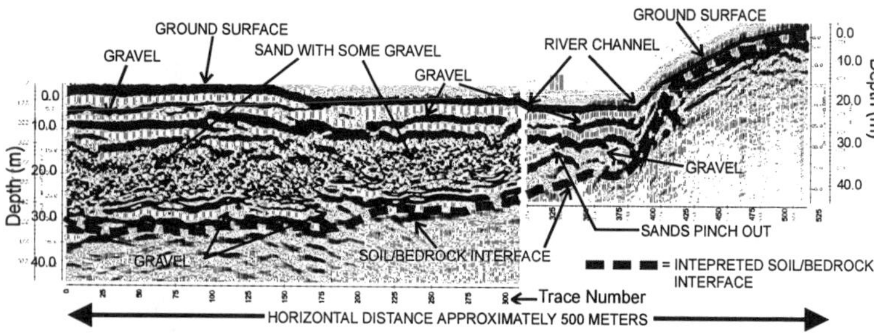

Figure 3. Ground penetrating radar data profile acquired at the Toutle River crossing with interpretation of soil stratigraphy and soil/bedrock interface overlain.

The stratigraphic interface between bedrock and overlying alluvial sediments is highlighted on the GPR data profile. Below this event the radar signals are observed to be severely attenuated which is interpreted to indicate a transition between alluvial sediments and bedrock. This event was then digitized, and an elevation profile was generated. The correlation between the depths calculated for the interpreted bedrock interface from the GPR data, and the bedrock depths encountered in the drill core results was to within 1.0 meter which enabled a reasonable confidence level to be placed in the overall bedrock interpretation.

The radar signal response observed above the interpreted bedrock interface contains multiple reflection events which indicate heterogeneous soil sediment conditions. All six drill cores encountered sand and gravel sequences overlying bedrock. High amplitude radar reflection events can be seen immediately below ground surface, and immediately above the bedrock contact. This data signature is interpreted to represent the radar's response to gravel sediments. Between the gravel sequences, the GPR data is observed to be chaotic, and signal reflections are laterally discontinuous. This region is interpreted to indicate the presence of sands with some gravel present. Note that within the river channel this sequence is interpreted to pinch out into gravel deposits. Figure 4. illustrates in cross section the final interpretation of the GPR results correlated with the drill core information.

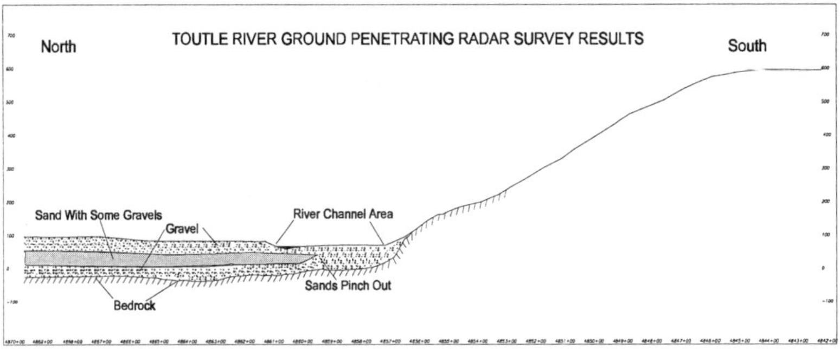

Figure 4. Interpreted ground penetrating radar results which illustrate soil stratigraphy and bedrock elevations along the proposed HDD alignment.

As mentioned earlier, a pilot hole attempt was made prior to the GPR survey which resulted in failure. The drilling contractor used the available drill core results in determining the HDD drill path geometry and presumed that the thick sequence of sand deposits identified by the core logs existed across the floodplain as well as below the river. Unexpectedly, gravel deposits were encountered within this region causing insurmountable complications.

The next attempt to drill the crossing would involve casing the entry hole down to bedrock on the floodplain side of the river, and maintaining a drill path within rock through to the exit point. Since drill core information was not available (or feasible) within the river channel or on the south shoreline of the river, the bedrock elevation profile from the GPR survey became of extreme importance in this area. The HDD effort would again be compromised if bedrock depths greater than 15.0 meters were observed on the south side of the river.

The interpreted results of the GPR survey revealed that bedrock elevations did not rise gradually across the river, but rather, remained relatively deep up to a location approximately 10.0 meters from the water's edge on the south bank of the river. From that point, bedrock elevations rose dramatically over a short distance to depths of less than 2.0 meters. Although this information indicated that maintaining a drill path geometry through bedrock within the river channel and onto the south bank was feasible, the drilling contractor was required to make a last minute adjustment to his previously desired drill path exit angle in order to avoid risking contact with unconsolidated alluvial sediments (i.e. coarse gravel) below the river. The subsequent drilling attempt was successful and the pipeline was installed without any further complications.

References

[i]Pipeline Crossings, ASCE Manuals and Reports on Engineering Practice No. 89, Prepared by the Task Committee on Pipeline Crossings of the Technical Committee on Pipeline Crossings of the Pipeline Division of the American Society of Civil Engineers, American Society of Civil Engineers 1996.

[ii] Jol. H.M. 1988. Seismic stratigraphic analysis of the southeastern Fraser River delta, British Columbia. M.Sc. thesis, Simon Fraser University, Burnaby, B.C.

[iii]Jol, H. M., Smith D.G., 1991 Ground Penetrating Radar of Northern Lacustrine Deltas, Canadian Journal of Earth Sciences, Vol. 28, pp. 1939-1947.

[iv]Killey, R. W. D., and Annan, A. P., 1989, Stratigraphic information from impulse radar profiling over unconsolidated sands, *in* Thomas, M. D., and Dix, D. F., Eds., Geophysical and related geoscientific research at Chalk River, Ontario: AECL-9085, Atomic Energy of Canada Ltd., 295-306

[v]Davis, J.L., and Annan, A.P., 1989 Ground Penetrating Radar for High-Resolution Mapping of Soil and Rock Stratigraphy, Geophysical Prospecting, 37 p. 531-551.

[vi]Morey, R. M., 1974. Detection of subsurface cavities by ground penetrating radar. Proceedings of the Highway Geological Symposium, 27: 28-30.

[vii]Unterberger, R. R., 1992 Ground penetrating radar finds disturbed earth over burials, Fourth International Conference on Ground Penetrating Radar, Geological Survey of Finland, Special Paper 16, 351-357.

[viii]Jol, H. M., Smith D.G., 1991 Ground Penetrating Radar of Northern Lacustrine Deltas, Canadian Journal of Earth Sciences, Vol. 28, pp. 1939-1947.

[ix]Annan, A. P., and Cosway, S. W., 1994, GPR Frequency Selection, Proceedings of the Fifth International Conference on Ground Penetrating Radar, Vol. 2. 747-760.

[x]Davis, J.L., and Annan, A.P., 1989 Ground Penetrating Radar for High-Resolution Mapping of Soil and Rock Stratigraphy, Geophysical Prospecting, 37 p. 531-551.

[xi]Ulriksen, C. P. F. 1982. Application of Impulse Radar to Civil Engineering. Ph.D. thesis, Lund University of Technology, Lund, Sweden.

[xii]Daniels, J.J. 1989. "Fundamentals of Ground Penetrating Radar" Proceedings of the Symposium on the Application of Geophysics to Engineering and Environmental Problems (SAGEEP), March 13-16, Society of Mining and Engineering Geophysics, Denver, CO, pp. 64-142.

Condition Assessment of Sewer Systems

Tom Iseley, P.E., Ph.D., M.ASCE[1]
Dulcy M. Abraham, Ph.D., AM. ASCE[2]
Sanjiv Gokhale, P.E., Ph.D., M. ASCE [3]

Abstract

Many municipal engineers are realizing that improvements in public health and safety as well as cost benefits can be achieved with a proactive sewer management system. The objective of such a system is not merely the collection of more data regarding the condition of the existing sewer system network, but also the use of this data to make intelligent, documentable, cost effective decisions regarding the rehabilitation, operation, and maintenance of this enormous and ever expanding network. This paper presents state-of-the art concepts and technology for sewer system data acquisition, data interpretation and utilization of the data for planning for intelligent renewal of underground sewer system.

Introduction

The U.S. has an estimated $20 trillion investment in civil infrastructure systems. But many of these systems are eroding due to aging, excessive demand, misuse, mismanagement and neglect (National Science Foundation 1993, 1995). Historically, infrastructure rehabilitation has been prompted primarily by repairing failures rather than preventing them (U.S. Congress 1991, World Bank 1994). Due to their lack of visibility, rehabilitation of underground utilities is frequently neglected until a catastrophic failure occurs, resulting in difficult and costly rehabilitation (WEF 1994).

Traditionally the methods used to assess and interpret status and inventory condition of underground sewer systems have been inadequate because they were often based on after-the-fact (or post-damage/deterioration) information. The need for adding tools to complement the current assessment environment is apparent. These tools will enable up-to-date evaluation and the development of maintenance

[1] Executive Vice-President, Iseley Enterprises, Inc., 12727 White Rock Court, Indianapolis, IN 46236.
[2] Assistant Professor, School of Civil Engineering, Purdue University, West Lafayette, IN 47906.
[3] Professor of Construction Technology, Purdue University, School of Engineering and Technology at Indianapolis, IUPUI, IN 46202-5160.

plans. Regular assessment is essential to determine the general health of a structure, and to determine its repair, retrofit, rehabilitation, life extension/replacement requirements. Current sewer condition information available to asset managers is subjective, resulting in handicapped financial justification of rehabilitation work, except for gross defects. (Campbell, et al 1995). One area that has been identified as an instrument for early detection of problems is integrating sensing mechanisms with current inspection technology (primarily TV technology) (Kuntze, et al. 1995).

State-of-the Art in Infrastructure Assessment Systems

Physical assessments must be made to establish the best strategy for maintaining and rehabilitating the underground infrastructure. These physical assessments include: inventory data defining quantities, types, location of system components, and condition data describing the physical state of a facility or component (e.g., cracking, deterioration, leakage, loss of strength, etc.). In addition, performance data describing the ability of a facility or component to perform its intended function (e.g., capacity, down time, safety, etc.) must be established. High speed noncontact sensory techniques such as video imaging, infrared thermography, optical character and code recognition, ground-penetrating radar, laser interferometry, and terrain conductivity have shown significant promise in other areas of inspection (Maser, 1988). However, most of these technologies were "not developed for civil infrastructure and hence there are some difficulties in their direct application to actual civil structures (Feng, et al 1995)." Additionally, in order to determine optimal specifications for a condition sensing system, it is necessary to understand the impact of the collected data on the decision process (Maser 1988).

A large number of automated data collection technologies such as photographic and video imaging, laser, radar and infrared thermography have made it possible to collect data in shorter time for the analysis of infrastructure performance. However, the accuracies and precisions of these technologies are not well understood, and infrastructure managers and planners have been reluctant to adopt them in practice.

Detection of interior defects is generally the first warning of problems that could occur within sewer lines. At present, the state-of-the-art inspection systems for sewer lines are dominated by Closed Circuit TV (CCTV) technology. Remotely controlled vehicles carry a TV camera through the sewer pipes. Diagnosis of failures or defects depends on the experience, the capability and the concentration of the operator, and the reliability of the TV picture. For example, serious cracks hidden by mud are not recognized at the monitor, and harmless glares are declared as damages. Thus, detection of subtle defects and assessment of the degree of deterioration are particularly error prone, so the method is adequate only for detecting gross defects.

To make high quality decisions that are needed for optimal operation and maintenance of a sewer collection system, the human decision maker must possess substantial experience and training to be able to accurately select the most efficient and cost effective rehabilitation solution with the specific set of pipe defects. The options for sewer system rehabilitation have multiplied tremendously (see Figure 1 for a classification system), making it more difficult for the decision maker to

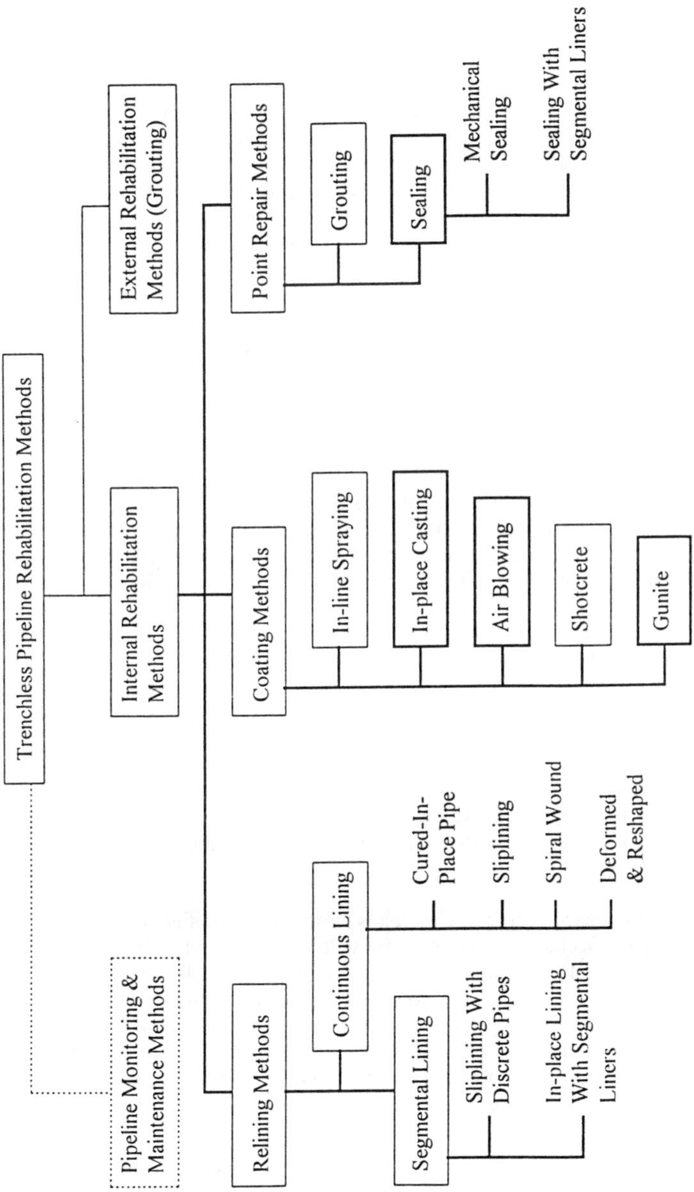

Figure 1: Sewerline Rehabilitation Methods

stay abreast of the evolving technologies and to select the appropriate technology for the identified defect.

Limitations of Current Sewer Inspection Technologies

While CCTV technology has advanced, the basic processes and results have remained unchanged. The technician has to identify pipe defects on a TV monitor located in a control vehicle (i.e., van, trailer, etc.) at the jobsite. He or she also attempts to classify each defect (i.e., corrosion, cracks, misaligned joints, root intrusion, etc.) according to the various levels of deterioration and risk. The data acquired from this process consists of video tape, photographs of specific defects, and record produced by the technician. The technician records can be produced manually or they can be automated. However, with either method, the quality of the data is dependent on the experience and skill of the technician. Typically, the field technician identifies and classifies each defect, and then submits these records to the responsible engineer, who evaluates the technician's records in the office to produce a recommendation regarding rehabilitation.

The potential weak link in this process is the judgment that the field technician must exercise when identifying and classifying pipe defects in the field. Many factors, such as experience, mental awareness, equipment capability, field distractions and interruptions, environment, etc., impact the judgment. Nevertheless, major decisions regarding rehabilitation methods hinge on this judgment (see Figure 1) Each decision has an associated cost. The owner deserves the highest quality decisions that provides him/her with efficient, accurate, and cost effective solutions.

Another limitation of the traditional CCTV technology is the difficulty in accurate establishment of budgets due to uncertainty in estimating field production. The time it takes to internally inspect a sewer pipe is dependent on the number and the degree of deterioration of the defects. At each defect, the operator must stop, assess, and record the condition. Therefore, it can be difficult to accurately estimate the time and cost to perform the CCTV work. Pressure to maximize the daily production demand can result in poor quality data being provided to the office engineer for decision-making, in which case the engineer will most often base his/her decision on a "worst case scenario." This can often result in costly and unnecessary fixes.

Inspection based on CCTV relies primarily on the use of a single sensing/assessment mechanism (namely, the video image). More reliable detection and diagnosis requires the introduction of several sensor signals which are "smartly" connected with each other. Recent work by a German industry-research institute team has resulted in a prototype for robot inspection system (Kuntze et. al 1995). The modular system concepts comprises a high resolution TV camera and a cluster of sensors which are capable of recognizing anomalies within the pipe during robot inspection. This system provides one basis for an automated assessment system. But the critical issue that governs the use of such systems in the field, is the ability to interpret the sensed information.

Sewer Scanner and Evaluation Technology (SSET)

The Sewer Scanner and Evaluation Technology (SSET) is a revolutionary break-through in pipeline inspection technology developed by TOA Grout, CORE Corp., and the Tokyo Metropolitan Government's Sewer Service (TGS) Company. This technology is the result of a three-year development process that began in 1994.

The Sewer Scanner and Evaluation Technology (SSET) is a flexible non-destructive evaluation data acquisition tool. The SSET removes the deficiencies of the CCTV to provide the engineer with more and higher quality information to make critical decisions regarding the most appropriate rehabilitation method. This is accomplished by utilizing scanner and gyroscope technology. The information produced provides the engineer the ability to see the total surface of the pipe from one end to the other. The scanned image is then digitized so that a color coded computer image can be printed out. Defects are illustrated by a designated color code. A written description of each defect is produced at the appropriate location along the pipeline. Figure 2 shows a plot of part of the distressed section of a sewer pipe. Horizontal and vertical deflections of the entire length of the line are also printed out. Such technology enhances the ability to automatically assess condition, eliminating subjectivity and inexperience. The system is being developed to suit the harsh environments and space constraints of live sewer lines.

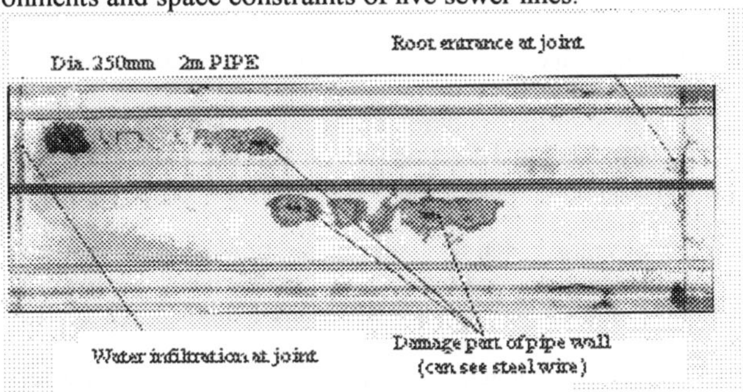

Figure 2: Sample image of sewer pipe

The survey machine houses three major components: (i) the scanner, (ii) the CCTV; and (iii) a three axis mechanical gyroscope. The current prototype has a diameter of 130 mm (5.2 in.) and a length of 850 mm (33.5 in) and weighs 25 kg (55 lb). The goal is to introduce into the market (in Spring/Summer 1997), a machine that is less than 90 mm (3.5 in) in diameter, 650 mm (25.6 in) in length and weighs less than 15 kg (33 lb). At present, the survey machine is equipped with three skids, and the body of the machine is rigid. It is expected that before the system is introduced into the market, the machine body will have two articulating joints to facilitate obstruction avoidance and traveling through bends (curved pipeline).

The mechanics of placing the SSET in the pipeline are similar to the CCTV inspection process. The SSET is designed to function with external winches to propel the machine through the pipe. Since the SSET utilizes state-of-the-art scanner technology, it can travel through the pipe at a uniform rate of speed. At present, this rate is 1.0 m (3.3 ft) per minute or 60m (197 ft) per hour. It is conceivable that a field crew could increase production up to four times current daily rates. Unlike conventional CCTV technology, the machine does not have to stop to allow the technician to assess each defect. All assessment will done by the engineer in the office. The technician is responsible only for proper equipment operation and acquiring the data, not interpreting the data.

Other components of the system consist of the necessary cable, control box, and monitors for the CCTV and scanner. The control box collects all data on a diskette. The information on the diskette is transferred to a central data processing location where it is processed, checked, printed, and developed into a professional report.

Benefits of the SSET

The major benefits of the SSET system over the current CCTV technology are:

a) The engineer will have more and higher quality data to make critical rehabilitation decisions. They will have the five levels of data to analyze on one printout. This will expedite the evaluation process and permit a more through comparative analysis.
b) Defects can be magnified, without the loss of resolution, to provide a clearer view.
c) Better documentation of the condition of the pipe is provided.
d) Horizontal and vertical pipe deflections are measured and illustrated graphically.
e) Field work is expedited because the technician is not required to assess the defects.
f) The defects can be color coded for better illustration and analysis.
g) Statistical data can be generated on the defects (i.e., the total area of pipe surface corroded can be determined and presented as a percentage of the total wall surface, etc.)

The costs of the SSET system will be comparable to that of the CCTV technology.

Automated Interpretation of Multi-sensory Data

Currently, the interpretation of the results of the scanned and videotaped images is done manually. Research is underway to automate the interpretation of the multi-sensory data. The interpretation system implements Artificial Intelligence techniques to automatically identify, classify and rate pipe defects. The task performed by the interpretation system is complex because pipe defects overlap and some complex surface geometry may be missing.

In addition to scanned and videotaped images, and data collected by the gyroscope, the following information will also be used in the assessment (Abraham, Iseley and Prasanth, 1996):

a) pre-diagnosis information such as intrinsic characteristics (age, sewer type, dimensions, depth),
b) working conditions (sewage nature, hydraulic regime),

c) environmental characteristics, including surface environment (traffic, works), soil environment (geology, hydrogeology)
d) history: construction mode (in trenches, trenchless),
e) difficulties during the construction.

Proposed Evaluation Process for the SSET

Reliable estimates claim that there are approximately 800,000 miles of underground sewer lines which are owned and operated by over 17,000 separate sewerage authorities. Additionally, there are at least 40,000 industrial facilities in the U.S. which have approximately 100,000 miles of sewerage and process water lines.

According to the 1996 Trenchless Technology Directory, there are 116 firms which offer CCTV services. Projections shown that there are over 1200 CCTV units operating in the USA. Each unit can televise approximately 50 to 70 miles per year. If one third of the units are actively in use everyday, then 24,500 miles of sewer are being televised every year. Some cities have adopted a program where they clean and televise all lines every five or ten years for preventive maintenance and infiltration reasons. Even assuming a very conservative plan of 22,500 miles of televised sewers per year as the expected annual USA market, the minimum CCTV contracts would amount to $148,500,000 per year (assuming a average cost of $1.25 per foot or $6,600 per mile).

Closed-circuit TV based assessment, which has been the norm, is prone to errors and subjective. Usually, such assessment has been used to detect large/gross defects, and not subtle ones. The SSET technology is capable of detecting finer defects, making the inspection data more reliable and objective. This will be very beneficial in the reliable prediction of system deterioration and the criticality of the deterioration. Before this technology is introduced in North America, it will undergo extensive evaluation.

The Trenchless Technology Center (TTC). has entered into a memorandum of understanding with the American Society of Civil Engineers (ASCE) Civil Engineering Research Foundation (CERF) to provide a cooperative mechanism for the evaluation of innovative trenchless technologies. In January 1996, CERF established CEITEC (Civil Engineering Innovative Technology Evaluation Center) to assist both the owners of new technology and the potential users of the technology. The SSET system will be the first technology to be evaluated by the ASCE-CERF/CEITEC process.

As part of the CEITEC process, CERF developed a proposal that was sent to over 100 municipalities inviting them to participate in a nationwide, comprehensive new technology evaluation program. Several have already sent commitments and others have expressed strong interest. The initial focus group will consist of 10-12 organizations around the U.S. Each participating municipality will contribute $20,000. For this fee, each municipality will get 10,000 LF of sewer lines inspected with the SSET system. A report will be generated for each municipality and distributed to the evaluation panel. Each municipality will have a representative on the evaluation panel.

The Trenchless Technology Center (TTC) will assist CERF with the evaluation process. TTC representatives will be responsible for conducting the evaluations at each site and producing a report . This formal, non-biased, third party evaluation process will aid and expedite the introduction of new and innovative technology. It will eliminate the need for each municipality having to conduct their own evaluation of new technology, and provides municipalities with a cost effective way of determining what works and what does not.

Conclusions

Current sewer condition information available to asset managers is generally subjective and often unreliable. In the current economic climate, most water and sewer authorities are seeking maximum efficiency, and reliable data to justify their maintenance programs. SSET is a state-of-the-art technology, that can provide quality data on sewer condition including: pipe geometry, vertical and horizontal deflection, structural deficiencies, location and extent of defects (both gross and minor), and an automated analysis and rating of the overall integrity of the sewer line. This information will assist engineers and asset managers in making a more reliable and cost effective determination of the repair and rehabilitation needs for the scanned pipeline. This will result in a more efficient allocation of scarce resources.

References

Abraham, D. M., Iseley, T., and Prasanth, R. K. (1996). "Integrating Sensing Technologies for Underground Utility Assessment," *Under preparation for presentation at the ASCE Conference on Infrastructure Condition Assessment*, Boston.

Campbell, G., Rogers, K. and Gilbert, J. (1995). "Pirat - A System for Quantitative Sewer Assessment," *International No-Dig*, Hamburg, Germany, September, 455-462.

Kuntze, H. B., Schmidt, D., Haffner, H., and Loh, M. (1995). "KARO - A Flexible Robot for Smart Sensor-Based Sewer Inspection," Proceedings of the International NO DIG '95 Conference, Dresden, Germany, September, 367-374.

Maser, K. R. (1988). "Inventory, Condition, and Performance Assessment in Infrastructure Facilities Management," *ASCE Journal of Professional Issues in Engineering*, 114(3), 271-280.

National Science Foundation (1993). *Civil Infrastructure Systems Research: Strategic Issues*, A Report of the Civil Infrastructure Systems Task Group, January 1993.

National Science Foundation (1995). *Civil Infrastructure Systems: An Integrative Research Program*, Program Announcement and Guidelines, National Science Foundation, March 1995.

U.S. Congress (1993) . Office of Technology Assessment, "Delivering the Goods: Public Works Technologies, Management and Finance," *OTA-SET-477*, U.S. Government Printing Office, Washington, D.C.

WEF (1994). *Existing Sewer Evaluation and Rehabilitation*, ASCE Manuals and Reports on Engineering Practice, No. 62, Alexandria, Virginia.

World Bank (1994). *World Development Report 1994: Infrastructure for Development*, Oxford University Press, New York, NY.

Valley Center Pipeline Tunnel

Jeff Moncrief [1] and Richard Trembath[2]

Abstract

This paper describes the San Diego County Water Authority's 165 meter long 1680 mm diameter pipeline tunnel under Interstate 15 in San Diego County as shown on the Location Map. Constraints imposed by the California Department of Transportation included the requirement for a trenchless crossing and minimal construction activity within the right of way. To reach competent material 12 meters beneath the freeway, a 10.5 meter vertical shaft was required on the west side of the freeway. Tunnel initial support consists of rockbolts, welded wire mesh and shotcrete, or ribs and lagging. The tunnel is expected to be generally in slightly weathered to fresh rock. Thickness of the steel pipe is 19 mm.

Introduction

The San Diego County Water Authority (Authority) currently operates two aqueduct systems importing water from Lake Skinner in Riverside County into San Diego County. The Authority supplies water on a wholesale basis to approximately 2.8 million people in San Diego County. The proposed Valley Center Pipeline will provide a 1680 mm diameter 6,520 meter long crossover pipeline between Pipeline 4 on the Second Aqueduct to Pipelines 1 & 2 on the First Aqueduct as shown on Fig 1. This pipeline will improve the operational flexibility of the system. The pipeline crosses Interstate 15 (I-15), a busy four lane freeway. The California Department of Transportation (CalTrans) required a trenchless crossing with minimal construction activity within their right of way. The main activities involved in the construction of the tunnel include excavation of the support of the portal and shaft, excavation of the

[1]Jeff Moncrief, Senior Civil Engineer, San Diego County Water Authority, 3211 Fifth Avenue, San Diego, CA 92103, (619) 682-4100, Fax (619) 692-9356.

[2]Richard Trembath, Senior Project Manager, Parsons Engineering Science, 9404 Genesee Avenue, Suite 140, San Diego, CA 92037 (619) 453-9650, Fax (619) 453-9652.

tunnel and installation of an initial support system, followed by installation of the steel pipe final lining in the tunnel.

Geotechical Investigation

Four wireline core boreholes were drilled to a depth of 12 to 23 meters along the tunnel alignment. Laboratory rock testing included unconfined compression and bulk density tests on selected rock core samples (Woodward-Clyde, 1992). In addition, a thin section petrographic analysis was performed on a core sample for investigation of corrosion potential. Also, three seismic refraction surveys were performed parallel to I-15 and one seismic line was made parallel to the tunnel alignment.

Site Geology

The tunnel penetrates the Cretaceous age granitic rocks of the southern California batholith. Surface outcrops of these rocks are exposed in man-made cuts near the alignment. The bore holes detected road embankment fills, highly weathered rock and slightly weathered to fresh rock. The portal on the east end and the shaft at the west end of the tunnel are expected to encounter minor amounts of fill, highly weathered rock, and slightly weathered to fresh rock. The tunnel is expected to be mainly within the slightly weathered to fresh rock except for approximately 60 meters at which point slightly to highly weathered rock is anticipated.

Regionally the granitic rocks in the tunnel vicinity have been classified as granodiorite to tonalite. The rock type encountered in the core borings and observed in the outcrops is primarily a tonalite. As observed in the recovered rock core samples, most of the rock mass discontinuities in the slightly weathered to fresh rock have narrow to very narrow joint apertures. The Rock Quality Designation (RQD) values below the weathered rock zone are generally greater than 50 percent, and for the most part are between 75 and 100 percent, an indication of good quality rock.

Portal and Shaft

To minimize the extent to which the tunnel which would be in weathered rock it was established that the crown of the tunnel should be approximately 12 meters beneath existing grade. This requires a 10.5 meter vertical shaft at the west end of the tunnel. At the east end, the portal was excavated to a depth of 10 meters from existing grade to the floor of the tunnel.

Adequate work room is required for equipment set up and to provide for the straight run of 12 m pipe lengths. The portal side slopes in soil are 1.5 to 1 and in rock vertical. In backfilling the portal, the existing ground slope cannot be replicated on a permanent basis and a geosynthetic reinforced embankment has been designed. The tunnel shaft at the west end of the tunnel connects to the tunnel 16 meters below existing grade. A vertical element of the pipeline 10.5 m long brings the pipeline up to an elevation to provide for cut and cover pipeline construction to the west. The shaft excavation will be conducted within a limited area within CalTrans right of way. A 45°

inclined shaft was bid as an alternative to the vertical shaft as shown in Fig 1. The construction contractor (Rasmussen) has elected to construct the vertical shaft.

Initial Liner and Tunnel Excavation

The tunnel will be constructed by conventional drill and blast methods for the majority of the alignment. At 2100 mm wide and 2300 mm high the tunnel will be of a size which will allow mucking by conventional load-haul-dump equipment. The initial support system consists of rockbolts with welded wire mesh and shotcrete. This system is shown on Fig 2. In addition, an initial liner system of steel ribs and lagging as shown on Fig 3 was designed and included in the construction drawings. This was included because there are portions of mixed face tunneling where crown bars or forepoling/spiling is required where slow to fast ravelling ground may be encountered.

Final Liner and Pipeline Installation

The final liner will be the 1680 mm diameter pipeline. The pipeline was designed on the basis of resisting all internal pressure in pipe hoop stress; no pressure load was carried beyond the pipe. The internal pressure head at the tunnel is 260 m of water column requiring 19 mm thick steel pipe (Parsons ES, 1991). All external loads are taken by the initial liner. The pipe is lined internally with cement mortar and externally with dielectric coating.

The pipes will be inserted into the tunnel in 12 m lengths and connected with an internal butt welded. Alternatively, the Contractor may elect to use bell and spigot joints with double fillet welds performed outside the tunnel, and then slide the pipeline into the tunnel on rollers as a unit. The space between the pipe and excavated surface will be filled with low density cellular concrete backfill.

References

[1)]Woodward-Clyde, Geotechnical Investigation Summary Report for San Diego County Water Authority's Interstate 15 Tunnel, July 1992

[2)]Parsons Engineering Science, Preliminary Design Report for Pipeline2A and Pump Station, October 1991

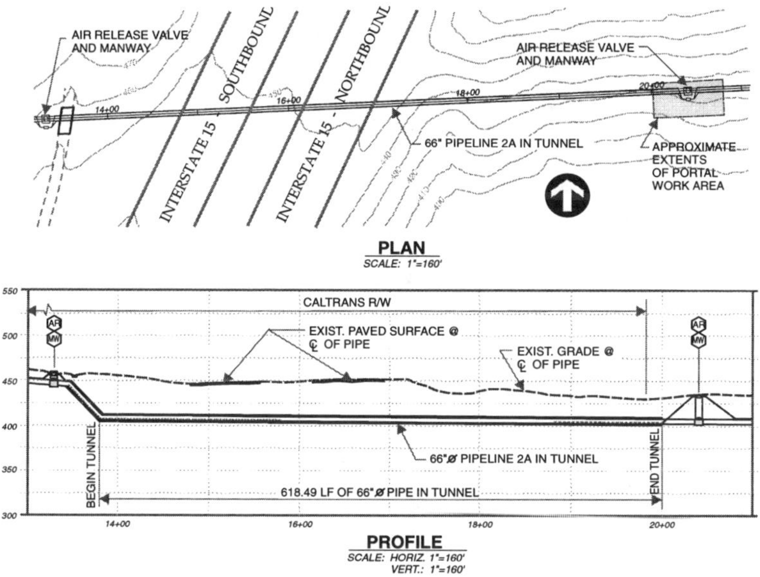

Figure 1
PLAN & PROFILE FOR VALLEY CENTER
PIPELINE TUNNEL

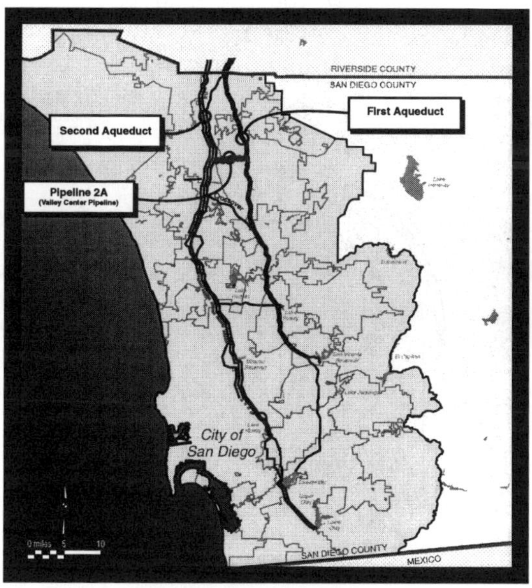

LOCATION MAP
VALLEY CENTER PIPELINE

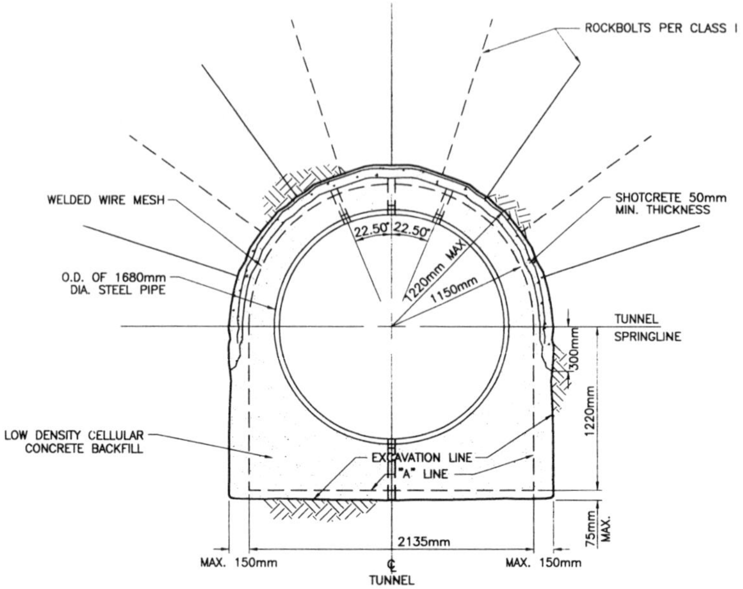

TUNNEL INITIAL SUPPORT
ROCKBOLTS W/WWM AND SHOTCRETE

FIGURE 2

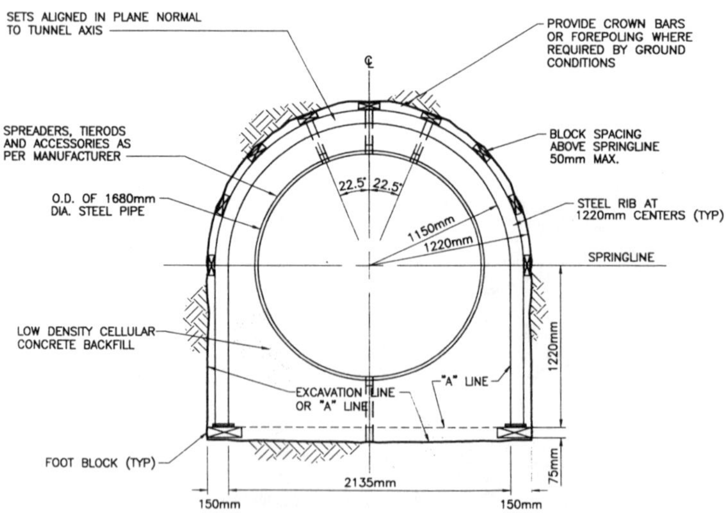

STEEL RIB OPTIONAL TUNNEL INITIAL SUPPORT

FIGURE 3

Trenchless Crossings for the MWRA Framingham Extension Relief Sewer Project

Robin Dill, P.E.[1]
Anthony Prata, P.E.[2]
Howard Goldberg, P.E.[3]

Abstract

The Massachusetts Water Resources Authority (MWRA) Framingham Extension Relief Sewer project involves installation of approximately 7.4km of 914mm diameter force main and 3.4km of 914 to 1524mm diameter gravity sewer pipeline through Boston's western suburbs. Installation of the pipe requires six (6) trenchless crossings beneath three roadways, a railroad, a 50 year old buried water aqueduct and an 810mm diameter white oak tree which has to be protected during construction. Steel sleeves, 1372mm in outside diameter, are being jacked to accommodate the 914mm diameter force main/gravity sewer pipelines in all cases except for the crossing under the oak tree, which requires an 1829mm diameter sleeve and will contain a 1219mm diameter gravity pipe. Crossings range in length from 18 to 60m. The jackings are primarily through saturated glaciolacustrine fine sands and silts. This paper will describe four out of six of the trenchless operations completed to date.

Introduction

The Framingham Extension Relief Sewer (FERS) facilities will provide needed additional sewer capacity and alleviate wastewater and storm overflow discharges to the Charles River. They will provide a critical link in Massachusetts Water Resources Authority's (MWRA) West Suburban facilities upgrade.

The FERS project includes five separate construction contracts consisting of a pumping station at the Framingham/Natick town line; a 7.4km long force main; a 3.4km long gravity sewer; and two existing sewer rehabilitation contracts.

---

[1] Vice President, Haley & Aldrich, Inc., 58 Charles Street, Cambridge, Massachusetts 02141

[2] Vice President, Anderson-Nichols & Company, Inc., 31 St. James Avenue, Boston, Massachusetts 02116

[3] Vice President, Bryant Associates, Inc., 77 North Washington Street, Boston, Massachusetts 02114

The design team is headed by Anderson-Nichols & Company, Inc. of Boston, with Bryant Associates, Inc. of Boston responsible for the gravity sewer contract, and Haley & Aldrich, Inc. of Cambridge responsible for geotechnical and hazardous materials investigations as well as underground engineering services for the entire project. All three firms also provided submittal review services and consultation to the MWRA throughout the construction phase. The MWRA is in charge of on-site inspection and construction management.

The design was completed in late 1995 and construction began in the summer of 1996. At present time, the construction is still in progress with a scheduled completion date of 1998. Four of the six trenchless crossings have been completed, all within the force main contract. This paper will focus on the four finished crossings, summarizing the subsurface conditions, explaining how the designers addressed various design and construction issues, and describing how construction was accomplished at each crossing.

Subsurface Conditions and Crossing Description

The force main contract consisted of installing 914mm diameter pipe with depths generally varying from 1.8 to 3.7m. The portion of the force main alignment indicating the location of the four crossings is shown on Figure 1. The pipeline passes through Framingham and Natick. The pipeline begins at a proposed pumping station in Framingham, crosses under active commuter/freight railroad tracks, then parallels the railroad alignment for approximately 3.8km before leaving the alignment and passing along roads through downtown Natick.

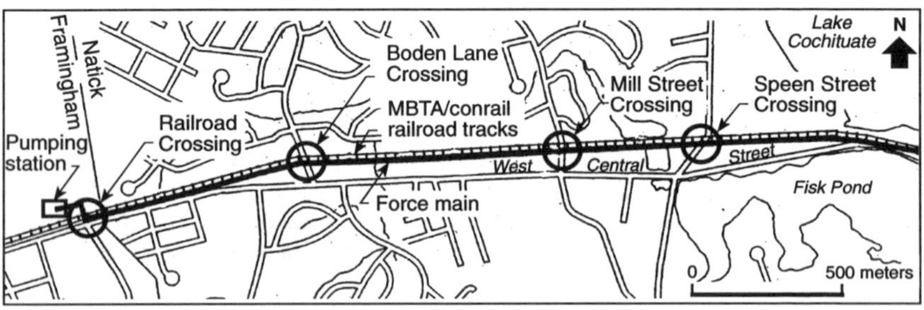

Figure 1. Project Location Plan

Routing the alignment along the railroad right-of-way had the advantage of avoiding traffic on existing streets, utilities and impact to adjacent businesses and properties. Three major town roads (Boden Lane, Mill Street and Speen Street) crossed over the railroad via bridges, with approach embankments to provide adequate vertical clearance over the railroad tracks.

To avoid the bridge abutments, it was necessary to divert the pipeline alignment away from the railroad tracks under the approach embankments. Since all three road embankments rose approximately 6 to 7m above surrounding grade and were major thoroughfares which could not be disrupted with open cut type construction, a trenchless crossing scheme had to be employed at each location.

The railroad is actively used by both Conrail and the Massachusetts Bay Transportation Authority (MBTA) for commuter traffic and freight hauling. The crossing under the tracks required a trenchless construction technique which had to adhere to requirements established by Conrail and the MBTA.

During the design phase a subsurface exploration program was performed to characterize soil and groundwater conditions along the alignment. The program integrated geotechnical and hazardous materials investigations to provide information which was used for both design and construction. At crossing locations a minimum of two borings, one on either side of the crossing, were completed. The subsurface investigations did not uncover any contamination within areas of proposed trenchless crossings. The geologic conditions along the railroad alignment were reasonably consistent for all four crossings. In general, three major soil strata were encountered below ground surface as described below.

Stratum	Typical Description
Fill	Loose to medium dense silty fine SAND with varying amounts of gravel.
Glaciolacustrine	Very loose to medium dense stratified fine SAND to medium stiff to stiff laminated SILT.
Glacial Till	Dense to very dense fine sandy SILT, to silty coarse to fine GRAVEL, some to trace coarse to fine sand, silt.

Groundwater levels were typically encountered at 0.8 to 2.4m below the general prevailing ground surface, and were found at or near the top of the glaciolacustrine stratum. The glacial till layer at crossing locations was found at depths below 6m, and hence did not impact either design or construction of the crossings.

Design Issues

Several different types of pipe were investigated for the project. For the pressure force main application, the two feasible options considered were ductile iron pipe (DIP) and prestressed concrete cylinder pipe (PCCP). Both required interior corrosion protection, with polyethylene and pvc liners determined to be the best suited for the DIP and PCCP, respectively. Since there was not any clear cut advantage of one system over the other, both systems were specified as Contractor's bid options, with the actual selection made by the Contractor after the bids were received.

After consideration for the difficulty of direct jacking DIP or PCCP without damaging the pipe or corrosion resistant liner, the design team decided that an oversized steel sleeve would be used at each crossing location. The carrier pipe would be installed within the sleeve. This proved not to be a significant issue since steel sleeve installation is a requirement by Conrail for crossings beneath their tracks. A 1372mm diameter steel pipe was selected as an appropriate size to accommodate either the DIP or PCCP options.

The horizontal alignment was established at the railroad crossing at approximately a 90 degree orientation to the tracks in order to minimize the jacking length, while also clearing the entire width of right-of-way as required by Conrail. For the road crossings, although there were no specific requirements established by local or state authorities, the casing/pipe locations were established so as to be far enough away from the bridge abutments to be out of the influence zones of the footings.

Figure 2 illustrates the generalized subsurface conditions at each crossing location.

Crossing	Meters				
	L	h_{min}	h_f	h_g	h_w
Boden Lane	50	2.6	6.4	2.0	1.8 - 2.7
Mill Street	49	1.5	7.3	0.9	1.2 - 2.0
Speen Street	52	1.7	7.8	0.2	1.7
Railroad	26	1.8	1.8	1.5	2.4 -4.0

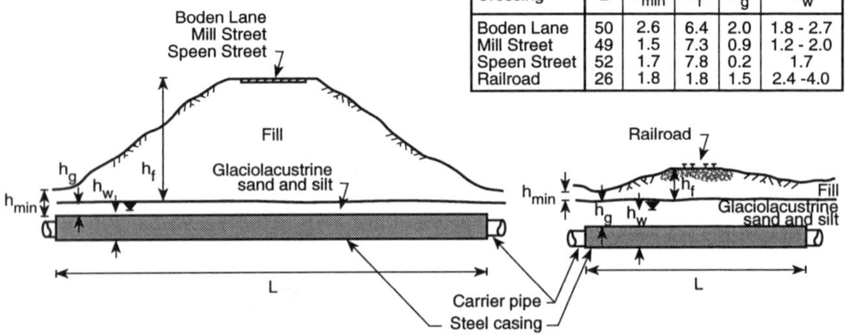

Figure 2. Subsurface Conditions at Crossing Locations

Average depth to design subgrade for the force main along the railroad alignment, independent of the crossings, was 2.1 to 3.7m below ground surface. This depth placed the pipe within the glaciolacustrine stratum and typically 0 to 2.4m below the groundwater table. During design, there was some flexibility to adjust the vertical alignment at crossing locations. Factors considered included satisfying a minimum cover criteria of 1.2m at jacking and receiving pit locations, and lowering the alignments where possible to place the sleeves in denser natural silt deposits and avoid potentially loose fill deposits. The sleeve alignments were established as shown on Figure 2 after some minor adjustments were made.

In all cases, the sleeves were anticipated to encounter saturated glaciolacustrine sand and silt. There was significant concern that a trenchless construction method would encounter difficulties in dewatering or otherwise stabilizing these soils if an open face jacking method were used. Consequently the geotechnical report and contract specifications focused on alerting the Contractor to the probable need for sophisticated predrainage dewatering methods and/or the use of chemical grouting of the fine sands. The contract documents also allowed the option to use a microtunneling or other closed face shield excavation method. Furthermore, the contract documents provided some minimum design requirements and required the Contractor to retain an experienced registered professional engineer to prepare detailed drawings and procedures for conducting the work.

Construction Phase

In April 1996, bids were submitted for the force main contract and after careful review, J. D'Amico Inc. was awarded the contract as low bidder. A specialty subcontractor, M&P Pipe Jacking Corporation from Newington, Connecticut, was chosen by J. D'Amico Inc. to perform all four of the trenchless crossings within the force main contract.

The Contractor's general approach to making the crossings consisted of jacking an open face shield and sections of steel Permalock casing, with excavation by clay spade. The custom-built jacking shield was designed to meet Cooper E-80 railroad loading and was further reinforced with 75mm by 25mm I.D. rings welded with gussets serving as brackets for steering jacks and bulkhead restraints. At the lead end of the shield a relief edge 12.5mm thick by 150mm wide was made part of the shield to facilitate steering and space for lubricant.

Figure 3. Lead End of Jacking Shield

The shield was also designed such that the excavated face could be breasted manually with wood lagging placed against bulkhead brackets as necessary if unstable soil conditions were encountered. The lead end was also fitted with a 25mm by 450mm horizontal breasting plate.

For stabilization of the soils the Contractor proposed to use sumps and/or deep well dewatering at jacking and receiving pits to draw down the groundwater table. Also, if unstable soils were encountered during jacking, the Contractor would implement chemical grouting using 25mm diameter grout pipes installed through a temporary wood bulkhead inside the shield. The pressure grouting would be performed ahead of the face in 3-m increments. The proposed grout mix consisted of sodium silicate, water and sodium bicarbonate.

Jacking pits were constructed using conventional interlocking steel sheetpiles and/or soldier piles and lagging, and interior bracing with sufficient space to accommodate the jacking frame and sections of Permalock casing. The proposed casing had a mechanical "press-fit" joint design and was supplied in sections typically 6m long and fitted with five grout ports per pipe section to allow bentonite lubrication during jacking.

Figure 4. Jacking Operation at Boden Lane

The use of a press fit joint was not in conformance with design specifications which required ASTM A 252 Grade 2 steel with beveled ends to be joined by continuous butt welding around the circumference. However, its use was allowed since the joint design was demonstrated to sustain tension loads and be leak proof when joints were sealed properly. Furthermore, the Contractor claimed a significant advantage in shortening the duration of each jacking cycle due to time savings in making press fit connections versus welded connections. Each joint was cleaned with a petroleum based solvent and treated with a silicone lubricant/sealant before the connection was made.

Steering of the shield was accomplished by adjusting each of the four 450 KN steering jacks as necessary within the shield. Line and grade measurements were obtained with a transit mounted near the back wall of the jacking pit, with monitoring for every 0.6m of sleeve advancement, or sooner, depending on the soil conditions.

Glaciolacustrine sands and silts were manually excavated at the face and loaded into a 0.4 cubic meter capacity cart with 125mm diameter rubber tires via a portable conveyor system. The cart with the excavated soil was pulled to the jacking pit with a hydraulically operated winch and then lifted out of the shaft with a large backhoe.

During jacking of the Permalock pipe, bentonite was injected around the exterior through 50mm diameter grout ports in order to reduce frictional forces so as not to exceed the jacking capacity of the four hydraulic jacks. At the completion of jacking each crossing, cement grout was pumped through the same ports to fill any potential voids. The grout pressures were monitored so as not to exceed surrounding overburden stress at the point of injection.

The receiving pits consisted of an open section of trench temporarily supported with a trench box on the sides, and a steel plate at the end of the trench where the jacked sleeve would be received. The steel plate was lifted when the sleeve reached its destination.

Figure 5. Mill Street Receiving Pit Showing Completed Sleeve and Backfilled Carrier Pipe

Once the entire steel sleeve was in place, the carrier pipe could be installed and the annulus backfilled with sand. This process started by welding steel braces onto the crown of the Permalock casing to prevent differential movement of the PCCP during backfilling. A 50mm diameter backfilling pipe was placed the full length of the sleeve near the crown, resting on the steel braces. Wood skids 0.9m long were then strapped onto the haunches of each section of PCCP as sections were added, to facilitate pushing the string of carrier pipe into the sleeve. Once the carrier pipe installation was complete, then brick bulkheads were constructed at each end of the sleeve. Next, sand backfill was blown into the annulus between the inside of the steel sleeve and exterior wall of the carrier pipe using the 50mm diameter pipe. This pipe was pulled back in 3m increments as the backfilling progressed until the entire annulus was filled.

Summary and Conclusions

Jacking and receiving pit construction began in July 1996 and M&P Pipe Jacking Corporation mobilized in August 1996 to begin jacking the steel sleeve. The actual sequence of construction was to first complete the Speen Street crossing, followed by Mill Street, Boden Lane and finally the Railroad crossing. A single shift operation was used for all crossings with the exception that a 24-hour weekend operation was performed at the Railroad crossing as required by Conrail. M&P completed the jacking, carrier pipe installation, sand backfilling and removal of equipment in 12 weeks. The advance rate for jacking steel casings averaged from about 0.6 to 1.1m per hour per crossing with a "best" production rate of 10.7m per 8-hour shift.

The jacking was completed successfully without any major incidents. Minor delays were experienced at only two of the crossings. At the Speen Street crossing, an unanticipated 300mm diameter water main was severed by the cutting edge of the shield, causing flooding of the casing and jacking pit, as well as a sink hole in front of the pit. After the main was shut off, the flooded excavation was pumped out and the line was relocated. Jacking was then successfully completed without further incident.At the Mill Street crossing, M&P encountered some boulders and cobbles in the fill, as well as some clay soils within the glaciolacustrine strata which slowed the excavation process. Some groundwater inflow also had to be dealt with during this jacking. Despite slow production, the trenchless crossing was made without any special means of soil stabilization being necessary.

The trenchless construction costs for the four crossings which had a combined length of 177 linear meters may be summarized as follows:

Item	Unit Cost	
	per linear foot	per linear meter
Steel Sleeve Installation (excluding pits and carrier pipe)	$458.00	$1502.00
Steel Sleeve Installation and Jacking/Receiving Pits (excluding carrier pipe)	$775.00	$2542.00
Complete Installation of Crossings-Steel Sleeves, PCCP and Jacking/Receiving Pits	$1013.00	$3323.00

In summary, the trenchless crossings were completed successfully and within schedule using fairly conventional open face pipe jacking methods chosen by the Contractor. In general the saturated fine sandy and silty soils along the sleeve alignment behaved somewhat better than anticipated by the designers and did not require any chemical grouting for stabilization during excavation, although dewatering was necessary at jacking and receiving pits. The costs for the trenchless crossings are considered very competitive in the New England marketplace.

Acknowledgments

The authors would like to acknowledge several people for their assistance with this paper. Maurice Plourde, president of M&P Pipe Jacking Corporation, provided detailed information on the construction aspects of the project. Edward Kinner and Tennyson Muindi of Haley & Aldrich, Inc. reviewed the draft manuscript and their useful comments were incorporated into the final publication.

Directional Drilling for Small Projects
A Resource for Meeting Schedule, Budget and Environmental Goals

Ms. Elizabeth Ellis, Member ASCE[1]
Mr. Tom Hargis, Member ASCE[2]
Mr. Iver Skavdal, Member ASCE[3]
Mr. Mike Wegener[4]

Abstract

Horizontal Directional Drilling (HDD) can be economical even for short drilled lengths of pipeline. The City of Petaluma, California, was required to relocate both the 20.3 cm (8-inch) sanitary sewer force main and water main as part of the demolition and subsequent reconstruction of the Payran Street Bridge over the Petaluma River. The relocation took place in late fall 1995 and was severely constrained by the fact that the sewer force main could only be taken out of service for 12 hours.

With the application of HDD, the two utilities were permanently relocated under the Petaluma River, which was flowing full, within seven days.

---

[1] Civil Department Manager, Winzler & Kelly Consulting Engineers, 495 Tesconi Circle, Santa Rosa, CA 95401

[2] City Engineer, City of Petaluma, 11 English Street Petaluma, CA 94952

[3] Region Manager, Winzler & Kelly Consulting Engineers, 495 Tesconi Circle, Santa Rosa, CA 95401

[4] Project Manager, Arrow Trenchless Technologies, 7307 Roseville Road, Unit 6, Sacramento, CA 95842

Actual tie-ins were completed in under eight (8) hours. Not only was the cost of this type of construction more economical than conventional cut and cover, but it allowed the City to comply with the requirements of the California State Department of Fish & Game. This Department does not allow excavation of any type in river channels from October 15 to April 1 in an effort to preserve water quality and biotic life in streams.

The decision to use the HDD method of trenchless technology took into consideration schedule, construction access, future river widening improvements, pipe diameter, traffic conditions, adjacent utilities and environmental concerns.

Project History

The Petaluma River HDD operation was the first successfully constructed segment of a $21 million Flood Control Project designed to mitigate frequent inundation of business and residential development in the City of Petaluma. The Flood Control Project, which took 15 years to develop is funded by a consortium of federal, state and local agencies, and consists of five distinct components:

- Replacement of the Payran Street Bridge
- Replacement of the Lakeville Street Bridge
- Demolition of two railroad bridges and replacement of one railroad bridge
- Construction of 244 m (800 ft) of flood control wall
- Construction of an on-site storm water management system for the water trapped on the dry side of the flood walls.

The HDD operation was the first phase of the Payran Street Bridge Replacement Project and evolved through a combination of technical, political, social and environmental concerns. While this was a small project, with the completed cost at $46,000, the use of trenchless technology proved to be practical and cost effective.

Technical Considerations

Petaluma's Payran Street Bridge Replacement Project required that the structure be replaced on its existing horizontal alignment. This necessitated complete demolition prior to beginning work on the new structure. The City street system allowed traffic to be diverted to other routes, alleviating the need to construct a temporary traffic crossing. However, the Payran Street Bridge also served as a crossing for a number of public and private utilities. From the City's viewpoint, the crossings of interest were the 20.3 cm (8-inch) water and sewer mains. Demolition of the Payran Street Bridge required at least temporary relocation of these utilities.

Three alternatives for relocation were analyzed.

- Temporary above ground river crossings, with permanent relocation in the new bridge structure
- Permanent below grade river crossings, constructed with conventional cut and cover technology
- Permanent below grade river crossings, constructed with directional drilling technology.

The temporary relocation option offered the advantage of permanent utility crossings within the new box girder bridge structure, which eased concerns about potential leaks (particularly of sewage) into the river and facilitated long term maintenance. The disadvantages included the relatively high cost of constructing both temporary and permanent relocations and, more importantly, the fact that the temporary crossing structures would impede the floodway of the river.

Permanent relocation using conventional construction techniques seemed to be the most feasible alternative. Both the water and sewer utilities are relatively small in diameter. The selected crossing was approximately 100 m (328 ft). If ductile iron pipe with concrete encasement was used, both pipes could have been placed in the same trench with little concern about leakage or cross contamination. The estimated cost of this type of construction was about \$490/m (\$150/ft). In addition, the technology is proven in this application, and the City has a considerable amount of experience with this type of construction.

Permanent relocation using HDD technology offered (the selected alternative) the advantage of locating all excavations outside of the river channel. This eliminated the need for dewatering and minimized water quality degradation related to construction. High density polyethylene (HDPE), SDR 11 pipe was selected because of its flexibility and durability. The HDPE pipe was installed in one continuous length with connections made onsite by butt fusion. This method of construction eliminates mechanical fittings and joining devices. The HDPE material met the leakage criteria established for the project and because of its flexibility and continuous construction, its long term performance will likely be superior to more rigid pipe with in-line joints. The actual construction cost for HDD pipe in Petaluma's application was $230m ($77/ft). This price included the entrance and exit pits for two separate drilled pipes of 100m (300 ft) each.

HDD technology also offered schedule advantages. Cut and cover construction in the active channel of the Petaluma River would have taken at least a month, with significant potential for delays associated with dewatering or unsuitable material. A directionally drilled crossing was installed in a week with utility tie-ins made in less than a day. This was especially important for the sanitary sewer force main, which could only be taken out of service for 12 hours without causing overflows in other areas of the collection system.

Social and Environmental Considerations

For all the potential benefits of directionally drilled utility crossings for the Petaluma River, the final decision was controlled by social and environmental considerations.

The Petaluma Flood Control Project took 15 years to develop and during this period the affected neighborhoods flooded four times. The final Environmental Impact Statement was approved in the fall of 1995. Upon receipt of this approval, the Petaluma City Council authorized immediate work on the project. The City in conjunction with its design consultants concluded that an effective early project would be the demolition of the Payran Street Bridge. The bridge formed a hydraulic constriction in the channel and its removal would provide an element of flood protection during the upcoming rainy season. A fast track design schedule allowed the utility relocations and bridge demolition project to take place prior to the end of 1995. In northern California, the majority of storm events take place from January to March so there was good reason to believe that we could enhance the community's protection against flooding if we made a fast track effort to design and construct the crossings and complete the bridge demolition.

The fast track option required negotiations with the California Department of Fish & Game which regulates construction in streams and rivers. The Department's standard practice prohibits construction in the river channels from October 15 through April 1. This requirement effectively eliminated the cut and cover construction option if we wished to consider fast tracking the project. HDD required bore and receiving pits on either side of the river but no work within the river channel. The trenchless construction technique allowed for work to be completed without disruption of the active channel and exemption from Fish & Game regulations and permit requirements.

Construction Experience

Construction of a directionally drilled pipeline begins with location of the entrance and exit pits. The entrance pit, also known as "rig-side," must be large enough to accommodate the drill rig which requires approximately 5 m (15 ft) of clear space. Entrance pits can be located in fairly constrained locations. Most of the work takes place from the exit pit, or "pipe-side," through which the pipe will be pulled. Ideally, there should be enough clear space around the exit pit for the construction crew to fuse the entire length of pipeline on the site. At a minimum there needs to be enough clear space to contain one and one-half lengths of pipeline. The minimum distance gives enough room for the contractor to make up pipe joints one by one as the pipe is pulled into the hole. For HDPE pipe, this minimum length is 12 m (40 ft).

The working space conditions of the pipe-side site offered the most challenges on the project. The "pipe-side" of the river crossing was located on a vacant 36.5m x 36.5m lot, in the middle of a residential block. The working area of the lot was constrained by the existing sanitary sewer pump station, in addition to underground gravity sewer pipes, water pipes and electric lines. To further complicate manners, the lot is also the future home of the Payran Storm Drain Pump Station.

The topography of both the rig-side and pipe-side sites is flat with only 0.3m different in elevation from the front of the lots to the back. The Petaluma River is approximately 55m wide at the point of the crossing for the sewer and water pipes. The river depth varies from 3.0m to 4.6m. Geotechnical conditions proved favorable for HDD pipes. Two borings were taken, one on each side of the river, and at the proposed depth of 5m, the soil conditions were medium dense, silty sand/sandy silt.

The contractor selected for the project Arrow Trenchless Technologies, from Sacramento, California, has considerable experience working in "tight spots." Their project manager, Mike Wegener, assessed the site and found that the pipe lengths would have to be fused together onsite, with materials stored out in the street of the adjacent neighborhood. The cooling time on the Petaluma pipes was approximately 45 minutes per joint, with space for two pipe sections at a time to be fused together before pulling.

Once the contractor had the equipment positioned, the drill rig completed the pilot hole. A tube behind the bit housed a transmitting device which was capable of sending a signal to a receiver that identified the depth, angle of inclination, and rotational position of the drill bit. The design consultant, Winzler & Kelly Consulting Engineers, had previously researched all record data in the area, in addition to future development in and along the Petaluma River in this reach, before laying out the proposed design profile of the two pipes. This advanced planning led to a required pipe depth of 5m in the center of the river with entrance and exit angles of approximately 15° for the pipes.

After the pilot hole was successfully drilled and the trace verified, the hole was back-reamed with a reamer of the same diameter as the final pipeline. Water and bentonite was used to lubricate the new hole. Drillers alternated back reaming with "pigging" (running a metal ball through the hole from the entrance to the exit pit) in order to condition the hole. Once conditioned, the final 20.3cm pipeline was pulled through the hole. No obstructions were met during the drilling process and the pipes paths had a vertical accuracy 0.2m and a horizontal accuracy of 0.4m. The pipes were then carefully mapped, giving accurate as-built information.

Once the pipelines were drilled, the actual tie-in locations were accurately located and constructed within a matter of hours. This eliminated the need for down time or bypass pumping during the utility relocation project.

Petaluma's project involved approximately 200 m (600 ft) of pipeline, the pipe was fused onsite. From start to finish, including mobilization, construction of entrance and exit pits, drilling of pilot hole and pulling of pipeline, the project took one week.

Summary

Trenchless Technology, as a whole, and HDD technology in particular, are frequently dismissed from consideration for smaller projects. Historically, these have been seen as expensive methods suitable only for large projects and/or difficult locations. The Petaluma River HDD Project illustrates that the technology has far broader application than historically assumed. Advances in materials, coupled with increased environmental restrictions on even small projects have combined to make trenchless pipeline construction a viable alternative for many projects. In order to evaluate the applicability of trenchless technology for smaller projects, engineers should consider the following items:

- Schedule. Trenchless applications frequently allow for accelerated construction, particularly during winter months.

- Tolerance for outages. The trenchless method can minimize, or eliminate, the need for utility outages.

- Overall Costs. When the costs of environmental mitigation, schedule delay and/or prolonged utility outages are taken into account, the full value provided by trenchless technology is realized.

- Social Considerations. Environmental, aesthetic or neighborhood concerns can frequently be mitigated by trenchless installation techniques that do not involve large scale cut and cover operations.

CASE STUDIES IN TRENCHLESS TECHNOLOGIES WATER MAIN

Kent A. Alms[1]

Abstract

This paper discusses the applications of trenchless technologies and alternative piping materials for water main replacement, specifically focusing on actual case studies of jobs performed and evaluated over the past two years at St. Louis County Water Company.

Experimental methods evaluated included pipe bursting, a cured-in-place pressure rated lining system, directional bore, and slip lining. The slip lining method was utilized on two jobs using different piping materials. On one job "bell-less", ductile iron pipe was slip lined through existing cast iron pipe. On the second job, high density polyethylene pipe (HDPE) was slip lined through existing cast iron pipe. Actual job experiences will be discussed in detail, including prior planning, construction practices and procedures, and job successes and failures.

Introduction

St. Louis County Water Company is a private investor-owned utility providing water service to 300,000 customers in St. Louis County and the surrounding area. St. Louis County Water Company operates and maintains approximately 6,437 km (4,000 miles) of main and 27,000 fire hydrants.

In 1994, an internal study was conducted addressing the growing concern over increasing main break occurrences and the need for replacement of obsolete cast iron water main. This report generally identified categories and vintages of pipe that are either in immediate need of replacement, or candidates for

---

[1] Manager, System Engineering, St. Louis County Water Company, 535 North New Ballas Road, St. Louis, MO 63141

replacement in the future. The vast majority of this pipe is located in older, established neighborhoods in St. Louis County. Properties in these neighborhoods can be characterized as being extensively landscaped and having many mature trees that must be addressed during construction. County Water is interested in developing applications for trenchless technologies for water main replacement in order to reduce associated costs and resultant property damage from conventional trenching.

Alternative Analysis

Numerous trenchless methods were discussed and evaluated in-house prior to job initiation to address identifiable issues. Cleaning and cement lining was discounted because County Water lime softens and produces a "scaling" water that produces a thin film of calcium carbonate that adheres to the inside of the water main. In addition, most of the water main to be replaced is cement lined, so in conjunction with the scaling effect of the water, the main does not require cleaning. Also, cement lining has no structural integrity of its own and relies on the host pipe, which in the present case has deteriorated to some degree.

Slip-lining, using "rolled-down" P.E. liners, although not completely disregarded, was not chosen during this phase of experimentation because of concerns over a thin lining withstanding high pressures and problems with reconnection of existing services. It was ultimately determined that County Water would experiment with pipe-bursting, slip-lining (with full size pipe), and directional bore trenchless techniques in 1996.

The last two major issues of concern were the use of alternative piping materials as a carrier pipe, and the NSF approval of these materials and their associated fittings. Prior experience with polyethylene service line piping indicated there may be problems associated with sloughing of accumulated lime scale with thermal expansion and contraction of the piping, with the potential of causing problems with clogged meters. These concerns were somewhat mollified by in-house plant experiments with the various piping materials. NSF approval was an issue that had to be addressed with the cured-in-place lining system.

Once it was determined these methods could be used, trenchless contractors were contacted to determine their interest in participating in the experiments. Since these types of technology had relatively little previous use in the water industry, the contractors were eager to gain experience by participating in the jobs.

Preliminary Engineering and Job Preparation

Project mains were selected from those identified as obsolete mains in need of replacement. Since these technologies were experimental in nature, the project mains were selected for ease of reverting back to conventional trenching and replacement of main if necessary. Preliminary engineering similar to that conducted for conventional replacement jobs was completed prior to job start up. Other utilities were located and plotted on job plans, existing service connections were identified, and transfers and reconnections were designed. A schedule of events and construction schedule also were developed and discussed with the trenchless contractors in pre-construction meetings. Finally, work order tickets were printed and the jobs were included in the overall construction schedule.

In addition, it was necessary to set up temporary service to customers along the main for the duration of the project. The types of trenchless technologies being evaluated, with the exception of directional bore, used the existing main either as a host for a new piping material, or as a pathway for installation of new pipe. Therefore, water service through the existing pipe was terminated during construction. In conventional trenching and laying, the existing pipe remains in service until the new main is installed and ready for use.

Pipe Bursting

During September and October of 1996, the pipe bursting method was used to replace water main along two different streets. Equipment rented from TT Technologies Inc. was used by County Water construction crews to burst and replace approximately 457 m (1500') of 150 mm (6") cast iron main. This method and equipment was used previously on a 1995 replacement job with some success. Problems encountered during the 1995 job were addressed prior to starting construction in 1996.

Briefly described, this method involves the pneumatic ramming of a cutter head through the existing cast iron main, opening a hole in the ground, and pulling through a casing pipe in one operation. After the installation of the casing pipe, standard ductile iron main is pushed through the casing pipe for restoration of service. On these two jobs, 150 mm (6") cast iron water main was burst and the hole was expanded large enough to accept a 250 mm (10") HDPE, Class SDR-17 casing pipe. After installation of the casing pipe, Class 52, rubber jointed 150 mm (6") ductile iron main was pushed through the HDPE pipe.

One of the problems encountered in the 1995 job was acquiring access for stringing out the HDPE casing pipe. For the 1996 job, a split ring butt fusion heater was used to connect the polyethylene pipe in individual lengths as the project progressed.

A second problem encountered in 1995 was the splitting of repair sleeves. The cutter head was successful in splitting a ductile iron four bolt split sleeve, however, it could not split a ductile iron solid sleeve. This was still the case during the 1996 job. Therefore, when solid sleeves were encountered, they had to be excavated and split with a pipe saw.

A build up of friction between the HDPE casing pipe and the ground as the polyethylene pipe was being pulled through the hole caused problems in 1995. The friction caused the HDPE pipe to pull loose from the cutter head during the burst and pull. In 1996, a larger hole was made through the ground to overcome this problem. In addition, a bentonite slurry was provided for added lubrication if necessary.

The fourth concern was the noise and vibration caused by the pneumatic hammer. During the 1996 job, County Water contracted with a vibration analysis consultant to record and analyze readings on homes along the construction route. The consultant determined that the vibration from the process, although audibly and tangibly perceptible, would cause no damage to the foundations of neighboring houses.

A lesson learned in 1996 was that butt fusing individual lengths of HDPE pipe as the job progressed added a considerable amount of time to the construction process. Actual costs for these two jobs totaled approximately $229,000, compared to estimated costs of $219,800 for conventional trenching and replacement of main. Estimated costs included all clean up and property restoration costs.

Directional Bore

In addition to pipe bursting, two other pipe laying methods new to County Water were used on the above job. In one experiment, a directional bore of 73 m (240') length was used to minimize excavation across landscaped properties and driveways. A 54 mm (2-1/8") diameter bore was made followed by a 250 mm (10") back ream pulling 250 mm (10") HDPE casing pipe. Again, standard 150 mm (6") ductile iron pipe was pushed through the 250 mm (10") casing pipe.

This method was utilized again on a second job later in 1996, however it was modified to eliminate the need for PE casing pipe. On that job, approximately 305 m (1000') of 150 mm (6") polywrapped ductile iron main pipe was pulled in behind the 250 mm (10") back ream of the directional bore. The pipe was restrained from coming apart by using "Field-Lok" gaskets. The path of the directional bore was approximately parallel to the existing pipe in the street. The one downside to this project was the necessity to go deep with the vertical dimension of the bore path in order to avoid sewer laterals and other utilities. This increased the depth of the excavations and added shoring requirements for restoring the service connections to the new main.

Pipe Ramming

The second new method tried was a pipe ramming experiment utilizing the pneumatic ramming head of the pipe bursting equipment. In this experiment, a cutting head was attached to the front of the casing pipe. The casing pipe was then rammed through the ground under a street for a side street connection and under pavement to make an end connection at the end of a cul-de-sac. The bentonite slurry was used as a lubricant during the two rams. The dirt that had accumulated in the casing pipe was augured out in one ram, and was blown out in the second ram using air pressure from the air compressor. These methods were tried instead of conventional auger boring.

Cured-In-Place Lining System

In August of 1996, St. Louis County Water Company, in conjunction with Insituform Technologies Inc., performed the first lining of a water main in the United States utilizing the Paltem WHT cured-in-place pipe lining system. The lining system was developed in Japan by Ashimore Industries for use in the gas and water industries to provide a pressure rated liner in the event of pipeline breakage after an earthquake. This system recently had received NSF approval, and therefore was certified for use in potable water systems.

The project involved the lining of 128 m (420') of 150 mm (6") cast iron main under the asphalt pavement of a parking lot. Service connections to the main consisted of two 19 mm (3/4") corps and a 100 mm (4") patent tap for a fire line. The resin impregnated, 3.2 mm (1/8") fabric liner was installed using an inversion method. An inversion "pot" was used with pressurized air over water to allow for better control of the inversion rate. Installation of the liner took five days from the beginning of setting up temporary service for existing customers to the end of making the final end connections. This included one full day of curing time for the liner using hot water.

Insituform had not yet developed methods for re-establishment of service connections when this project was executed. Therefore, the newly lined pipe was cut off short of the service connections. Standard ductile iron pipe was then connected to the end of the lined cast iron pipe to conventionally reestablish existing service. The lined piece of pipe with the installed service connections was given to Insituform for experimentation in developing new methods of tapping the liner.

Because of the composition of the resin used for impregnating the liner, it was considered prudent to monitor for the presence of Volatile Organic Compounds (VOC's) in the water after flushing and before returning the main to service. All compounds identified were found in quantities less than regulatory limits. A representative from NSF Laboratories visited the job site and supervised the extraction of four samples of the liner for their experimental purposes.

Actual costs for this job totaled approximately $57,400.00, which compares to estimated costs of $52,500.00 for conventional trenching and replacement of main. If cost effective methods can be developed for reestablishing services and side connections, this cured-in-place lining system may become a viable alternative to be considered when designing future water main replacement projects.

Slip Lining

Two projects were completed in 1996 using the slip lining process. Simply described, the process involves pulling (or pushing) a smaller diameter pipe inside the original obsolete pipe. The two jobs completed in 1996 were similar, except for the type of material used in each project.

In one project, 50 m (164') of 100 mm (4") MT bell-less pipe, manufactured by American Cast Iron Pipe Company, was inserted inside of an existing 150 mm (6") cast iron main. The 100 mm (4") pipe was inserted inside the existing 150 mm (6") pipe by pushing it through with a backhoe. Some minor difficulties were encountered during this job. First, the wrong type of lubricant was supplied by the manufacturer. The lubricant supplied was too viscous and resulted in the gasket tearing when the pipe was homed. This problem was overcome by making a lubricant composed of paraffin wax and linseed oil from a recipe found in the ACIPCO pipe installation manual. Another problem was encountered at the end of the job while pushing in the last couple lengths of pipe. Resistance was met when the lead pipe encountered an offset joint. The bell-less pipe could not navigate the deflection in the existing joint, and therefore had to be unearthed to continue the push. One of the lead joints came out of home during this process.

Actual costs for this job totaled approximately $39,000, which compared to estimated costs of $30,000 for conventional trenching and replacement of main. This type of material has limited application in that there is little deflection that can be achieved at the joint. Before this job was attempted, an elevation profile of the main was established to determine job feasibility. This process may be practical for replacing obsolete mains that were bored under railroad crossings or highways where the bore was essentially level.

The second job used HDPE pipe as the slip lining material. In the second project, County Water first attempted to insert 192 mm (630') of 125 mm (5") SDR-11 HDPE pipe inside an existing 150 mm (6") cast iron main.

The first attempt failed due to restricted clearance between the inside diameter of the existing cast iron main, and the outside diameter of the HDPE pipe. At this point, it was decided to go down one pipe size in diameter to 100 mm (4") HDPE pipe. The ID of the new polyethylene pipe is 92 mm (3.63"). In this installation, the smaller diameter polyethylene pipe was inserted easily by pushing it by hand. Prior to insertion, the entire length of HDPE pipe was butt fused together and strung out along a neighboring side street. This is the first installation where HDPE was used as the carrier pipe, and therefore the first project to experiment with side wall fusion of a polyethylene corp for reconnection of services. This polyethylene corp was then reconnected to the copper service line by means of a transition fitting.

An obvious concern of the slip lining process is the loss of hydraulic carrying capacity of the existing pipe. The slip lining project involving the 100 mm (4") bell-less pipe was done on a cul-de-sac with no fire hydrants, and therefore supplied only domestic service. The project utilizing the 100 mm (4") polyethylene pipe was on a street with an existing fire hydrant. Prior to construction, registered fire flow on this main was 4,542 liters/minute (1200 gpm) at 207 kPa (30 psi) residual. After the project was completed, fire flows on the same main were registered at 2460 liters/minute (650 gpm) at 207 kPa (30 psi) residual. When considering the slip lining process as a trenchless main replacement alternative, hydraulic considerations must be taken into account.

Actual costs for the job utilizing the 100 mm (4") HDPE pipe totaled approximately $67,000, which compares to estimated costs of $89,500 for conventional trenching and replacement.

Conclusion

Water main replacement is a relatively new arena for application of trenchless technologies. Additional concerns unique to the water industry provide engineering challenges for applying such technologies. Public health concerns, maintaining service during construction, and reconnection of services that will withstand pressure are issues that must be addressed when applying these technologies in the water field. Public health concerns regarding alternative piping materials were addressed by third party NSF certification. Public health concerns during construction were addressed by installing temporary service in conjunction with a thorough flushing and sampling program. While public health concerns were addressed with relative ease, setting up temporary service added time and costs to the trenchless jobs. On jobs where standard ductile iron pipe was pulled through HDPE casing pipe, services were reconnected by conventional tapping methods. On the slip lining job where HDPE was used as a carrier pipe, side wall fusion of polyethylene corps was used for reconnecting services. In the case of the Insituform cured-in-place lining system, methods for reconnection of services are still being developed.

During the slip lining job using ACIPCO "MT" bell-less pipe, negotiating joint deflection in the existing pipe proved to be a problem. This will limit future application of this material to level and relatively short lengths of replacement. In addition, when considering the slip lining alternative, hydraulic concerns have to be taken into account due to the reduction in hydraulic carrying capacity of the pipe.

It should be noted that in most of these projects, the actual costs were greater than the estimated costs for conventional trenching and replacement of pipe. This was expected due to the experimental nature of these jobs and the first time application in water main replacement. The HDPE slip lining project came in under the estimated amount because of large cost savings realized in pavement replacement. Various new ideas were tried as the jobs progressed in order to learn as much as possible about the procedures. With increased application and more experience in using these technologies, these costs would be expected to decrease.

St. Louis County Water Company will continue to investigate and evaluate new and innovative methods for replacing obsolete water mains. Clean up costs and pavement repair costs are major components of overall project costs in fully developed and mature neighborhoods where most of these mains are located. Implementation of technologies that can reduce these associated costs and the higher labor costs associated with tedious installation by conventional trenching methods will be necessary for future efficient operations. County Water is planning to develop a "Menu" of technologies from which to choose when designing water main replacement projects in order to cope with the inherent idiosyncrasies and uniqueness of each individual job.

Extending Life and Adding Control -- The Rehabilitation of a 70 Year Old Water Transmission Main

Mark Nankivil, P.E.[1] and Daniel J. Nichols, P.E.[2]

Abstract
The City of St. Louis Water Division and its consultant, Sverdrup Civil, Inc. jointly developed a system to evaluate the risks, consequences, costs and benefits of alternative strategies to rehabilitate Conduit No. 1, a 1.5 m-(60-inch), 25.6 km-(16-mile) water transmission main. The rigorous analysis proved the benefits of reinforced mortar lining, which the City is installing to prolong the life of the 70-year old water transmission main. The "lessons learned" from the analysis, design and construction support the value of this method of trenchless construction.

The importance of analysis

Built many decades ago, the water transmission mains that supply our cities are aging while the budgets to repair them are shrinking. For most utilities, the entire transmission and distribution system competes with other municipal services for limited funding. The process of setting priorities demands rigorous analysis to support the capital costs of rehabilitation and repair. When the City of St. Louis Water Division began to experience leakage problems in the oldest of its three transmission mains from the Howard Bend Plant on the Missouri River, the Water Division staff began a systematic evaluation of the risks, consequences, costs and benefits of alternative rehabilitation strategies. The analytical process that they

---

[1]Assistant Division Engineer, Transmission and Distribution Section, Water Division, City of St. Louis, 4600 McRee, St. Louis, MO 63110

[2]Project Manager, Sverdrup Civil, Inc., 13723 Riverport Dr., Maryland Heights, MO 63043

developed in partnership with Sverdrup Civil, Inc. provides some practical guidance for utilities that need:

- a systematic means of identifying the most critical pipeline sections;
- an objective process for evaluating rehabilitation alternatives and weighing social costs;
- a cost-effective approach to building in operational control.

A systematic means of identifying the most critical pipeline sections

Project background

Constructed in 1926, the history of Conduit No. 1 is well documented because it was an impressive public works achievement for its time. Conduit No. 1 is the oldest of three parallel water transmission mains that convey finished drinking water from the Howard Bend Water Treatment Plant to the Stacy Park Reservoir, and from Stacy Park Reservoir to the City of St. Louis.

The 1.5 m-(60-inch) inch riveted steel pipe was originally installed with an asphaltic external coating and interior lining, but without cathodic protection. When it was installed, most of the property along its 25.6 km- (16 mile) length was rural or low density residential. Today approximately 19.2 km (12 miles) are moderate to high density residential and commercial property.

When the pipe began to experience leakage problems, the Water Division selected Sverdrup to help them evaluate and design a cost effective solution. The Water Division set three objectives for the rehabilitation of Conduit No. 1:

- systematically identify the pipeline segments most at risk
- evaluate the alternatives and their actual and social costs
- design a rehabilitation solution.

Pipeline Inspection -- identifying a representative sample

To analyze the condition of the pipeline and identify those sections with the greatest risk of failure, the Water Division and Sverdrup developed a plan to guide the field work. The team began with a review of the pipeline records -- including the original plans and specifications, construction records, repair records, leak history, as well as the plans and specifications for the unreinforced mortar lining added to some sections of the pipeline in the 1940's and 1950's. Then Sverdrup staff walked the entire right-of-way to document the preconstruction conditions with a camera, and to identify street crossings and other surface conditions that might effect the condition of Conduit No. 1.

Using the records as a guideline, the team began the process of identifying sections that might serve as a representative sample of the condition of the entire pipeline. While it was important to check on high risk sections (for example, sections that were unlined, in corrosive soil, or reporting a history of leaks); the

Water Division also planned internal and external inspections for a broader sampling to reflect the varied conditions prevalent in the main. The Water Division and Sverdrup judiciously selected sites which would represent a reasonable compromise between statistical reliability and affordability. A total of approximately 5-6% of the 25.6 km- (16 mile) pipeline was internally inspected.

The Water Division selected sites for internal inspection based on several factors:

- moderate to aggressive soil conditions
- high risk areas in terms of the consequences of pipeline break or failure
- sections with and without reinforced mortar lining
- sections with and without significant external repairs
- areas of high external loading.

As the Water Division and Sverdrup staff internally inspected the pipeline, they noted that most of the lining was in good condition, although external pipeline repair had delaminated the lining in some areas. Measurements of the pipe showed ovality in some areas -- caused by traffic loading or problems with compaction in the original installation.

Water Division crews also excavated the pipeline at ten locations to reveal the external condition of the pipe. Most of the external sites were aligned with the points of internal inspection. The external asphaltic coating had bonded to the soil in some areas, making the coating ineffective and causing some pitting from the springline to the bottom of the pipe. Soil samples were also taken at three sites to evaluate the effect of corrosive soil conditions on the external condition of the pipe. Ultrasonic measurements of the thickness of the pipe were inconsistent and were not useful in evaluating the condition of the line. However, as Sverdrup and the Water Division compared the external conditions of the line with the results of the internal inspection and the soil evaluation, some of the more critical pipeline areas began to emerge.

Planning for construction

With a preliminary identification of the most critical pipeline sections, the Water Division began to plan for the rehabilitation work ahead. Although Conduit No. 1 lies in a 30-m (100-foot) city right-of-way, it borders or travels through six cities and crosses several state, city and county roads. During the pipeline evaluation phase, the project team identified and notified all public agencies that could be effected by the construction. The early planning created important benefits. Planned highway resurfacing was rescheduled until the pipeline project was complete. The early communication also streamlined the permitting, contracting and construction process. As the Water Division and Sverdrup worked with the affected agencies and communities, the areas with the greatest risks and consequences of pipeline failure became more apparent.

An objective process for evaluating alternatives and weighing social costs

Evaluating rehabilitation alternatives

The Water Division staff set two important parameters to guide the process of evaluating rehabilitation options. First, they wanted to consider the widest possible range of alternatives, so that this and future projects could benefit from emerging trenchless technologies. Working with Jason Consultants, an international trenchless engineering firm, Sverdrup and the Water Division studied over a dozen options, including some that have been effective for similar projects in Europe, but are not widely used in the United States. Secondly, the Water Division was interested in effective long term solutions -- not quick fixes. They preferred a technology that would provide a complete structural solution.

To determine the viability of the full range of trenchless techniques, Sverdrup and the Water Division analyzed the successful uses of each technique in pipelines similar to Conduit No. 1. Each technique was evaluated based on its appropriateness to the project, including:

1. Size -- appropriate to use on a 1.5 m- (60-inch) diameter pipeline;

2. Pressure -- appropriate to use at 150 psi;

3. Potable water use - acceptance by the National Sanitation Foundation for use in a potable water system;

4. Full structural potential -- the capacity of the technique to provide complete structural security for the pipeline;

5. Reduction in hydraulic capacity -- system redundancy and normal operation at less than full capacity made this less important for Conduit No. 1.

The evaluation also considered elements important to the St. Louis location, including:

1. Contractor availability -- the availability of materials and the capability of most contractors to perform this type of work

2. Construction access and disruption -- excavation, site storage and use of existing access points

3. Cost per square foot for each option.

The final analysis evaluated mortar repair, reinforced mortar or concrete lining insitu ferrocement and gunite lining, cured in place pipe, sliplining with glass-reinforced-plastic, polyethylene, ductile iron and steel pipe, open cut construction and pipe-jacking.

Assessing Risks and Consequences

To begin the assessment process, the Water Division and Sverdrup defined several important concepts. The *risk* of failure of the pipe (low, medium or high) would be assigned based on the likelihood of failure before and after rehabilitation. Historical records, soil tests and inspection information determined the risk value assigned. The *consequences* would be assessed based on the direct costs and the social costs of pipeline failure (impact on nearby residences, businesses and traffic).

To evaluate the consequences accurately, the pipeline was divided into six sections: section one is rural, sections two and three are moderate density residential, section four is high density residential, section five is highway and section six is a business district. (See the benefit analysis table)

Studying the records and pipeline inspection information, the Water Division and Sverdrup assigned a risk of failure to each section. To evaluate the consequences, the Water Division examined the direct costs of recent pipeline failures. The direct costs of repairing two recent pipeline failures were $125,000 and $145,000. Jason Consultants provided an extensive study of the social costs associated with traffic delays, disruption and loss of business when pipelines fail in an urban environment. The study concluded that the ratio of social costs to construction costs is one or greater. In densely populated or high traffic areas, the study suggests that the total cost consequence (direct and social costs) of a pipeline failure can escalate to three or four times the direct cost of repairs.

Analyzing costs and benefits

Each pipeline section received a low to high assessment of the risks and consequences of pipeline failure. Then Sverdrup developed preliminary costs per foot for each rehabilitation option. Comparing the cost of each option with the likely direct and social costs of future repairs, the advantages of reinforced mortar lining rehabilitation became clear (see table below). Those sections with the highest risks and consequences were recommended for rehabilitation in Phase One.

BENEFIT ANALYSIS

Section	Solution	Cost of Rehabilitation			Cost of Future Repairs Over 50 Years ($ x 1000)*		Benefit
		Rate $/ft	Approx. Length ft.	Total Cost (x $1000)	No Rehabilitation	Rehabilitated	(x $1000)
1-rural	Mortar Repair	30	15,000	450	18,750	3,750	14,550
1	Reinforced Mortar Lining	250	15,000	3,750	18,750	Nil	15,000
1	Pipeline replacement	300	15,000	4,500	18,750	Nil	14,250
1	Sliplining	400	15,000	6,000	18,750	Nil	12,750
2-resid.	Mortar Repair	30	12,300	370	30,750	6,150	24,970
2	Reinforced Mortar Lining	250	12,300	3,080	30,750	Nil	27,670
2	Pipeline Replacement	350	12,300	4,310	30,750	Nil	26,440
2	Sliplining	400	12,300	4,920	30,750	Nil	25,830
3-resid.	Mortar Repair	30	19,500	590	48,750	9,750	38,410
3	Reinforced Mortar Lining	250	19,500	4,880	48,750	Nil	43,870
3	Pipeline Replacement	350	19,500	6,830	48,750	Nil	41,920
3	Sliplining	400	19,500	7,800	48,750	Nil	40,950
4-resid	Mortar Repair	30	9,200	280	34,500	6,900	27,320
4	Reinforced Mortar Lining	250	9,200	2,300	34,500	Nil	32,200
4	Pipeline Replacement	360	9,200	3,300	34,500	Nil	31,200
4	Sliplining	400	9,200	3,680	34,500	Nil	30,820

BENEFIT ANALYSIS, continued

		Cost of Rehabilitation			Cost of Future Repairs Over 50 Years (x $1000)*		Benefit
Section	Solution	Rate $/ft	Approx. Length ft.	Total Cost (x $1000)	No Rehabilitation	Rehabilitated	(x $1000)
5-highway	Mortar Repair	30	9,300	280	23,250	6,900	16,000
5	Reinforced Mortar Lining	250	9,300	2,330	23,250	Nil	20,920
5	Sliplining	400	9,300	3,720	23,250	Nil	19,530
6-business	Mortar Repair	30	12,100	360	90,750	30,250	60,140
6	Reinforced Mortar Lining	250	12,100	3,030	90,750	Nil	87,720
6	Sliplining	400	12,100	4,840	90,750	Nil	85,910

*Calculated as (50 years)x(length/10,000)x(No. of incidents corresponding to risk of leakage)x(Cost consequence of repair)

A cost effective approach to building in operational control

The planned rehabilitation of Conduit No.1 gave the Water Division an opportunity to add operational control improvements to the water transmission system. The final design includes transmission main interconnection and flow control stations and line valves to improve operational and maintenance flexibility. To enhance the monitoring of flow through the three conduits, flow monitors and pressure transmitters will be installed at the new interconnect and control stations.

By monitoring flow and pressure and integrating the information into a supervisory control and data acquisition system (SCADA), a major leak can be detected quickly. Operators can respond quickly and reduce the risk of property damage or service interruption. Fiber optic cable (installed in part by guided boring) will connect the pressure sensors and flow indicators to the SCADA system. Sverdrup and the Water Division packaged the construction for maximum efficiency in cost and schedule.

Construction Issues

With construction of Phase I related work approximately 50% complete at this writing, the Water Division can identify several lessons learned from the process.

With construction along the full length of the right of way, careful planning and coordination was vital. Early in the project Sverdrup staff identified all the jurisdictions that would be effected by the rehabilitation of Conduit No. 1. The Missouri Department of Transportation, local utilities, and six local governments were all notified in writing of the impending project. This "external prework" helped to plan the construction schedule with minimal disruption. Because trenchless construction is new to some areas, water utilities planning this type of project may also want to provide the local authorities with a drawing or photo of the type of street opening that is typical for trenchless projects.

"Internal prework" is also important in scheduling construction start-up. Utilities planning this type of work should schedule the approval of the design and construction packaging so that the vault structures are in place before winter. Application of the lining can be done during any season.

The ovality in some sections of the pipe also created some initial problems in the application of the lining. The steel reinforcing cage originally specified proved to be too inflexible for the pipe, and the lining was applied unevenly. Working together, the design and construction teams redesigned the reinforcement using two layers of smaller gauge steel. The lining is applied in the following sequence: lining, reinforcement, lining, reinforcement, lining.

Conclusions

Careful analysis of the costs and benefits of alternative rehabilitation strategies begins with an effective field inspection program. City records, internal and external inspections allowed the city to identify representative sections of the line so that its condition could be accurately characterized. Then the City could accurately prioritize the sections most in need of repair. The department's extensive cost-benefit analysis incorporated social costs and validated the need for the project. A complete evaluation of the alternatives -- from pipeline replacement to trenchless alternatives proved the value of reinforced mortar lining. Although the construction process is incomplete at this writing, it has already established the importance of accounting for ovality in older transmission mains.

Design and Rehabilitation of an Urban Gravity Interceptor Using Pipebursting in Norfolk, Virginia

Aaron B. Small, P.E. Assoc. Member, ASCE, Jeffrey S. Gidley, P.E., Member, ASCE and Daniel A. Riley[1]

Abstract

The City of Norfolk, Virginia, Department of Utilities recently completed 1,270 feet (387 m) of 20 inch (508 mm) gravity sanitary sewer pipe rehabilitation by pipebursting an existing 18 inch (457 mm) clay and concrete gravity interceptor sewer. Initial television inspections of the pipe showed areas of partial collapse and severe deterioration which prompted Baker and the City to recommend that the 72 year old sewer pipe be replaced or rehabilitated. Major issues encountered during the construction phase were the time constraints, coordination with other contractors in the construction corridor, traffic control, bypassing of the sewage flow, and the location of pipe layout/staging areas. Pipebursting enabled the project to be completed within the time constraints while minimizing disturbance to the downtown streets and other construction activities.

Project Description

In July, 1994, the City of Norfolk Virginia, Department of Utilities was asked what would be required to upgrade its sanitary sewer lying under City Hall Avenue as part of an urban revitalization project undertaken by the Norfolk Redevelopment and Housing Authority (NRHA). The overall project (see Figure 1) was to build an urban mall in the downtown area of the city. In order to have a project that the City could use to attract tourist and businesses, the NRHA set a goal for all the utility owners to not disturb the streets in the blocks surrounding the mall for twenty years after completion of the mall. This noble goal required the Department of Utilities, which administers water and sewer service throughout the city, to review their existing sanitary sewer

---

[1]Authors are, respectively, Project Engineer and Project Manager, Michael Baker Jr., Inc., 770 Lynnhaven Parkway, Virginia Beach, Virginia 23452 and Civil Engineer, Department of Utilities, City of Norfolk, 400 Granby Street, Norfolk, Virginia 23501.

mains. Michael Baker Jr., Inc. was hired as a consultant to review the sewer and recommend rehabilitation or replacement options.

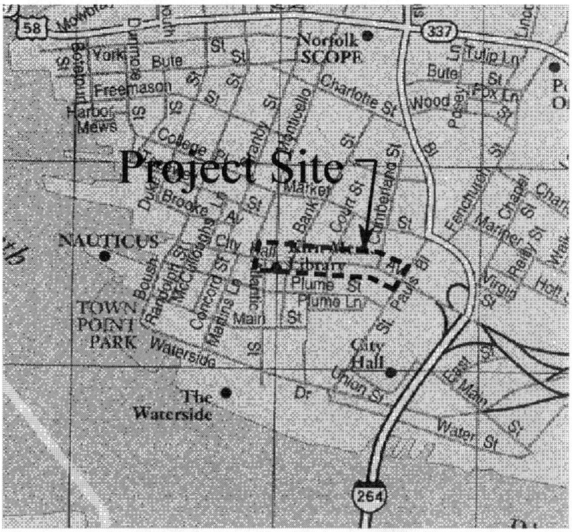

Figure. 1. Project Location

Design Considerations

Initial television inspection of the 72 year old pipe showed many areas where the partial collapses of the 450 mm (18") vitrified clay pipe were evident. Much of the pipe being investigated was not able to be televised due to obstruction in the sewer line. Inflow and infiltration was quite prevalent along most of the sections televised. The pipe served as a gravity interceptor for "high-line" systems before the effluent reached the pump station near a corner of the mall project area. The depth of the sewer varies from 5.0 to 5.5 m (16 to 18 feet) below City Hall Avenue, four-lane boulevard in the center of the downtown district. Mid-way along the street exists the former city hall building which now houses the General Douglas MacArthur Memorial and Library. On the east end of the street lies historic St. Pauls Episcopal Church, while at the west end is the main branch of the Kirn Memorial Library.

Baker and the City determined that the only options available to repair the pipe were open cut replacement or rehabilitation by pipebursting. Due to the urban vicinity and the depth of the sewer, the idea of open cut replacement was severely impaired. Open cut replacement was further handicapped by the minimal time allotted to replace the sewer and the disruption caused by open trenching the full length of City Hall Avenue. Other trenchless rehabilitation methods such as fold and form and cured-in-place were ruled out due to the eminent collapse and jagged edges from the existing broken pipe (see Figure 2). In addition the mall project required the line to be up sized from 450 mm (18 inch) to a minimum 500 mm (20 inch) diameter pipe which left only the pipebursting and open cut trenching as viable replacement/rehabilitation options. After analyzing costs, Baker concluded that the cost of replacement by open cutting was slightly more expensive than pipebursting due to the depth and problems expected with open cut trenching. Based on the above considerations, Baker recommended that the city proceed with the design and construction of the pipebursting option.

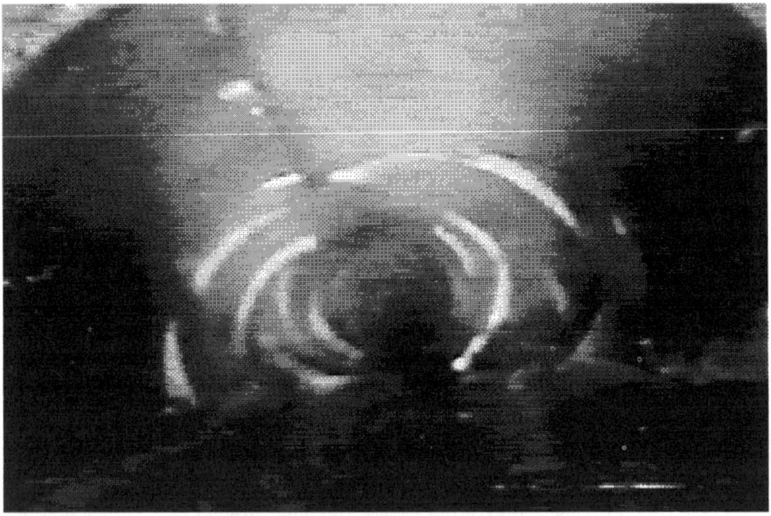

Figure. 2. Existing Pipe Collapse

Bid Process

Due to deadlines imposed by the mall project a "window of opportunity" became available of approximately 3 months where the pipebursting could be performed with minimal imposition on other contractors working in the construction corridor. Bids for the project were opened on June 4, 1996 and Vico Construction Corporation of Chesapeake, Virginia was awarded the project. Vico subcontracted TTS Northwest of Seattle, Washington to provide equipment and on the job training to their workers for performing the actual pipebursting. TTS uses a static bursting process which uses a 200 ton hydraulic winch within a pit to pull a chain which is attached to the pulling head (see Figure 3).

Figure. 3. Hydraulic Winch

Construction

Construction on the project began immediately after the award of the contract. Many constraints were placed on Vico in order to comply with City and other requirements. The locations of insertion and pulling pits become an issue because of the many improper locations of pits. Some of the locations included the area in front of the fire station that had to maintain emergency ingress and egress, street intersections which could not be blocked in order to maintain traffic, and areas where existing underground utilities would imped excavation of the pits. Due to the depth of some of the pits, maintaining structural integrity of adjacent structures was paramount. The pit located at the east terminus of the project had to be located adjacent to a brick wall surrounding the grounds and cemetery of historic St. Pauls church. Vico placed gauges on the wall in order detect any possible movement due to construction vibrations. The gauges allowed the contractor to monitor possible movement so that corrective action could be taken prior to possible collapse.

In addition to the "static" areas where pits could not be located, Virginia Power was in the process of relocating an electrical substation within the construction zone. This relocation required the construction of large underground electrical duct banks directly over the sanitary sewer to be pipeburst. The Virginia Power construction work within the City Hall Avenue Corridor caused further problems with maintenance of traffic and required coordination and cooperation of both contractors to meet all requirements. Staging of the new HDPE pipe strings (i.e., approximately 400 L.F. each) was thought to be an initial problem, however, after discussion with the mall contractor, Vico was able to obtain a portion of the mall site for their staging activities.

Another pit location that caused unforseen problems was at the extreme west end of the rehabilitation project. The manhole at the western terminus as indicated on the plans allowed only a minimum space for a pit. In addition a 600 mm (24 inch) sanitary force main was found to be imbedded in the side of the brick manhole where the pit was to be located. However, the actual connection of the 450mm (18 inch) gravity sewer to the manhole was found to be located at a wye connection approximately 4.6m (15 feet) upstream of the manhole. The connection lies directly underneath a 900 mm (36 inch) storm drain. Later television inspection showed the last section of pipe to be in fairly good condition and therefore Vico, Baker and the City concluded that the best course of action was to not attempt to dig a pit at the terminus location but to relocated the center of the pit 15.25m (50 ft.) upstream of the original location with the installation of a manhole to connect the existing pipe to the new sewer pipe.

The contractor was required to provide an initial sewer cleaning and television inspection prior to pipebursting in order to locate possible lateral connections not noted on the preliminary television tapes. The contractor elected to bypass the sewage for the entire construction period to alleviate future problems during construction while trying to maintain sewage flows. This bypass required additional coordination with the fire station because the line had to cross their entrance. Crossing of active side streets and the mall site construction entrance was also required. Vico used an HDPE pipeline which provided the flexibility to be buried at the fire station entrance and cross streets as well as to have the ability to be moved easily if required by other contractors working within the corridor.

The overall sequence of construction proposed by the contractor was to setup the sewage bypass and traffic control, dig all of the pits, burst the pipe, reinstate all services, and finally perform site restoration. The majority of the construction time was spent on the digging of the pits. Numerous abandoned underground utilities which were not indicated on city or private utility records were encountered. The Norfolk area was also hit by heavy rainfalls from Hurricanes Bertha, Edouard and Fran which caused slight delays in the pit excavations. Vico anticipated that three pulls would be required to burst the sewer. Each pull was between 113 to 130 m (370 to 425 ft.). Vico intended to pull the middle section first, the east end second and the west end third. The east end pull was to occur in the middle of the night because St. Paul's Boulevard, a major city thoroughfare, had to be closed for the pipeline insertion.

The first pull began at 3:00 PM on September 17, 1996. The majority of pipe in this section was broken and partially collapsed. The first 100m (330 ft.) of the burst proceeded at a pace of approximately 450mm per minute (18 inches per minute) without incident. However, at approximately 7:00 P.M., the pull hit resistance. The next stroke of the hydraulic cylinder separated the chain from the pulling head. An immediate point repair was required to reconnect the chain to the pull head. It was discovered that the resistance was caused by concrete encasement around the existing clay pipe which prevented the bursting head from advancing. The concrete was between 150 to 200mm (6 to 8 inches) around the top and sides of the pipe. Fortunately, the last 21 m (70 ft.)

of the pull was located in the median of the roadway and other underground utilities were not in conflict, therefore open cut trenching was possible to complete the installation of this section.

The second pull began on the morning of September 25, 1996. Much of the morning hours and the previous days activities were spent laying out the pipe to be inserted and pulling the chain through the existing pipe to attach to the winch. The second section of pipebursting occurred without any problems in about 4 hours from the time of insertion to removing the winch from the pulling pit. See Figure 4.

Figure. 4 Pipebursting section 3.

The concrete encasement was expected to continue throughout the eastern pull, and due to the problems encountered on the end of the first pull, the method of rehabilitation for this section needed to be reevaluated. Vico, Baker and the City came up with three options for rehabilitation of this section of pipeline. Fortunately the additional flows from the proposed mall were designed to flow into the manhole where concrete was encountered, therefore, upsizing upstream of this manhole was not required, although initially desirable. The options discussed were fold and form polyvinyl chloride or deformed polyethylene, conventional open cut trenching, or sliplining.

Open cut of the third portion would be very expensive compared to the other options and would severely impede the work at the Virginia Power substation site not to mention conflict with the fire station ingress and egress. Existing underground utilities, including a 50 year old 900 mm (36 inch) reinforced concrete sanitary force main, would also have to be crossed with an open trench. The fold and form/deformed pipe option was not recommended by the contractor due to the many in-line irregularities in this section of the pipe. The irregularities included unused hubbed-in laterals and sections of jagged and broken pipe.

The recommended solution was a combined slipline/pipebursting method which would use a 480 mm (16 inch), SDR 26 HDPE liner pipe pulled with the pipebursting equipment. Total reduction in flow capacity was less than 15% with this liner. This method would also use the existing pit locations that had already been excavated. The bursting head was modified for use as a mandrel to clear the protruding lateral connections and broken sections of pipe from the path of the liner pipe. The total pull time was less than 4 hours and occurred without incident.

Conclusions

The construction of the City Hall Avenue sanitary sewer rehabilitation project has provided an excellent learning tool for all parties involved. The project has also demonstrated that pipebursting is a reliable and cost effective solution when faced with the many problems of rehabilitating sanitary sewers in urban areas, where the disruption of the construction can cause more problems than could be caused by trenchless methods . Pipebursting enabled the project to be completed within the time constraints while minimizing disturbance to the downtown district and other construction activities.

Protecting the Environment - Directional Bore Crosses Hudson Bayou

Bryan T. Veith, EI[1]
&
Douglas H. Taylor[2]

ABSTRACT

The City of Sarasota, located in Southwest Florida, is expanding its Urban Reclaimed Water Reuse System to provide for more flexibility in its effluent disposal options. The important completion of the South Loop pipeline helped to close the inner loop of the City's Reclaimed Water Reuse Distribution System. The South Loop pipeline project included an 18-inch crossing of environmentally sensitive Hudson Bayou, which is a tributary of Sarasota Bay. As a prestigious honor, Sarasota Bay was chosen as a National Estuary Program waterbody in 1988 by the U.S. Environmental Protection Agency (EPA).

An alternative analysis was performed during the preliminary design phase of the South Loop project before the directional bore method was selected for the crossing. The pipeline was analyzed based on several important criteria including permitability, constructability, aesthetics, estimated cost of construction, estimated future maintenance/replacement costs, and navigability of waterway during construction. Construction related issues during the directional bore included limited equipment staging area, maintenance of traffic, soil conditions, and the depth of the existing bridge wing walls. This paper will present the alternative analysis along with the specific construction details that enabled the Contractor to complete the City's largest directional bore to date.

INTRODUCTION

The length of a directional bore, without examining all of the other variables, can be deceiving in judging the complexity of the bore. A short directional bore does not necessarily signify a simple design or uncomplicated construction, which was the case with the City's recently completed 226 foot directional bore across Hudson Bayou.

----

[1]Boyle Engineering Corporation, 5580 Bee Ridge Rd., Sarasota, FL 34233

[2]Environmental Resources Manager, City of Sarasota Public Works Dept., 1750 12th Street, Sarasota, FL 34236

The City authorized Boyle Engineering to design the South Loop pipeline which included the environmentally sensitive crossing of Hudson Bayou. The South Loop pipeline crosses Hudson Bayou at Osprey Avenue. (Figure 1)

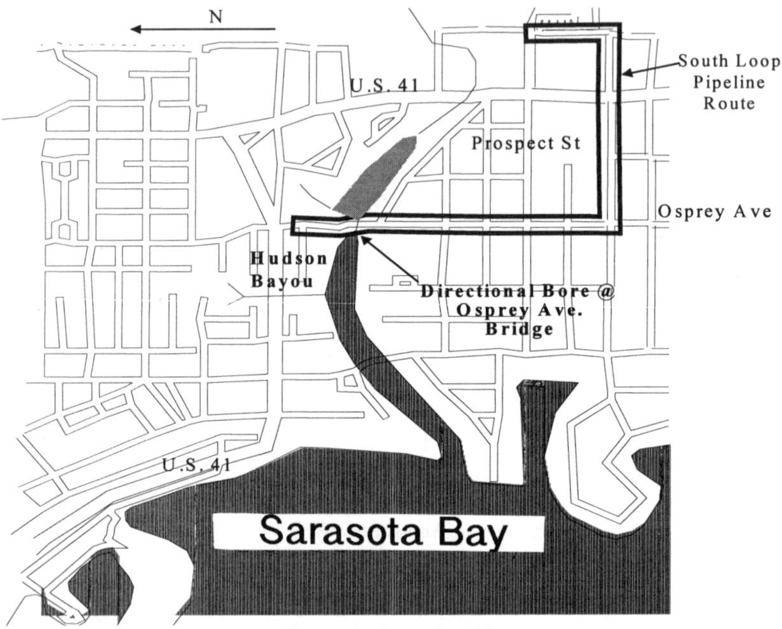

Figure 1 - Location Map

The existing Osprey Avenue bridge, which spans about 60 feet over Hudson Bayou, is an old two lane bridge which the City may need to replace in the future. On the eastern side of the bridge, wing walls extend into the bayou. Dense vegetation including mangroves are located immediately north of the northern bridge wingwall. During the original site visit to the area, numerous cracks were noticed that had formed in a rock retaining wall that is located on the north east side of the approach to the bridge. The bridge currently has five different utilities attached to it, leaving virtually no room for an additional utility pipeline. Boyle performed an alternative analysis for the Hudson Bayou crossing in the Urban Reuse South Loop Preliminary Design Memorandum for the City. This analysis is presented in the following section.

ALTERNATIVE ANALYSIS

The amount of right of way available on the western side of the bridge was very limited, so the eastern portion of the bridge was chosen for the pipeline crossing alternative analysis. (Figure 2)

Figurc 2 - Hudson Bayou on east side of Osprey Avenue facing North

Five pipeline crossing alternatives were considered with the analysis based on the following criteria: constructability, permitability, aesthetics, estimated cost of construction, estimated future maintenance/replacement costs, and navigability of waterway during construction. A brief discussion of each alternative is presented below.

1. *Directional Bore (Drilling)*

This alternative consisted of directional boring to install an 18-inch pipe underneath the bottom of Hudson Bayou then recovering to minimum cover within a relatively short distance on either side of the Osprey Avenue Bridge. Boyle developed a conceptual plan for the directional bore and presented it to several directional bore contractors during their respective site visits in order to obtain their opinion of the bore's constructability. The conceptual plan included sending and receiving pit locations, equipment requirements, pipeline cover requirements, and proposed boring angles. The directional bore contractors that visited the site concurred that the bore could be accomplished within the allowable bore limits for vertical and horizontal deflections.

Since navigation, aquatic wildlife and the channel bottom would be unaffected by the directional bore construction, the permitting requirements for this directional bore were minimal. The Army Corps of Engineers (ACOE) required a general permit, but the bore was exempt from permitting from the Florida Department of Environmental Protection (FDEP) and the U.S. Coast Guard.

2. *Jack and Bore*

This alternative consisted of jack and boring an oversized casing pipe underneath the Bayou along with installing a 16-inch carrier pipe inside the casing pipe. In order for the bore to be completed underneath the Bayou and bridge wingwall, the necessary 20 foot depth of the sending and receiving pits would be extreme. Due to the required dewatering, trenching and shoring requirements, limited right of way area for the pits, and the rock layer that would be encountered with the bore, this alternative was dismissed from further consideration.

3. *Pipe Bridge*

The pipe bridge alternative can be separated into two variations. The first variation comprises spanning the length of the Bayou with a 33 ft. by 8 ft. double tee prestressed concrete utility bridge that would safely carry the load from three pipes, the 16-inch reclaimed water line along with two future pipes, a 14-inch force main and a 10-inch water main. Support piers for the double tee would be placed landward of the existing bridge wingwalls so as not to disrupt navigation through and aquatic life within the Bayou. To avoid disruption of navigational traffic, the bottom of the double tee would be set at elevation 10.2 feet NGVD, which matches the clearance of the existing bridge. The maximum top elevation of the proposed pipes would be approximately four inches below the elevation of the top of the bridge's concrete guard rails. Dense vegetation including mangroves is located immediately north of the northern bridge wingwall and its removal is required during the pipeline construction. Removal of the mangroves and other water vegetation would require mitigation measures. The permitting requirements for this utility bridge will be greater than the directional bore, but not as extensive as open cut construction. It was anticipated that permits would be required from the ACOE, FDEP, and the Coast Guard. (Boyle, 7)

The second variation of this alternative involves using a self-supporting pipe bridge. Preliminary structural analysis indicated that the self-supporting pipes would not be capable of spanning the length of the Bayou, so intermediate supports would be proposed to be constructed within the Bayou to shorten the pipe span. Placement of the support piles would require coordination with the U.S. Coast Guard. Permitting efforts for this option would increase due to the construction and placement of the pipe bridge supports within the Bayou and their impact on navigation and the environment. For these reasons, this variation was not recommended. (Boyle, 7)

Steep side slopes and dense vegetation (mangroves) just north of the northern bridge wing wall would most likely cause an increase in construction time and construction costs for the pipe bridge alternative. (Figure 2) Due to the deteriorating condition of the northern rock retaining wall, discussions would need to occur between the City Engineer, the City's Utilities Engineer, and Boyle regarding the rock retaining wall and the impacts of proposed construction activities on the wall.

The historic Osprey Avenue Bridge enhances the picturesque view from adjacent condominiums overlooking Hudson Bayou. (Figure 3) The proposed addition of the pipe bridge would disrupt the view of the Osprey Bridge and Bayou and therefore aesthetics becomes an issue in selecting the pipeline crossing alternative. (Boyle, 8)

Figure 3 - View of Hudson Bayou & Osprey Ave. Bridge from Central Park Condos

4. *Conventional Open Cut Construction*

This alternative involves using conventional open cut trench construction to bury the 18-inch reclaimed water pipe underneath the bottom of the Bayou. The depth of the pipe was determined by permitting requirements. In order for the pipe to remain within the City's right-of-way on each side of the crossing, the pipe must pass either through or over the bridge wing walls. To preserve the structural integrity of the existing bridge walls, it was proposed that the pipe not pass through the walls. As discussed earlier, dense vegetation and steep side slopes will increase construction time and costs. Preliminary discussions with the permitting agencies and previous experiences on similar projects indicate that this alternative would be the most difficult to permit, especially with FDEP's new Environmental Resource Permit which became effective on October 3, 1995. (Boyle, 9)

5. *Utilities on Existing Bridge*

The existing Osprey Avenue Bridge has five different utilities attached to it including a 6-inch gas main, a 10-inch water main, a 14-inch force main, three 4-inch telephone conduits, and an unknown utility conduit. These utilities leave little room for adding the proposed reclaimed water main. With the congestion of the utilities and the possibility of future replacement of the Osprey Avenue Bridge, this alternative was dismissed from further consideration.

Based on alternative analysis presented above using the selection criteria previously outlined, the directional bore and prestressed double tee utility bridge were considered viable alternatives for the Hudson Bayou pipeline crossing. The City selected the directional bore for the Hudson Bayou crossing primarily based on its permittability and because it does not impact the area's pleasing landscape.

DESIGN CONSIDERATIONS

Before the final design of the directional bore was started, two very important design parameters needed to be identified: the depth of the concrete bridge wingwalls and soil characteristics within the proposed bore path. At the request of Boyle and the City Utilities Department, the City's geotechnical consultant, Ardaman & Associates, conducted two Standard Penetration Test (SPT) Borings near the proposed sending/receiving areas and two jet-wash borings at the existing wingwalls. The specific purpose of these borings was to determine the subsurface soil characteristics, the depth to the rock layer, and the approximate depth of the bridge wingwalls. The SPT boring results are summarized in Table 2 below.

Table 2 - Soil Boring Information

Soil Description	Boring #1 Elev (ft)[1]	Boring #2 Elev. (ft)[1]
Grass (Ground Surface)	10.84	11.17
Brown to gray fine sand	6.34	6.17
Brown to gray fine sand, some shell	4.34	3.17
Water Table (March 8, 1996)	2.34	1.67
Gray fine sand. trace roots	-2.16	-1.83
Gray slightly silty fine sand	-7.66	-7.83
Gray Cemented Silt (Rock)	Start @ -7.66	Start @ -7.83

[1]Bottom Elevation of Soil Layer
Source: Table 26-2, *Ardaman Report*

The approximate bottom elevation of the bridge wingwalls was determined to be -5.7 NGVD and -6.7 NGVD for the north and south wingwalls, respectively. (Ardaman, 2) Due to the depth of the bridge wing walls and the relatively short horizontal distance from the sending area to the walls, a flexible pipe material was required to make this vertical deflection. A high density polyethylene pipe (HDPE) SDR 11 was specified due to its small bending radius. The HDPE is very thick for the SDR 11 pressure rating and therefore the pipe was upsized to 18 inches to equal the flow area of the connecting 16 inch PVC pipe.

Due to the nature of the soil conditions, the directional bore contractor was provided with two separate bid items, a 6-inch pilot bore and the 18-inch directional bore. In order to avoid possible delays in the completion of the entire south loop pipeline project, the contractor was required to complete the 6-inch pilot bore within 30 days of the notice to proceed. This requirement would have enabled the City and Boyle to reevaluate and proceed with other crossing options if the pilot bore had not been successful.

CONSTRUCTION

Proper planning was the first important step by the contractor in successfully completing the directional bore in a timely manner with minimal complaints by the general public. The contractor, Communications by Poire', chose the southern side of the Osprey Bridge for the drill equipment (sending area) and the northern side of the bridge for pipe joining and pull back operations. There was adequate distance available north of the Bridge for stringing out the pipe without adversely affecting traffic. Using the north side instead of the south side for the pipe area enabled the contractor to heat fuse the HDPE joints together ahead of completing the directional bore which helped reduce delays during pipe pull back.

The contractor selected a Vermeer D24a Navigator™ with 17,900 pounds of thrust and 23,800 pounds of pullback for boring and backreaming operations. For underground location of the bore, a SPOT D TECK II™ bore locator was utilized which enabled the contractor to continuously monitor bore location, depth, pitch and roll. Hydraul-EZ®, a specially treated bentonite-based drilling fluid designed for directional drilling, was used to maintain borehole integrity while reducing fluid loss. Due to the bore's close proximity to environmentally sensitive Hudson Bayou, the contractor employed a septic tank company to provide tanker trucks onsite to collect excess drilling mud that overflowed from the pits.

The contractor began with the 6-inch pilot bore, penetrated through the rock layer, and bored under the Bayou with little effort exerted on the drill machine. However, a major problem occurred when the contractor's bore attempted to return through the rock layer on the other side of the north wing wall on its way to the north receiving pit. The contractor surmised that a soft clay layer was underlying the rock and the bore was taking the path of least resistance through the clay instead of cutting up through the rock layer. After many attempts to drill through the rock, the contractor decided to bring in a second drilling machine, a D7x11 Navigator™. Using the SPOT D TECK II™ bore locator, the contractor was able to puncture the rock layer with the D7x11 from the north side of the bridge and enter into the bore hole drilled by the D24a. The contractor used the D7x11 to push the D24a drill bit/string back to the original sending pit. The two independent drill strings from the D7x11 and D24a were connected at the south sending pit and the D24a pushed the D7x11 drill string back through the bore hole. From this point forward the backreamer was attached to the D24a drill string and the D7x11 drill string was connected to the backreamer through the swivel. This left drill string in the hole at all times in order to avoid losing the path of the original 6-inch pilot bore. The contractor used four passes with a 9-inch, 12-inch, 18-inch and 24-inch backreamer to construct the bore hole for the 18-inch pipe. The 18-inch pipe was pulled through on the fifth pass with the 24-inch backreamer. The bore was successfully completed in five construction days. The total cost of the south loop pipeline project was $944,000. Table 3 below summarizes the cost data for the bore only.

Table 3 - Partial Summary of Cost Data from Original Bid

Description	Quantity	Low Bidder	High Bidder	Average
Mobilization Pilot Bore	L.S.	$43,370	$29,000	$32,680
6" Pilot Bore	203 L.F.	$6,750	$62,524	$42,693
Subtotal (Pilot Bore)		*$50,120*	*$91,524*	*$75,373*
Mobilization Directional Bore	L.S.	$52,840	$29,000	$32,460
18" Bore	203 L.F.	$37,555	$102,312	$81,149
Subtotal (Directional Bore)		*$90,395*	*$131,312*	*$113,609*
Total		**$140,515**	**$222,836**	**$188,982**

CONCLUSION

The completion of the directional bore across Hudson Bayou in conjunction with the South Loop pipeline project completes the inner loop of the City's Urban Reclaimed Water Reuse System. The successful completion of the Hudson Bayou crossing with the directional bore can be attributed to proper planning by the City, Boyle, and the Contractor.

A manatee, which is on the endangered species list, was present in the Bayou in the vicinity of the Osprey Bridge during the directional bore construction. This encounter with the manatee confirmed the wisdom of using the directional bore method as the manatee and other aquatic life may have been affected by the other previously discussed construction methods for the Bayou crossing.

Another underground crossing on the South Loop project, which was originally designed and bid as a jack and bore crossing of US 41, was also constructed using the directional bore method. This method was chosen by the City and the Contractor largely due to the success of the directional bore crossing of Hudson Bayou.

REFERENCES

Ardaman & Associates, "Subsurface Soil Exploration, Analysis and Recommendations for proposed 'Hudson Bayou Reuse Pipeline Crossing @ Osprey Avenue, Sarasota, Florida'", prepared for City of Sarasota, March 1996.

Boyle Engineering Corporation, "Urban Reuse South Loop Preliminary Design Memorandum", prepared for City of Sarasota, January 1996.

Urban Reuse South Loop Contract Documents - Conformed Copy, Sept. 4, 1996.

PREDICTING I/I REDUCTION FOR PLANNING SEWER REHABILITATION

George E. Kurz[1] , P.E., DEE, ASCE Member

ABSTRACT

This paper will discuss measured I/I reduction following trenchless lining work conducted as part of Nashville's sewer rehabilitation program. Eight previously unreported projects consisting of 20,600 m (67,600 ft.) of sewer lining work are evaluated. The level of lining intensity within each of the projects varied from 1% to 46%. The goal for each project was to improve capacity and reduce (I/I) to a level where the peak flow resulting from a storm having a five-year recurrence probability could be carried by the sewers. As a result of these projects, an estimated 3.0 x $10^6 m^3$ (800 million gallons) of I/I is excluded from the collection system annually. This represents nearly a 30% reduction. Additionally, peak hour flow rates were reduced by nearly 43%.

Targeting I/I producing defects and selecting appropriate pipe segments and groups of segments for rehabilitation, appear to be the greatest variables for planning rehabilitation to effectively remove I/I. The combined results from these projects and others previously reported for Nashville, indicate that it is reasonable to predict removal rates of 14,700 m^3/year/100m of lining (12 million gallons/year/1000 ft of lining) for deteriorated sewers. This success rate depends on proper selection and on sufficient lining intensity in an affected mini basin to forestall easy migration of groundwater.

INTRODUCTION

Despite the growing number of sewer rehabilitation processes and products, few articles are available in the literature about their effectiveness in I/I removal. The objective of this paper is to identify a basis for planning and designing I/I reduction projects based on actual project performance. In an effort to develop a data base of objective measures of effectiveness for sewer rehabilitation projects, Kurz, et al (1994) analyzed eight projects conducted in

----

Consoer Townsend Envirodyne Engineers, Inc., 545 Mainstream Drive, Suite 200, Nashville, TN 37228 (615) 244-8864

the Nashville, Tennessee Overflow Abatement Program (OAP) which included 16,830m (55,220 ft) of sewer lining or replacement. The goal of the Nashville program was to improve capacity and/or reduce I/I to a level where the peak flow resulting from a storm having a five-year recurrence probability could be carried by the sewers without overflows. Linear regression analysis was used to predict I/I resulting from a five-year, 24-hour storm of 114mm (4.5 in). Some projects showed sufficient I/I reduction to achieve a five-year payback in treatment costs. The eight projects reviewed resulted in an annual I/I reduction of 2.12×10^6 m^3 (560 million gallons).

Since 1991, the City of Nashville, has required the rehabilitation of all service laterals contiguous to rehabilitated public sewers. Vogel (1994) also reported on the significance of service lateral rehabilitation in North Olmsted, Ohio. As a result of smoke testing, 215 laterals were identified with defects and were repaired by epoxy resin (Amkrete) injection. Additionally, lining (1.3%) and grouting (37.5%) rehabilitation work was conducted in 12,000m (400,000 ft) of the sewer system under study. As a result, the peak flow rate at the wastewater treatment plant in North Olmsted was reduced about 60%.

Projects ranging in length from 1,158m (3,800 ft) to 11,277m (37,000 ft) in Broadview, Mattesen, and Melrose Park in Illinois were rehabilitated by CIPP (Insituform) as reported by Goumas (1995). In one case, Melrose Park estimated annual savings between $2,570,000 and $3,030,000 due to I/I reduction. Flow monitoring documented successful reduction of 49% to 82% of I/I rates.

I/I contributions from deteriorated manholes may be significant, as reported by Croxton (1995) and rehabilitation of these structures may result in significant flow reduction. A 97% reduction in annual total normalized inflow volume was projected after 88 manholes were rehabilitated in a pilot project in Newport, Texas. Additionally, flow monitoring documented a 60% reduction in average daily wet weather flow and a 28% reduction in average daily dry weather flow.

Sewer Rehabilitation

Sewer lining has been the principal rehabilitation technique used in the OAP and has been the major factor in I/I and overflow reduction. From 1991 to 1996, lining projects with a construction value of $30.4 million have been installed. Almost 35,000m (114,500 ft) of cured-in-place pipe (CIPP) installed by Insituform and 26,000m (85,100 ft) of fold-and-form pipe (FF) installed by U-Liner have been the principal lining processes used in the program. CIPP by InLiner has been tested in part of the Brookwood project and installed in another project area not yet evaluated. FF by NuPipe has also been included in project specifications as an acceptable alternate.

Over 1,000 service laterals have been renewed in Nashville projects. This work represents about a tenth of the rehabilitation contract cost. Assuming removal rates similar to levels achieved in the Oak Valley project (Bible, 1991), then this results in an incremental I/I removal rate of 20% to 25% for 10% of the construction cost. In many cases, these renewals have been accomplished using a proprietary, trenchless CIPP process (Insituform), to avoid street cuts and other disruption. A small pit is usually required to install a clean-out for lateral liner insertion. Traditional renewal by trench excavation is allowed in unpaved, easement areas and in some paved areas where new paving is planned as part of regular street maintenance. Work to date has demonstrated that quality control of the service reinstatement is very operator dependent. An air pressure test after service renewal is required to verify a water-tight renewal.

With fold and form pipe lining systems, achieving a leak-proof seal at the manhole and eliminating I/I "tracking" from the annular space to the manhole has been very difficult. Additional measures such as installation of hydrophilic rubber seals or chemical seals must be used. Extended warranties and follow-up visual manhole inspection are administrative tools to help verify seal integrity. Cured-in-place products have not been as likely to have this problem, but construction conditions can cause manhole/line leakage. The OAP currently requires seals on all linings to minimize this problem.

METHODOLOGY

Demonstration of Rehabilitation Effectiveness by Flow Monitoring

The measure of effectiveness of rehabilitation is directly related to the original goal to reduce I/I and/or eliminate overflows. Flow monitoring and rain measurement must be conducted to collect enough data to predict peak flow conditions for the design return interval storm (Sullivan and Ewing, 1985). The ideal monitoring system would incorporate hardware and software for flows to be automatically calculated using both the continuity equation and a calibrated Mannings equation. A network of permanent and temporary flow meters and rain gauges in the MWS system supplied by ADS Environmental Services Inc. (Huntsville, AL) satisfies these requirements. Comparing simultaneous hydrographs calculated by each equation would immediately identify periods of backed-up flow (non-Mannings free flow).

Following rehabilitation, temporary flow monitors are installed in the same sites as used in the original investigation (or logical alternative sites are selected if the sewer system has been altered by new construction). Generally, a two month wet period during winter or spring has been sufficient to capture data from several storms for flow projection. Storms between 25.4 mm (1-inch) and 63.5 mm (2.5-inches) have produced satisfactory results.

Selection of storm events must include consideration of soil antecedent moisture conditions (AMC). This may be a major source of variability in the regression analysis. Generally, in the Nashville program, it has been possible to minimize variability with intelligent selection of storm events based on observation of previous rainfall. Merrill (1994) proposed an algorithm to predict long term levels of AMC if a more precise analysis is required. Project areas are televised as completely as possible by Metro crews. All defects are noted and given a rating based on the severity of the defect. Development of the rating system and its defects criteria application has been described by Dillard (1993). The City of Edmonton also has reported using a sewer defect rating system (Smith, 1996). This system also enhances the uniformity of information from each TV operator. These criteria are used to select line segments for rehabilitation.

RESULTS

As shown in **Table 1**, rehabilitation by lining and replacement of 20,600 m (67,600 ft) of sewer resulted in a projected annual reduction of 3.012 x 10^6 m^3 (795.8 million gallons) of I/I. As a gross average, annual I/I is projected to be reduced by 29.6% from lining 8% of the length of the monitored project areas. As discussed earlier, a portion of the I/I reduction is attributable to the MWS policy to renew all services contiguous to the rehabilitated public sewers. The projects were listed in ascending order of length of rehabilitated pipe. As expected, there is a concurrent trend of increasing annual I/I reduction (with some notable exceptions). The Paragon Mills projects were conducted in areas that did not produce large amounts of annual I/I before rehabilitation. Rehabilitation was conducted in those areas to correct significant capacity problems with those sewers. The success of those projects is readily seen in the reduction of the peak hour flow rate projected for the five year storm as shown in **Table 2**. The projects are listed in ascending order of % rehabilitation. Overall, there has been a 42.7% reduction of the peak flow for these eight projects.

TABLE 1 ANNUAL I/I REDUCTION

PROJECTS	Rehabilitated Sewer	Annual I/I Reduction		
	Length (m)	Percentage	Volume (m^3x10^3)	Percentage
* Lower East Nashville (5)	195.0	12	35.2	38
* Anderson Road (PR3)	351.0	<23	26.9	34.2
Collier Avenue (NR1)	410.0	12.1	162.5	58.1
*Lower East Nashville (6)	521.0	24.9	84.0	68
* Smith Springs (SS3)	693.0	33	6.0	28.6
Kenner Avenue	751.0	45	344.4	88
Oak Valley	1,341.0	41	253.6	76
* Seven Mile IDU45	1,583.0	1.1	552.2	18.7
* Paragon Mills (PM1)	2,474.0	21	19.6	32.6
* Paragon Mills (PM2)	2,522.0	45.8	2.5	30.6
Hopedale	4,902.0	21	1,093.9	73
West Linden	6,402.0	93	193.0	45
* Sugartree	12,261.0	15.1	2,285.6	33.5
GRAND TOTAL	34,406	11.9	5,059.5	38.6
* NEW PROJECT TOTAL	20,600	8.0	3,012	29.6

* Indicates projects not previously reviewed.

DISCUSSION AND CONCLUSIONS

One of the main problems confronting municipalities with excessive I/I and wet-weather overflows is, "How to develop a rehabilitation strategy?" Generally, communities don't consider rehabilitation or replacement of all sewers. Rather, it is assumed that rehabilitation will be conducted until "enough" I/I has been removed to eliminate overflows and backups. Until recently, sufficient information has not been readily available to provide a basis for projecting the level of rehabilitation effort required to eliminate overflows (or reduce them to a frequency accepted by the community). Overall, the cumulative performance

TABLE 2 REDUCTION OF PEAK HOUR FLOW RATE

		DESIGN STORM PEAK HOUR PROJECTIONS		
PROJECTS	**Rehabilitated Sewer Percentage**	**Before Rehab (m^3/sec)**	**After Rehab (m^3/sec)**	**Percentage Reduction**
Seven Mile IDU45	**1.1**	**0.645**	**0.648**	**-0-**
Lower East Nashville (5)	**12**	**0.023**	**0.013**	**43**
Sugartree	**15.1**	**2.175**	**1.061**	**51.2**
Paragon Mills (PM1)	**21**	**0.263**	**0.036**	**86.4**
Anderson Road (PR3)	**<23**	**0.079**	**0.047**	**40.2**
Lower East Nashville (6)	**24.9**	**0.136**	**0.110**	**19**
Smith Springs (SS3)	**33**	**0.075**	**0.044**	**41**
Paragon Mills (PM2)	**45.8**	**0.036**	**0.007**	**80.6**

of these Nashville projects suggests that 14,700 m^3/year/100 m of lining (12 million gallon/year/1000 ft of lining) is a conservative basis for planning I/I reduction in badly deteriorated sewers.

This performance rate includes the apparent poor performance from the West Linden project. However, as previously discussed (Kurz, 1994) a significant amount of I/I remaining in this area came from sump pumps and roof drainage. If this project is removed from the analysis, then a performance rate of 17,400 m^3/year/100 m of lining (14 million gallons/year/1000ft of lining) may be calculated. A more refined approach might calculate lining effort in terms of mm diameter - km (inch-miles). However, since most of the Nashville projects lined 200 mm (8 inch) diameter pipe, this should not affect the final result significantly.

Figure 1 illustrates the individual performance of each of the 13 Nashville projects with a calculated straight line regression plot. For comparison, the average performance rate calculated above (without West Linden) is also shown.

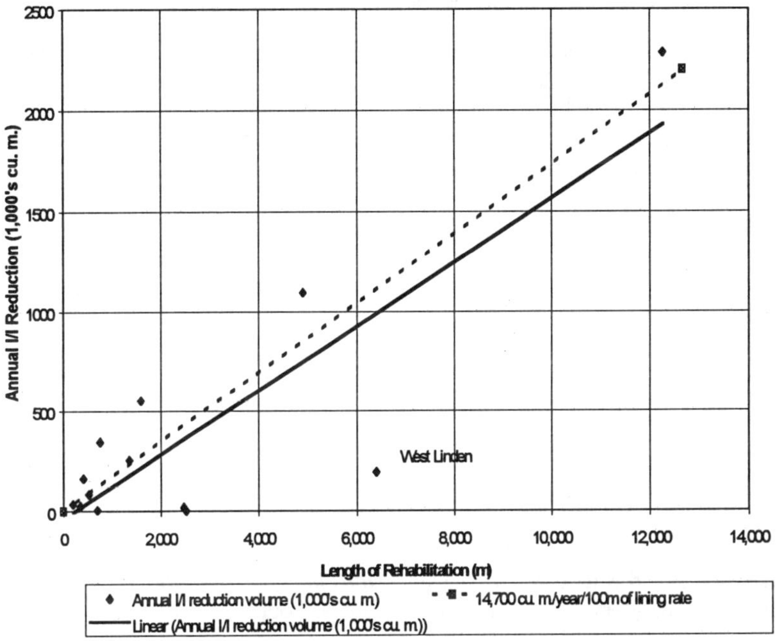

Figure 1. Rehabilitation Effectiveness as Measured by Annual I/I Reduction

ACKNOWLEDGMENT

The projects in this paper were funded by The Department of Water and Sewerage Services of the Metropolitan Government of Nashville and Davidson County, Tennessee. Cyrus Q. Toosi, P.E. of that Agency was instrumental in developing the flow monitoring plan. Also, the linear regression figures (graphics) included in this paper were adapted from an analysis program developed by Michael A. Burgett, P.E. and Jason W. Griffin of Gresham Smith and Partners. Their assistance is greatly appreciated and acknowledged by the author.

REFERENCES

Bible, D., *"Results from Sewer Rehabilitation Test Areas".* Water Pollution Control Federation 64th Conference, (1991).

Croxton, R., *"Manhole Sealing Stops Severe I/I at Houston - area Residential Development."* Trenchless Technology, Vol. 4, No. 3, (1995).

Dillard, C.W., G.E. Kurz, and P.A. Stonecipher, *"Management of Sewer System Rehabilitation for the Overflow Abatement Program in Nashville, Tennessee"* Pipeline Infrastructure II, American Society of Civil Engineers, (1993).

Goumas, J., *"Tri-Villages of Greater Chicago Reduce I/I with CIPP."* Trenchless Technology, Vol. 4, No. 3. (1995).

Kurz, G., Dillard, C., Stonecipher, P., Ballard, G., Scott, L., and Toosi, C. *"Successful Sewer System Rehabilitation,"* WEFTEC - 94 Chicago, Illinois, (1994).

Merrill, M. S. and Butler, R., *"New Dimensions in Infiltration/Inflow Analysis,"* WEF - A Global Perspective for Reducing CSOs, Louisville, Kentucky (1994).

Smith, G. *"Evaluation of Sewers for Trenchless Rehabilitation,"* No-Dig Engineering, Vol. 3, No. 3, (1996)

Stevens, P., *"Basin Size is the Most Effective Variable an I/I Engineer Can Control."* Water Environment Federation 66th Conference, (1993).

Sullivan, R. and J. Ewing, *"Selected Topics Related to Infiltration and Inflow in Sewer Systems."* EPA/600/2-85/020, NTIS No. PB 85-177202, U.S. EPA, Water Engineering Research Laboratory, Cincinnati, Ohio, (1985).

Vogel, M., *"Rehab Saves Flooded Basements,"* Trenchless Technology, Vol. 3, No. 6, (1994).

The Houston Experience: Recipe For A Successful Demonstration Project

A.D. "Art" Kidder, Jr.[1]
William Sukenik, P.E.[2]
T. Lynn Chance, P.E.[3]

Abstract

Houston is engaged in a $7.5-million sanitary sewer rehabilitation demonstration project with the following objectives:

- demonstrate the effectiveness in the reduction of rainfall dependent inflow and infiltration (RDI/I) and peak RDI/I flow rates achieved by two levels of rehabilitation;
- evaluate the comparative costs of the two levels of rehabilitation; and,
- characterize the contribution of defective private services and illegal inflow sources on private property to the overall experienced RDI/I.

Four sewer basins, each with less than 7,010 m (23,000 feet) of main lines have been completely or partially rehabilitated. The preliminary analysis of flow data indicates twice the reduction in RDI/I with the higher level of rehabilitation.

Introduction

In the 1980s, it was evident that Houston was experiencing an inordinate amount of sanitary sewer failures, stoppages and overflows. Deteriorating sewers were contributing to the erosion of sewer service quality and customer confidence, while inconveniencing citizens and visitors. Furthermore, structural failures and stoppages were threatening water quality and public health, and causing damage to public and private property. Only an intensive sewer system rehabilitation program focused on the pipes in the worst structural condition could turn the situation around.

In its fourth year of implementation, the structural rehabilitation program is now paying dividends as corrective maintenance work orders show significant decline while customer satisfaction improves. Houston has initially committed to spend over $300 million on its structural rehabilitation program in a five-year period starting in 1992. It is anticipated that this program will continue beyond this period with an average annual expenditure of $50 million or more.

However, Houston recognized that its rehabilitation efforts were focused on repairing the structural deterioration of its 9,640-kilometer (6,000-mile) sewer system without a thorough understanding of any concurrent benefits to the reduction of rainfall dependent infiltration and inflow (RDI/I), and more specifically, peak RDI/I flows.

[1] City of Houston, 1100 Louisiana, Houston, Texas
[2] Greater Houston Wastewater Program, 1100 Louisiana, Houston, Texas
[3] Camp Dresser & McKee, 1800 West Loop South, Houston, Texas 77027

Houston is under an EPA administrative order requiring the control of sanitary sewer overflows by the end of December 1997. The need for rehabilitation plus the terms of the EPA order prompted Houston's interest in identifying how it might modify its structural rehabilitation program to cost effectively maximize its contribution to a reduction of RDI/

Project Goals. As a measure to fill this information gap specific to the sewer conditions and environmental characteristics such as soil types and climate of the Houston area, the city embarked on a four year, $7.5-million Sanitary Sewer Rehabilitation Demonstration Project ("the project") with the following objectives:

1. Demonstrate effectiveness in reduction of peak RDI/I flows of two levels of rehabilitation:
   - "City Method" - rehabilitation of only those mainlines and laterals in the public rights-of-way or easements meeting certain city-specified criteria for structural condition based on television inspection data; and,
   - "Comprehensive Method" - rehabilitation of virtually all mainlines, laterals and services, sometimes referred to as upper laterals, on private property.
2. Evaluate comparative costs of two methods of rehabilitation; and,
3. Characterize contribution of defective private services and illegal inflow sources on private property to overall experienced RDI/I.

The objective of this presentation is to share the key experiences of Houston in the implementation of the project so others planning similar projects may use this information to assist in the design of the most effective demonstration to suit their respective objectives.

Selection of Test Basins

Since the essence of the project is the comparison of pre- and post-rehabilitation sewer flow conditions, the physical condition of the test sewer basins should be similar. This includes similar percentages of differing pipe materials, pipe age, land use, type of storm drainage systems, soil permeability and other physical characteristics. Additionally, relative close proximity of each basin to one another is important as a measure to minimize some of the inherent variability in intensity and coverage of rainfall events commonly experienced across a region.

The design of the project included four test basins with two additional control basins. As with other scientific investigations, the project used areas in which no rehabilitation was performed as control for comparisons of pre- and post- rehabilitation conditions. It is important to monitor basins not rehabilitated under similar rainfall conditions to allow valid comparisons and to account for seasonal RD/I variations.

The City Method was implemented in two of the four test basins while the Comprehensive Method was implemented in another two basins. Basin selection was highly dependent on physical similarities and sewer system response to wet weather flows. The following criteria were used in screening 14 basins (Camp Dresser & McKee, 1994) initially nominated for participation:

1. Land Use. To minimize flow variability, the most desirable basins for investigation are those that serve predominantly residential users. Diurnal flow patterns are much more consistent in these basins. An optimal basin had at least two-thirds residential land use with limited commercial/ industrial land uses.

2. Basin System Age. The most desirable basins are at least 20 years old which generally assures a variety of pipe materials in use, and are likely more subject to I/I.
3. Area Size. The objective was to use a basin with approximately 6,096 m (20,000 feet) of mainline pipe that would assure sufficient tributary flows to allow flow monitoring without extraordinary measures like installing flumes or weirs, but small enough for a cost-effective demonstration project.
4. System Layout and Location of Sewers. To have ready access to all mainline sewers, basins with the majority, if not all, of its mainlines in public rights-of-way were desired. Furthermore, gravity mainlines were desired with no interbasin cross-connections.
5. Flow Characteristics. A basin must exhibit normally expected diurnal flow variations without surcharging the system at peak dry weather flow periods. An acceptable basin should also return to normal flow patterns within 24 hours of a storm, and should exhibit relatively uniform response to subsequent storms of similar intensity and duration.
6. Pipe Conditions and I/I Sources. The desirable basins are those that have not been rehabilitated and have minimal numbers of illegal inflow sources on private and public property.

The application of the above criteria resulted in a group of six basins subjected to further screening criteria. These included:

1. Storm Drainage System. Basins selected were characterized by open ditch drainage, which provided the greatest potential for leaks from the drainage system into the sanitary sewer system via breaks in laterals within and under ditches.
2. Accessibility and Suitability for Flow Monitoring. The availability of readily accessible manholes with adequate hydraulic characteristics to support relatively accurate flow monitoring was confirmed.
3.Design and Maintenance Status. Plans of ongoing and future infrastructure improvements and maintenance were reviewed to make sure that conflicts with other programs would not materially affect any selected basin prior to completion of the project.
4. Geographic Location. It was desired to have all test and control basins in relatively close proximity to one another to optimize storm effect over the entire group on a simultaneous basis as much as possible.

Table 1 summarizes the key physical attributes of the project's test basins located in Houston's Northside II Service Area, tributary to Houston's 69th Street Wastewater Treatment Plant, the largest plant in the system.

Table 1. Summary Description of Test Basins

Basin Number	Rehabilitation Method	Mainline (m)	Diameter (cm)	Area (ha)	Service Connections
II 013	Control	5,888	15.2-30.5	60	650
II 017	Comprehensive	6,893	15.2-25.4	53	650
II 021	Comprehensive	4,253	15.2-30.5	26	341
II 109	City	6,496	15.2-38.1	49	360
II 149	City	4,267	15.2-20.3	40	550
II 186	Control	6,517	15.2-38.1	77	356

Physical Inspection And Flow Monitoring

The initial step in any physical inspection and flow monitoring project, or Sewer System Evaluation Survey (SSES), is to establish the scope of the SSES work. This was one of the most critical steps in the project's success. Issues such as location and number of flow meters and rain gages, extent of smoke testing and closed circuit television inspection (CCTV), and the procedure for locating "hidden" manholes, were decided before the contractual arrangements were initiated. Alternatively, the SSES can be performed by in-house forces. This was not feasible for the project. In either case, an administrative authority must determine the types of physical inspection tests needed to establish the appropriate rehabilitation approach, followed by the formulation of an SSES work plan.

Rain Gages and Flow Meters. The first step in implementing the SSES plan was to install rain gages and flow meters, since this portion of the data gathering required the greatest amount of time. The desired area coverage for rainfall monitoring was determined before site selection and installation began. Normally, the number and location of rain gages is dependent on the shape and area of each basin being monitored, method being used for calculating the total rainfall, and nature of the storms in the geographical area. Soil moisture conditions may affect RDI/I quantity; and rain gages should be installed first for at least 20 days before the flow meter. This was not possible on the project due to time constraints at the beginning of the pre-rehabilitation monitoring period.

The general nature of rain storms and size of the demonstration project area should be used to determine the rain gage density. Experience shows limiting the number of rain gages used to monitor the rainfall in a basin to conserve costs is counterproductive. If the prevailing storm pattern in each basin area is fairly uniform and the individual basins are reasonably close together, fewer rain gages can be used to monitor rainfall. Under ideal conditions, the rain gages in adjacent basins can be used to establish the average rainfall pattern, and the minimum number of rain gages would be one gage located near the center of each basin with a sufficient number of rain gages around the perimeter of the aggregate basin area to adequately define the storm pattern.

Where rain storms are highly variable, such as periods of thunderstorm activity, rain gages at the center of each basin should be supplemented with additional gages in a pattern around the basin's periphery. Irregularly shaped basins, instead of the desired square or circular shape, will require additional rain gages. The project was fortunate in that Houston has a system of permanent rain gages with 15 of them used to supplement the gages located in each basin. In addition, five of the six project basins were close enough to use the data from the adjacent test basin's gage to calculate average rainfall amounts.

Monitoring rainfall by the use of Doppler radar data was investigated. This method establishes rainfall intensities during a particular storm event by "calibrating" the Doppler intensity bands from real-time data acquired by rain gages in the area being monitored. This system could result in more accurate rainfall data generated with fewer numbers of rain gages. However, this method is a broad coverage monitoring system. The minimum unit size of Doppler data is 0.75 sq. km (185 acres). Therefore, if the basin areas are not close to this size, this method is not suitable for rainfall monitoring. This method could not be used in Houston because the basins were smaller.

Physical Inspection. Once rain gage and flow monitor locations have been determined and installation begun, other SSES work should commence. Smoke testing of all sewers in the study area should be performed first to determine the sequence and number of additional tests needed. Where smoke testing indicates that there may be connections between the storm drain system and the sanitary sewer system, dyed-water flood testing with concurrent CCTV inspection should be performed to verify the cross connection. Direct, manmade connections need to be verified and eliminated as soon as possible, since they are usually very large sources of rainfall dependent inflow. Naturally occurring cross connections via soil fissures between defects in the storm drains and the sanitary sewers are also large contributors of rainfall dependent sewer flows.

The performance of smoke tests on all sewers in the study area cannot be overemphasized. During the project's kickoff meeting, smoke testing of sewers in the control basins was eliminated as a cost saving measure. Subsequently, flow anomalies were identified in Basin II 013, which led to verification of a sanitary sewer cross connection with the adjacent Basin II 012. Had smoke testing been performed, the cross connection would have been identified on a more timely basis. Corrective action could have been taken to eliminate the cross connection, or to select a new control basin. Instead, remedial measures were taken to salvage the use of Basin II 013 as a control basin with additional flow monitors and an extended monitoring period. This increased project costs.

During 1989 and 1990, Houston had performed a major, comprehensive SSES program to inspect over 60 percent of the city's sewers. At that time, guidelines were set up to perform supplemental SSES inspections. The project's mandate, therefore, was to perform inspection of all sewers and manholes that were not previously inspected, in addition to those sewers identified during the project's smoke testing work as requiring dyed-water flood testing with concurrent CCTV inspection. Reinspection by smoke testing and CCTV inspection allowed comparison of the sewers' condition between the 1989/90 SSES period and the present. The project inspection work found:

- general worsening of the sewers' conditions,
- only 25 percent repeatability of leaks previously identified by smoke testing; and,
- an unexpected amount of silt deposition in sewers cleaned only four years previous.

One conclusion is that it is not necessary to CCTV inspect all of the sewers at the initiation of a major SSES program. Sanitary sewers suspected to be structurally unsound due to incidents of blockages, overflows and other operational problems, should be CCTV inspected, while such inspection of remaining sewers should be on a random basis to determine the mean condition of each basin. Experience in other communities has shown that CCTV inspection of approximately 10 percent of the sewers chosen at random is generally sufficient to establish the mean condition of the sewer basin. Since CCTV inspection is the most expensive of SSES work and conditions in the sewers change with time, Houston has found that CCTV inspection of the remainder of each basin should be performed immediately before undertaking the rehabilitation design to assure that design is based on the most up-to-date information. If design is to immediately follow the SSES, as with the case of the project work, then all sewers, irrespective of planned rehabilitation approach, should be CCTV inspected.

Inspection of all existing manholes not previously inspected in the test basins was also performed. Manhole inspection is generally much less expensive than CCTV inspection. The manhole inspection data should include a lamp and mirror inspection of the first few meters of sewer upstream and downstream of the manhole. On other projects where a random portion of the sewers are to be CCTV inspected, data gathered during manhole inspections can be used to extrapolate the data gathered during the random CCTV inspection to other line sections, as well as determining the need for manhole rehabilitation.

Selecting Rehabilitation Technologies

The data gathered during the SSES phase was used as the basis for evaluating the condition of the sewers. This also supplied information for selecting appropriate rehabilitation methods for both the Comprehensive Method and City Method designs. Assuming that all available rehabilitation technologies for pipelines are equally efficient in reducing RDI/I, one primary project objective was to determine the relative effectiveness of these two rehabilitation approaches in reducing peak rate and volume RDI/I. The project compared two rehabilitation design approaches. Houston established the City Method to address sewers which have less than minimum structural integrity with limited regard to concurrently reducing RDI/I. The project team developed a comprehensive rehabilitation design that addressed all defects in every sewer in the basin.

City Method. Sewers scheduled for City Method rehabilitation are selected on the basis of the structural condition of each sewer segment with an average manhole-to-manhole length of 100 m (330 feet), which is calculated on a numerical rating system. Sewers that have a condition rating of 12 or higher are scheduled for rehabilitation. Sewers that have a rating between 8 and 11 are reevaluated during the rehabilitation design phase to assure that their structural condition is adequate to remain in service without rehabilitation. Sewers with a condition rating of less than 8 are not considered for rehabilitation. The City Method also addresses rehabilitation of a maximum of 2.5 m (8 feet) of the service lateral immediately upstream of the rehabilitated sewer main. The City Method rehabilitation approach currently assumes a 15 percent reduction in peak RDI/I rates. The decision as to which rehabilitation technique is used for each segment of sewer is based on a "decision/logic tree" that limits the maximum number of point repairs that can be performed on any pipe segment and takes into account the relative capacity losses impacted by each technology.

Comprehensive Method. In contrast to the City Method, the Comprehensive Method addresses defects in all sewers in the project area. Comprehensive rehabilitation uses the same pipeline rehabilitation techniques as the City Method but addresses the rehabilitation of all sewer pipes in the project area, regardless of their condition rating. The sewer service lateral located within the publicly controlled right-of-way or easement is completely rehabilitated, which usually means replacement by either open cut installation or pipebursting. Comprehensive rehabilitation addresses the condition of privately owned portions of the sewer system when possible. Comprehensive rehabilitation has been shown to achieve reduction of RDI/I volumes as high as 70 percent and RDI/I peak flow rate as high as 60 percent (Townsend and Miles, 1995).

The Comprehensive Method design team developed a data analysis technique to choose the most appropriate rehabilitation technique for each sewer segment based on total life-cycle cost to rehabilitate the segment. This approach does not limit the number of point repairs that can be performed on a sewer segment. However, should the number and extent of point repairs for support of other rehabilitation work become excessive, cost for that rehabilitation method will be greater than the cost for other rehabilitation techniques and will be eliminated on the basis of cost only. In this manner, a greater number of point repairs can be allowed on shallower pipes where the unit price for point repairs is lower and still be the most cost effective rehabilitation method.

Analysis of Pre- and Post-rehabilitation Flow Data

One method of evaluating RDI/I is to calculate it as a percentage of total rainfall. By using weighted averages of adjacent rain gauges to determine the amount of total rainfall that falls in the basin area and then using the flow monitoring data to determine the amount of RDI/I that enters the sanitary sewer system, the RDI/I percentage can be calculated. Calculation of the RDI/I in this manner facilitates comparison of the effects of different rain storm intensities under varying soil moisture conditions.

The effectiveness of a rehabilitation project can be measured by determining the reduction in both the peak rate and volume of RDI/I entering the sanitary sewer system, and then comparing the pre- and post-rehabilitation flows. A common analysis of storm data is focused on the reduction of the volume of RDI/I reduction. However, the project is quantifying the volume of RDI/I reduction, and estimating the reduction in the peak rate of RDI/I.

Usable Storms and Flow Monitoring Data. The collection of the pre- and post-rehabilitation flow data is critical to the success of determining the effectiveness of the rehabilitation plan. However, data from all storms are not usable. Some storms are too large and cause system surcharge, reducing the amount of RDI/I that can enter the sanitary sewer system. Other storms are too small and do not provide a measurable response in increased flow in the sanitary sewer system. Thunderstorms often do not distribute uniform rainfall over the basin. Therefore, the total rainfall falling on the basin cannot always be determined accurately. To effectively assess I/I reductions, an adequate number of storms with high quality data needs to be monitored.

Antecedent Rainfall Relationship. Soil moisture content may also affect the amount of RDI/I entering the sanitary sewer system. The database needs to include storms occurring during periods of both wet and dry soil moisture conditions. Storms that do not meet the quality level for complete analysis of RDI/I can still be used to correlate between antecedent rainfall relationships and the maximum rate of RDI/I that enters the sanitary sewer.

Houston experiences both desert- and monsoon-like climatic events. The Houston experience teaches us once again that storms providing high quality data come few and far between, and monitoring equipment must be kept in proper working order at all times so that an opportunity to capture unique data is not lost. This was learned during the pre-rehabilitation flow and rainfall monitoring phase, and additional measures to reduce the possibility of lost data due to equipment malfunction have been implemented. These measures include:

- Installing two rain gages at each site for redundancy;
- Performing reconnaissance and meter checks at the beginning of rain storms or when high probabilities of rainfall are forecast;
- Gathering data from both flow monitors and rain gages at an increased frequency by telemetric polling; and,
- Performing equipment maintenance on a more frequent schedule.

Flow Monitoring Data. A goal of the flow and rainfall monitoring program is to obtain a total of at least six high quality storms during each of the pre- and post-rehabilitation monitoring periods. Three storms should occur during wet soil conditions where near maximum RDI/I is contributed, and three during dry soil conditions. If adequate high quality data cannot be obtained, then data from a greater number of lower quality storms or historical data, when available, can be considered, as was the case for the pre-rehabilitation monitoring for the project. Although this can result in conclusions of less reliability, this approach is better than not having any information with which a rehabilitation plan, partly focused on the reduction of RDI/I, can be developed. Concurrently and beyond the life of the project, flow and rainfall monitoring will be continued in sewer basins not part of the project but part of Houston's ambitious rehabilitation program. This will allow ongoing accumulation of data to assess overall effectiveness of the City Method and any modifications to it as a result of the implementation of project recommendations.

Project Costs and Results

Estimated capital costs including design, inspection, and administration summed to $5.7 million, while the estimated costs of project management, planning and flow monitoring of the four-year project totaled $1.8 million. Preliminary evaluation of flow data using several analytical methods suggests 23 to 36 percent and 30 to 51 percent ranges in peak RDI/I rate reduction for the City Method and comprehensive rehabilitation (not including private services), respectively. The results are still under review, and a correction factor to adjust for the drought conditions encountered during the post-rehabilitation flow monitoring period is likely warranted. This could significantly decrease the calculated peak RDI/I rate reductions.

Summary and Conclusions - Successful Project Implementation

There are many issues that can affect the success of a sewer rehabilitation demonstration project. They must be addressed early in the planning stage, when possible, and throughout the project as they arise. Some of these key issues include:

- Availability of financial resources to cover all project expenditures;
- Ability to expend public funds on private property improvements;
- Ability to identify potential construction conflicts of other ongoing and planned capital improvements (e.g., drainage, paving and sidewalks) when selecting test sewer basins;
- Quantification of the needed intensity, duration and number of rainfall events to adequately establish pre-/post-rehabilitation sewer flow conditions;
- Accessibility of private property to evaluate and rehabilitate private services;
- Cooperation of property owners in eliminating illegal inflow sources.

References

Camp Dresser & McKee (1994) City of Houston, Greater Houston Wastewater Program, sanitary sewer rehabilitation demonstration project update of Task E technical memorandum.

Townsend, J.A. and Miles, S.W. (1995) Charlotte's facility plan to mitigate sanitary sewer overflows. EPA National Conference on Sanitary Sewer Overflows, April 24-26, 1995.

THE NORTH METRO INTERCEPTOR SEWER

A Trenchless Pipeline Project

Mike Faramarzi, PE[1]

Abstract

This major City of San Diego Metropolitan Wastewater Department project involves design and construction of a key 2750 mm (108-in dia) parallel interceptor sewer extending 3,060 m (10,036 ft) through a highly congested urban area and a 2900 mm (114-in dia) two-pass tunnel pipeline continuing another 1,440 m (4,725 ft) under a hazardous landfill. The project, entailing over 2,600 m (8,550 ft) of trenchless pipe installation, is designed to convey a peak flow of 11.8 m^3/s (294 mgd). The parallel sewer section, completed late last year, conveys over one-half of all flow within the Metro area. The two-pass tunnel portion, now under construction, will be completed by mid-year next. The budget for project construction is approximately $45 million.

Introduction

The existing North Metro Interceptor (NMI) is a 2450 mm (96-in dia) gravity sewer built in the early 1960s of reinforced concrete pipe (RCP). Running 4.0 km (13,200 ft) from north to south, the pipeline provides only 400 cm of fall (.001 slope). The interceptor begins at the San Diego River, follows numerous narrow local streets and easements, then passes through the Marine Corps Recruit Depot (MCRD) and Naval Training Center (NTC) property, terminating at the City's Pump Station No. 2 (PS2). The pipeline is entirely below the groundwater table and located in San Diego Bay mud deposits subject to liqucfaction.

The NMI sewer conveys wastewater generated within the northern, eastern and central portions of the San Diego Metropolitan Sewerage System, accounting for 55 percent of all flow from the 1,170 km^2 (450 sq mi) service area. A South Metro Interceptor (SMI) conveys the remaining 45 percent to PS2 where the entire flow is conveyed through two tunnels to the Point Loma Wastewater Treatment Plant for processing and deep ocean discharge. Average dry weather flow in the NMI is presently metered at approximately 4 m^3/s (100 mgd). During major storm occurrences, peak flows have approached the pipe's design capacity of 5.88 m^3/s (147 mgd). The old RCP sewer, manufactured with calcarious concrete to withstand hydrogen sulfide attack, has experienced a loss of 5 to 15 cm (2-6 in) in crown lining extending the length of the pipe. Due to its deteriorated state and limited capacity to convey expected flow, upgrading was necessary.

---

[1] Project Manager, City of San Diego, Metropolitan Wastewater Department
600 B Street, Suite 500 San Diego, CA 92101

Predesign and Preliminary Investigations

The existing NMI sewer extends below a hazardous landfill site located on NTC property. The landfill, which began operations in 1940, was abandoned following inception of the Clean Water Act (CWA) in 1972. Designated as a Comprehensive Environmental Response Compensation Liability Act site by EPA, the landfill incorporates NMI right-of-way for approximately 365 m (1,200 ft). The NMI pipeline alignment also traverses a protected *Least Tern* nesting site located on the northern portion of the MCRD/NTC property.

A Predesign Report[(a)] evaluated four alternative pipeline alignments for increasing capacity, reducing the rate of corrosion, improving structural integrity and avoiding landfill excavation. The report also identified major archeological and historical sites, the presence of hazardous wastes, subsurface conditions, permits and environmental restrictions, and methods of construction. The preferred project included rehabilitation of the existing sewer line north of Barnett Avenue as well as construction of a new parallel pipeline and large tunnel under the MCRD/NTC site. The new interceptor was specified as PVC-lined reinforced concrete pipe (PLRCP).

Proposed Alignment and Construction Methods

The NMI improvement project is included as part of a federal Consent Decree order mandated by the EPA. The initial project was divided into two phases; 1) the parallel pipeline and 2) the tunnel portion, as a result of protracted negotiations with the MCRD/NTC regarding necessary easements and construction techniques. An Environmental Impact Report was required for Phase 1 while an Environmental Assessment was necessary for Phase 2, including preparation of a comprehensive Waste Management and Assessment Plan.

The Phase 1 pipeline extends from the Friars Road/Napa Street intersection just north of the San Diego River south along Taylor Street, Pacific Highway, Enterprise Street and west along Barnett Avenue to the property line of the MCRD/NTC site. The new 3.1 km (10,040 ft) alignment roughly parallels the old NMI sewer with the exception of the last 400 m (1,310 ft) segment along Barnett Avenue, as shown in Figure 1.

To overcome multiple surface restrictions at four locations, a 3.35 m (11-ft dia) tunnel boring machine (TBM), using jacking for mobilization, was deployed. Tunneling was designed for a crucial 150 m (500 ft) section under Interstate-8, as well as another 305 m (1,000 ft)section under Interstate-5 and the Southern Pacific Railroad. Although not mandatory, tunneling was selected to avoid multiple utility relocations, traffic congestion, and community disruption at two additional locations including a 600 m (1,975 ft) segment under a heavily congested section of Taylor Street through Old Town, a popular tourist area, and a 115 m (380 ft) section passing under a busy three-way intersection near the main Post Office on Midway Drive. In total, 38 percent (1,170 m) of the Phase 1 pipeline was designed to be tunneled and the remainder to be installed using conventional open trench methods.

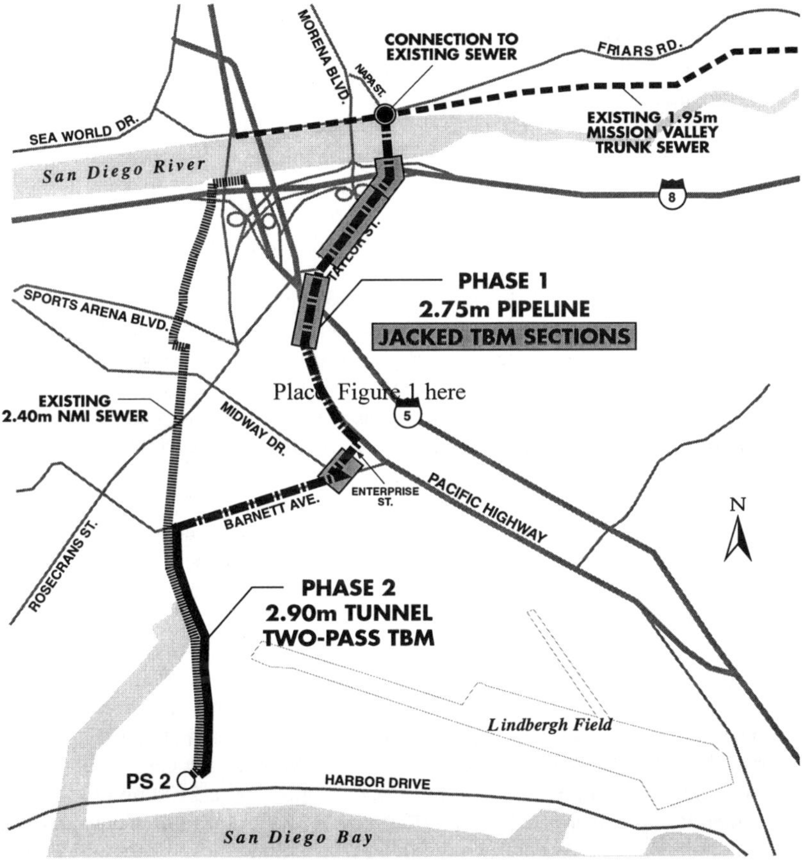

Figure 1
North Metro Interceptor Project

The Phase 2 two-pass tunnel, providing a finished pipe diameter of 2900 mm, extends 1,440 m (4,750 ft) south from the terminus of the Phase I pipeline, under the NTC landfill to PS2 situated along Harbor Drive just to the west of Lindbergh Field. This tunnel, currently under construction, deploys a new, larger 4.60 meter (15 ft dia) TBM.

Phase I Design and Permitting/Subsurface Conditions

Geotechnical investigations[b] for the new NMI revealed extensive artificial fill and bay mud deposits to depths of 4.5 to 6.0 m (15-20 ft). The groundwater table was recorded at a depth of 2.4 m (8 ft) but fluctuated considerably due to tidal action. Cover depths above the proposed pipeline are in the range of 2.0 to 7.5 m (7-25 ft). The tunneling methods considered included 'one pass' (jack/bore) and 'two-pass' (initial liner plate support followed by interceptor pipeline insertion and grout backfill). Both methods required that all voids created by the TBM overcut be pressure grouted via contact grout holes. The contract documents also required chemical grouting to stabilize soils around the perimeter of the pipeline (a minimum 180 cm reach) in areas where there might be an interface between alluvium and the bay mud deposits.

Issues addressed during design, including disposal of contaminated soils, dewatering, unstable bay mud formations, liquefaction potential, and traffic control, resulted in a firm recommendation to tunnel extensive portions of the project. Various pipe materials were evaluated based on the observed poor soil conditions. PLRCP double gasket joints were selected to allow for maximum deflection without leaking. Minimum pipe design loading for a single pass jacking operation was specified as '3000 D'. Special transition and junction structures had to be designed and constructed to accommodate the new pipeline. Access and exit shafts, designed by the contractor, were required for each tunnel.

Instrumentation and monitoring to document both existing conditions and post-construction activities were also required by the contract. Surface settlement, lateral shaft deformations and groundwater levels had to be monitored and duly recorded by the contractor.

The Taylor Street alignment, near its terminus at the railway crossing, was changed to avoid a sharp 90-degree turn and probable closure of one side of the street for a four-month period. The revised pipe alignment was designed through State Park property where the Metropolitan Transit Development Board (MTDB) was constructing major improvements including a new light-rail trolley station and a 2100 mm (84-in dia) replacement to an existing 1950 mm (78-in dia) RCP storm drain. Tunneling was selected for the pipeline portion passing through the parking lot and continuing under I-5. During design, critical construction activities and milestones were agreed to and carefully coordinated with the MTDB to accommodate the contractors without significant delay to either project. The MTDB agreed to remove part of the existing storm drain under the railway line to avoid obstructing the TBM operation.

A large 2,000 m^3 (2,500 CY) concrete 'Conflict Structure' was needed on State Park property to accommodate the replacement storm drain at its intersection with the new NMI pipeline. A computer model was developed to study alternative layouts and the complex hydraulics associated with this structure.

Phase 1 Trenchless Pipeline Construction

Over 20 local, State and federal permits were required for construction of the Phase 1 pipeline. A special permit was required from the California Department of Transportation (Caltrans) for crossing Interstate-8 and -5 rights-of -way. Caltrans requested a two-pass tunnel (or pipe encasement) for the crossings rather than direct jacking to avoid potential leakage from the NMI that might undermine the freeway embankments. This would have required a larger diameter boring machine than deployed thus increasing project costs and resulting in significant delays. Negotiations lasting over six months eventually resulted in an agreement to jack a reinforced concrete cylindrical pipe (PLRCCP) including a steel canister to contain possible leakage within the right-of-way. In addition, follow-up Soil Reports© indicated that the I-8 embankment, placed over 50 years ago, was unstable and susceptible to liquefaction, requiring upgrading improvements as well as pressure grouting before tunneling.

The contractor began operations with tunneling 150 meters of the Phase 1 pipeline under I-5. Different jacking equipment was required for PLRCP than needed for the PLRCCP. The contractor elected to tunnel through the Conflict Structure by providing a siphon to the storm drain. Sections of the jacked pipe were later removed from the site where the structure would subsequently be built using open cut excavation.

Some old concrete block pylons, used to anchor blimps during World War II, were reported buried in the vicinity of the project, however, records of specific locations could not be found. Several of these large blocks were encountered as the tunnel boring operations proceeded. The cutting head of the TBM received some damage and later needed to be replaced.

As tunneling continued, the TBM also encountered old steel sheeting of the existing storm drain in the vicinity of the Conflict Structure site. An open cut had to be made immediately adjacent to the railway tracks so that both the sheeting and portions of the storm drain could be removed. Several recently placed MTDB improvements were also broken out causing some delay. Although the TBM's cutting head sustained more damage, tunneling proceeded and was completed on schedule. On removal of the TBM from the exit shaft it was discovered that the trailing pipe exhibited excessive dip, causing separation of some of the joints. These were repaired by using internal stainless steel sleeves with (Waco Seal) gaskets.

The second tunnel was constructed using the TBM under Taylor Street for a distance of approximately 600 m (1,975 ft). This 2450 mm diameter PLRCP sewer was one of the first to be jacked this length in North America. Tunneling went smoothly with the exception of the last 30 m (100 ft) section when the TBM was pushed at above normal rates while excavated material removal was reduced. This action resulted in some ground-heave that damaged a nearby water main.

The third trenchless operation was constructed from the same entrance shaft as the Taylor Street tunnel but proceeded 150 m (500 ft) in the opposite direction under the I-8 Caltrans right-of-way. Dependable geotechnical data could not be found for the freeway embankment that was constructed over 50 years ago of mixed sand, gravel, clay and boulders. The TBM encountered an extensive gravel fill in this embankment that induced a substantial amount of additional friction on the pipe. As the contractor had decided against installing an intermediate jacking station, the increased friction resulted in the entrance shaft jacks blowing up, crumbling the 'dead man' and temporarily shutting down TBM operations. New, stronger jacks were ordered for delivery by ground transport from Texas over a weekend and the TBM was put back in service. Upon completion, the body of the machine was found to have sustained additional damage due to encountering boulders through this section.

The fourth and final 115 meter (375 ft) pipe section, termed 'The Midway Tunnel' passed under a highly congested (both traffic and utilities) three-way intersection at Barnett Avenue and Midway Drive. The TBM operations went smoothly on this relatively short section.

Phase 2 Tunnel Design and Subsurface Conditions

The 1.4 km Phase 2 tunnel generally parallels, at a distance of 10 to 30 m (30-100 ft) east, the existing NMI sewer. Invert elevations of the tunnel were determined by the existing sewer elevations at both ends. These range from approximately -6.42 m (-21.40 ft below msl) in the south to -5.10 m (-17.00 ft msl) in the north, resulting in a depth from ground surface to crown of between 6.0 and 7.5 m (20-25 ft). Tunneling is proceeding from the entrance shaft near PS2 northward to an exit shaft at Barnett Avenue.

The existing NTC landfill and its contents have been studied by the US Navy for nearly a decade. The Navy is in the process of obtaining regulatory agency approval to cap the site. Landfill excavation and exposure of potential hazardous material were of major concern to both the Navy and the regulatory agencies. Tunneling appeared to be the only viable option for the new interceptor. Authorization to conduct a preliminary geotechnical investigation and site assessment during predesign was difficult since the landfill was classified 'untouchable'. The landfill depth varies from 2.5 to 3.5 m (8-12 ft) while the proposed tunnel crown would be approximately 6 m (20 ft) below the surface.

Numerous discussions were held with the Navy and the regulatory agencies to receive approval for construction. Extensive testing for contaminants is required during tunnel and shaft excavation although contaminants have not been detected at the tunnel depths.

The MCRD also expressed major concerns regarding site-specific issues including surface excavation, noise and vibration, surface settlement, damage to several historical buildings, liquefaction, and overall disruption from construction activities. The MCRD insisted on maintaining the new easement as close as possible to the existing sewer, which contained several large curvatures. Several regulatory agencies were also concerned about the potential for adverse impact of vibrations from the TBM operation on the *Least Tern* site and required that additional mitigation measures be addressed.

Subsurface conditions along the tunnel alignment consist of hydraulic fill, bay mud and terrace deposits. Tunnel construction is expected to be predominately through terrace deposits, with some segments excavated partially or entirely through bay mud. There are no portions of the pipeline that are to be tunneled through fill material or the landfill. Groundwater has been encountered in the initial boring and cone penetrometer explorations at depths of 2.25 to 3 m (7.5-10 ft) below the surface. The close proximity of the project to San Diego Bay results in evident tidal fluctuation of the groundwater table.

Specified tunneling methods are derived from a combination of ground conditions, project restrictions, and available tunneling technologies. Based on ground conditions, consisting primarily of sand and silty sand with no or little cohesion below the water table, much of the material is defined by the *Tunnelman's Ground Classification* as 'Flowing Ground'. Jacking would have required an intermediate shaft every 450 m (1500 ft) with attendant surface excavation and dewatering. Due to presence of the landfill and *Least Tern* nesting site this approach was unacceptable. The only tunneling method found acceptable was a two-pass 'earth pressure balance' technique. The earth pressure balance provides for a positive external pressure which reduces the possibility of leaching hazardous materials from the landfill and also eliminates the need for dewatering. Although a 300 meter (1,000-ft) radius is required to accommodate the existing NMI curve, dewatering will be unnecessary.

The two-pass tunnel construction consists of excavation and installation of an initial support lining followed by installation of the PLRCP. Due to the shallow depth, the linings are designed to experience full overburden and water loads. The initial water tight lining consists of gasketed and bolted concrete segments installed within the tunnel shield. To limit ground movement, the annulus outside the lining is grouted immediately. Although the PLRCP line will be operated as non-pressure pipe under most conditions, it can be subjected to as much as 10.7 m (35 ft) of head for limited periods.

A contractor's staging area and entrance shaft was identified at the south end of the project on MCRD/NTC property. This 2.4 ha (6 ac) site was selected because of its adequate space for stockpiling excavated material and for testing of possible contaminants prior to disposal.

Conclusions

This project is notable in several respects. Phase 1 involves the successful jacking of 2750 mm (108-in dia) pipe over 1,150 m (3,700 ft) in aggregate length employing earth pressure balance single-pass TBM methodology. Construction of the Phase 2 tunnel, now in progress, offers the only viable means for installing the remaining 1.4 km section of pipe. The lessons learned regarding trenchless pipeline installation during the first phase are expected to pay off handsomely in execution the Phase 2 work. Minimizing disruption to traffic, utility interfacing and agency approvals, combined with less risk of excavating contaminated soils and reduced dewatering represent the more salient advantages associated with tunneling extensive segments of this interceptor sewer.

References

(a) James M. Montgomery, *Predesign Report for the North Metro Interceptor Sewer*, City of San Diego, October 1991.

(b) James M. Montgomery, *North Metro Interceptor Sewer NTC/MCRD Tunnel Concept Study* (*Preliminary Geotechnical Evaluation*), Woodward-Clyde Consultants, City of San Diego Clean Water Program, November 1992.

(c) Woodward-Clyde, *Geotechnical Investigation North Metro Interceptor Tunnels San Diego, California*, Montgomery Watson, September 1994.

Acknowledgments

The author wishes to acknowledge the assistance of Fred Le Febvre, Jr. with HYA Consulting Engineers, Terrance Hoeper, Montgomery Watson and Greg Sherry of Haley & Aldrich in preparing this paper.

HORIZONTAL DIRECTIONAL DRILLING FOR BYPASS TUNNELS

by Michael J. Robison[1], Tracy J. Lyman[2], Larry Erwin[3] and Steve Wolfman[4]

ABSTRACT

Horizontal directional drilling was used to construct a 0.91 m diameter tunnel 409 m long and a 0.71 m diameter tunnel 627 m long through weak sedimentary rock and hard crystalline rock. The tunnels were lined with high density polyethylene pipe, which was contact grouted in-place. This paper documents the drilling, lining and contact grouting of the two bypass tunnels.

INTRODUCTION

In 1928, the City of Santa Cruz sited a landfill in two steep-walled canyons. The landfill blocks the natural flow of a perennial creek (Lombardi Creek) and a seasonal tributary to Lombardi Creek. As the landfill developed, pipelines were constructed beneath it to convey the creek downstream. However, the pipelines failed to properly separate creek water from leachate and contaminated water flowed down Lombardi Creek.

In 1986, the pipelines were plugged at their upper ends in the east and west canyons and a temporary pumping system was constructed. This system collected flow in the creeks and pumped the flow up and over the landfill. In 1994, the California Regional Water Quality Control Board (RWQCB) adopted a Cease and Desist Order requiring the landfill to maintain continuous flow in Lombardi Creek, prevent groundwater from entering the landfill, and reduce ponding against the landfill during storms.

To satisfy the requirements of the RWQCB and to provide a reliable, low-maintenance solution, the City decided to construct two tunnels to bypass the landfill. The Upper Tunnel would convey water from Lombardi Creek to the tributary creek in the West Canyon. The Lower Tunnel would convey water from the West Canyon to the downstream end of Lombardi Creek.

This paper presents a case history of tunnel construction on this project, known as the Freshwater Bypass Tunnels Project. It describes the use of horizontal directional

[1]Senior Project Manager, Jordan, Jones & Goulding, Inc., Charleston, SC; [2]Vice President, Haley & Aldrich, Inc., Denver, CO; [3]Director of Public Works, City of Santa Cruz, CA; [4]Associate Civil Engineer, City of Santa Cruz, CA

drilling (HDD) through both weak sedimentary rock and hard crystalline rock to create the tunnels and the installation and contact grouting of high-density polyethylene (HDPE) pipe as the final lining.

PROJECT DESCRIPTION

The Freshwater Bypass Tunnels Project consists of two tunnels. The 409 m long Upper Tunnel transports flow from the East Canyon to the West Canyon. The 627 m long Lower Tunnel transports flow from the West Canyon to Lombardi Creek, which flows to the Pacific Ocean. A plan of the landfill and the two tunnels is shown in Figure 1.

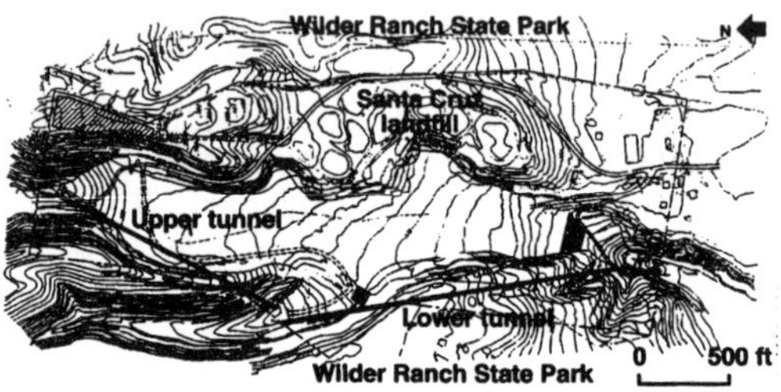

Fig. 1: Plan of Freshwater Bypass Tunnels

Both tunnels were designed with constant grades. The Upper Tunnel goes from Elev. 232 ft (71 m) at the outlet in the West Canyon to Elev. 277 ft (84 m) at the inlet in the East Canyon, for a grade of 3.4 percent. The Lower Tunnel goes from Elev. 111 ft (34 m) at the outlet in Lombardi Creek to Elev. 225 ft (69 m) at the inlet in the West Canyon, for a grade of 5.5 percent.

The tunnels were designed with several options for final lining, depending upon the method of construction. If microtunneling or a tunnel boring machine was used, the final lining could be either 0.91 m diameter glass fiber reinforced pipe or 1.07 m diameter reinforced concrete pipe. If horizontal directional drilling was used, then 0.91 m diameter high-density polyethylene pipe could be used.

SITE CONDITIONS

Geologic Setting and Ground Conditions

The City of Santa Cruz Landfill is located within the Coast Range geomorphic province. The Coast Range is characterized by steep, rugged terrain with northwest-trending mountain ranges, broad basins, and narrow valleys that tend to parallel major geologic features. The landfill lies in one of the many canyons that have been eroded into the gently sloping marine terraces that extend up to several hundred meters above sea level along the coastline.

Surficial sediments in the vicinity of the landfill are derived from Santa Cruz Mudstone and Santa Margarita Sandstone. These sediments consist of mixed rock fragments, sand, silt and clay. The underlying Santa Cruz Mudstone consists of medium to thick-bedded, faintly laminated, blocky-weathering, pale-yellowish-brown, siliceous organic mudstone. The Santa Margarita Sandstone consists of thick-bedded to massive, well-sorted, yellowish-gray to white, medium to fine grained arkosic sandstone. The older, plutonic rock at the site consists of quartz-diorite, granodiorite and granite.

Upper Tunnel

The Upper Tunnel was expected to be driven primarily in the fine-grained sedimentary rock units, which consisted of mudstone, siltstone and fine sandstone. Excavation would begin in the West Canyon in an area of landfill material/alluvium with shallow cover, pass quickly into the mudstone/siltstone/sandstone beneath the canyon wall with cover of up to 55 m, and terminate in an area of landfill material/alluvium with shallow cover in the East Canyon. The groundwater level was expected to be at or slightly above the tunnel invert elevation. Groundwater inflows were expected at up to 3.2 liters per second (lps) for 2 hours duration for open-face tunneling equipment.

Lower Tunnel

The Lower Tunnel was expected to be driven through ground conditions that were expected to be more difficult than the Upper Tunnel. The upper 427 m of the Lower Tunnel was expected to be driven through the mudstone/siltstone/sandstone beneath the canyon wall with cover of up to 64 m. The Lower Tunnel would then encounter 122 m of harder, more abrasive quartz-diorite. The lower end of the Lower Tunnel was expected to encounter schist and quartz-diorite before ending in a 81 m long section of cut and cover through quartz-diorite and sandstone in Lombardi Creek. Groundwater infiltration rates were expected to reach 0.6 lps.

CONSTRUCTION CASE HISTORY

Constructability Review

The design of the Freshwater Bypass Tunnels was controlled by several criteria. First, drilling and blasting would not be permitted for creating the tunnels, due to the concern that blast vibrations could fracture the host rock and create flow paths for groundwater, possibly contaminated with leachate from the landfill, to enter Lombardi Creek. Second, the use of a steel pipe lining would not be permitted, due to the concern for corrosion and damage to the protective coating during installation. Third, contact grouting would be required around the final pipe lining, to stabilize the opening around the pipe and to prevent groundwater from reaching the outside wall of the pipe and traveling to the outlet structure.

The first task of the construction manager was to review the design at the 90 percent stage. At the 90 percent stage, the design called for the tunnels to be constructed as microtunnels, i.e. using remote controlled tunneling equipment to excavate the tunnels. However, because of the great depth at which the tunnels would be driven, it was probable that microtunneling contractors would have high contingency costs in their bids to cover the cost of any recovery shafts that may be needed. Based on our review of the project, the costs anticipated for construction using microtunneling, and the desire of the owner for a low cost project, the construction manager suggested that the specifications be rewritten to allow for tunnel construction using horizontal directional drilling or a tunnel boring machine (TBM). That suggestion was adopted, and the specifications were rewritten to allow prospective contractors to use microtunneling, HDD or TBM tunneling techniques.

Alternative Technology/Contracting Options

The specification requirements for tunneling required the use of a small diameter tunnel boring machine capable of excavating and installing pipe from the jacking shaft. The requirements for microtunneling required a mechanical type earth or slurry pressure counter-balanced system with a reversible cutterhead drive system to minimize rotation of the pipe during installation. Tunneling and microtunneling envisioned that either glass fiber reinforced pipe (manufactured by Hobas) or reinforced concrete pipe would be used to jack the tunnel boring machine. With either of these pipes, the specifications required that the pipe be contact grouted in place, following installation.

The specifications for horizontal directional drilling required that the tunnels be drilled using a pilot hole followed by enlargement of the hole to accommodate that product line. The product line specified for the directional drilling method was high-density polyethylene (HDPE) pipe. The joints between pipe sections would be fused in the field, prior to pulling the pipe sections into the tunnel.

All three tunneling methods required that the departure from line or grade, as measured from the design centerline of the tunnel, remain within a 0.76 m diameter target centered on the design centerline of the tunnel. The TBM and microtunneling specifications required the use of lasers projected onto targets to provide guidance for the excavation. The HDD specification required the use of a steering tool and a tracking system for guidance.

During the pre-bid phase of the project, interest was shown by tunneling, microtunneling and horizontal directional drilling contractors. At bid opening, three horizontal directional drillers and one microtunneler submitted bids. The low bid for the project came in at $4,125,632, with the contractor planning to use HDD. The second low bid was $4,153,000, with that contractor planning to use microtunneling. The other two bids were higher, with both contractors planning to use HDD. The project was awarded to the low-bid, HDD contractor.

Upper Tunnel Case History

Horizontal directional drilling began with the Upper Tunnel. The drilling equipment was mobilized to the West Canyon, where the drilling rack was set in a pit dug at the location of the outlet structure. There, the drill was set to grade and drilling began. By drilling uphill, cuttings from the drilling operation flowed back to the pit. From within the pit, the larger cuttings were removed using a clamshell bucket while the drilling mud was recirculated through a series of shaker screens and centrifuges to remove sand and silt size particles.

The pilot bore was drilled using a 0.25 m diameter roller bit, which was pushed from the drilling pit to the East Canyon. Following punch-out of the pilot bore, a series of backreams was performed using barrel reamers. The barrel reamer consists of a steel cylinder with a series of teeth mounted to its outside edge. The barrel reamers used in the reaming process were pulled back the entire length of the bore, from the East Canyon to the drilling rig in the West Canyon. The reams were performed sequentially to enlarge the bore from 0.25 m to 0.44 m to 0.76 m to 1.17 m.

The drilling operation was successful in creating the opening for the Upper Tunnel. The ground conditions were as anticipated and groundwater inflows were relatively minor, with the completed bore producing approximately 0.1 lps. The pilot bore was drilled over a period of four working days, followed by the backreaming operation, which was accomplished in eight working days. During drilling of the pilot bore, line and grade was checked at the end of each 9.1 m push, using the Sharewell navigation instrumentation in the drilling string and the Tru-Tracker surface tracking device.

The final lining consisted of 0.91 m diameter, SDR 21, HDPE pipe, which was installed in the bore following completion of the last ream. Pipe sections 15.2 m in length were delivered to the West Canyon. From there, the leading pipe section was attached to the pulling head and subsequent 15.2 m sections were fused to the pipe string as it was pulled into the bore. A series of 50 mm diameter steel pipes were attached to the pulling head and pulled into the bore along with the final lining, for use in the contact grouting operation. Pipe installation was accomplished over a period of two working days.

Contact grouting was performed through the 50 mm pipe, starting at the downstream end of the line in the West Canyon and proceeding in a continuous manner to the upstream end in the East Canyon. The grout was a mixture of one part cement to eight parts flyash, combined with water. Prior to grouting, the pipe was filled with water. During the grouting operation, the grout pipe was withdrawn from the tunnel, although the end of the pipe was kept buried in the grout. At the completion of the grouting operation, groundwater that had collected between the pipe and the tunnel was displaced out the upstream end. Inspection of the interior of the pipe following the grouting operation indicated that the pipe installation and contact grouting had been successful.

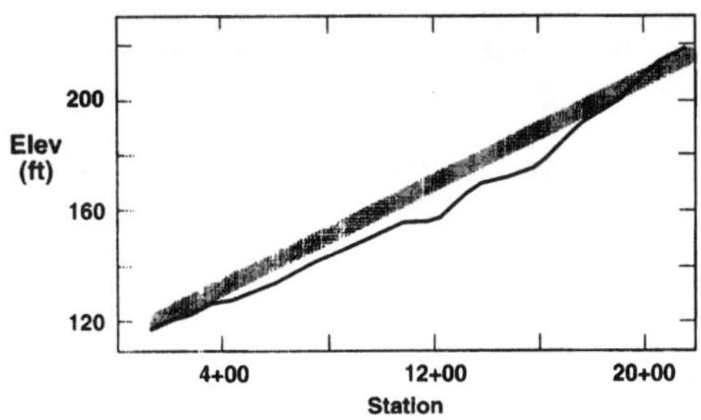

Fig. 2: Upper Tunnel Profile

A final, as-built survey of the Upper Tunnel was performed from within the pipe following completion of the contact grouting. The results of the survey indicated that the maximum departure from grade varied from 2.0 m low to 0.6 m high. The survey also indicated that the slope of the tunnel flattened around Sta. 29+00 ft. Despite not meeting the specification requirements, this final tunnel alignment was considered acceptable because the slope of the pipe maintained a positive pitch throughout its entire length. A plot of the as-built grade of the Upper Tunnel is shown in Figure 2.

Lower Tunnel Case History

Shortly after award of the contract, the contractor proposed that the lower 61 m section of the Lower Tunnel be constructed using HDD, rather than using cut and cover. The proposal resulted in elimination of the 81 m long cut and cover section and reconfiguration of the outlet structure, with a cost saving of $61,000 to the City.

The Lower Tunnel was begun following completion of pipe installation at the Upper Tunnel. The drilling equipment was mobilized to Lombardi Creek, where the drilling rack was placed into a pit excavated in the sandstone. Similar to the Upper Tunnel, the Lower Tunnel was drilled uphill, in order for cuttings to flow back into the drilling pit.

The pilot bore was drilled using the 0.25 m diameter roller bit. On this tunnel, the first ream was attempted using a 0.44 m barrel reamer, backreaming from the West Canyon towards the drill. However, the barrel reamer advanced approximately 427 m until it reached the harder quartz-diorite. The barrel reamer was unable to advance through the quartz-diorite. Consequently, the barrel reamer was removed and a 0.44 m diameter "hole-opener" was pushed from the drilling pit. A hole-opener is a reaming tool designed for use in hard rock, which uses a series of roller cones to cut the rock. The hole-opener was successful in reaming through the quartz-diorite. The second ream was made using a 0.71 m diameter hole-opener pushed from the drilling pit. The next two reams were performed as backreams, using 0.91 m and 1.07 m hole-openers pulled by the drill.

The drilling operation was successful in creating the opening for the Upper Tunnel. The ground conditions were as anticipated and groundwater inflows were less than anticipated, at approximately 0.1 lps. The pilot bore was drilled over a period of five working days, followed by the backreaming operation, which was accomplished in 15 working days.

The final lining for this bore was identical 0.91 m diameter, SDR 21, HDPE pipe. It was installed in the bore following completion of the last ream. The 15.2 m pipe sections were made up into 30.5 m lengths in the field prior to installation into the tunnel. As with the Upper Tunnel, a series of 50 mm diameter steel pipes were attached to the pulling head and pulled back through the bore for use in the contact grouting operation.

Pipe installation was more difficult on the Lower Tunnel than the Upper Tunnel. On the Lower Tunnel, the pipe had to be pulled back through a horizontal curve, whereas on the Upper Tunnel, the pipe was pulled in on a tangent. The 50 mm grout pipe that extended the entire length from the West Canyon through the tunnel to Lombardi Creek was found to have broken in the location of the horizontal curve. However, the grout pipe that extended 213 m up the tunnel from Lombardi Creek was usable.

The contractor had planned to grout the Lower Tunnel in the same manner as the Upper Tunnel. The contractor revised the method of contact grouting as a result of the broken grout pipe. The revised grouting method began with filling the 0.91 m HDPE pipe with water, grouting the annular space around the lower 213 m of tunnel through a short pipe in the downstream bulkhead and pumping until grout reached the 213 m pipe. The next day, grouting was planned to be performed through the 213 m pipe until it could be seen at the upstream end of the tunnel. However, because of a supply problem, the grouting operation was stopped before the grout could be seen at the upstream end. The next day, the contractor resumed pumping grout through the 213 m grout pipe until grout was noted at the upstream end. On the final day of grouting, the upstream end of the tunnel was grouted by tremie through the upstream bulkhead. The grouting operation used the same mix of cement, flyash and water that had been used on the Upper Tunnel.

During inspection of the tunnel following completion of the contact grouting operation, a large blockage was found in the pipe, extending approximately 27 m in length, centered around the end of the 213 m grout pipe. Visual observation of the blockage indicated that the invert of the pipe had risen to the point of contact with the crown, as shown in Figure 3. The blockage was attributed to high grouting pressures that were applied to push the grout up the bore. Pressures in excess of 690 kPa were recorded at the pump.

Fig. 3: Photograph of Collapsed HDPE Pipe

The contractor considered his options at this point, which included drilling out the blockage and repairing the pipe, sinking a shaft and repairing the collapsed section out of the shaft, or drilling a complete new tunnel. The contractor considered the cost of drilling a new tunnel to be too expensive and it was not practical to sink a shaft at the location of the collapse because of the high ground cover in that location. Thus, the contractor decided to drill out the collapsed section of pipe. Based on discussions with the owner and designer, it was decided to slip line the entire length of the tunnel with a smaller size HDPE pipe and to contact grout the annular space between the pipes.

A horizontal auger drill rig was mobilized to the jobsite. The auger rig was set in the same pit as the directional drill rig and augering began. A 0.76 m diameter steel casing was installed as the augering proceeded, for use as a guide for the flights of auger. The casing was advanced a distance of 198 m, at which point the collapsed section was reached. During the course of advancing the casing, workers had to enter the pipe several times to help guide its advancement and to cut out sections of the original HDPE pipe that had been gouged by the casing. The lead section of auger was advanced slowly through the collapsed section. Numerous manned entries of the casing were made in order to remove cuttings and large pieces of cut HDPE pipe.

Following augering and removal of the collapsed section, the Upper Tunnel was relined with 0.71 m outside diameter, SDR 26, HDPE pipe. As with the original pipe, the new pipe was fused into 30.5 m lengths in the West Canyon. The new pipe was pulled into the original pipe without difficulty, using a wire rope and a tugger.

Contact grouting of the 0.71 m pipe was performed in stages using cellular concrete. A series of injection ports and gauge ports were installed within the pipe on 107 m centers. The gauges were used to measure the pressure that was developing as the head pressure of the grout increased. The maximum allowable pressure on the HDPE pipe was 55 kPa. At the Lombardi Creek end of the pipeline, the pressure gauge could be observed from within the pit. In order to observe the gauge pressure as grouting progressed up the line, a remote controlled video camera was positioned to observe the gauges. The grouting operation was performed over a period of five working days. Inspection of the interior of the pipe following the grouting operation indicated that the relining operation had been successful.

A final, as-built survey of the Lower Tunnel was performed from within the pipe following completion of the grouting. The results of the survey indicated that the maximum departure from grade ranged from 0.38 m high to 0.48 m low. A plot of the as-built grade of the Lower Tunnel is shown in Figure 4.

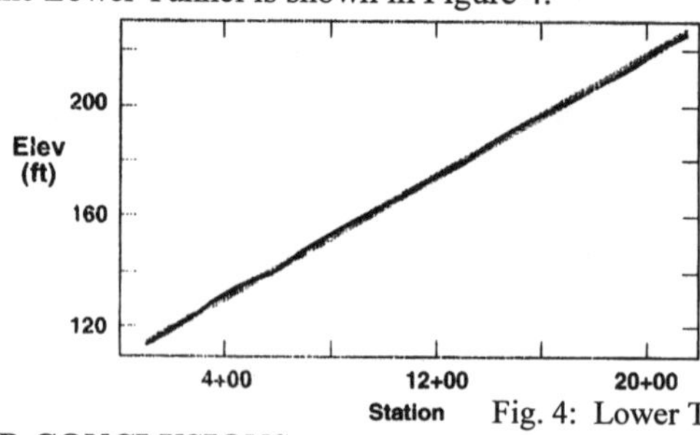

Fig. 4: Lower Tunnel Profile

SUMMARY AND CONCLUSIONS

The Freshwater Bypass Tunnels Project was a successful project, even with the difficulty experienced with the collapsed pipe section. Horizontal directional drilling proved to be a cost effective and efficient method of creating the tunnels for this project. The method worked well in both the soft sedimentary rock and in the harder igneous rock. HDD was also successful in drilling across the transition from the soft sedimentary rock to the hard quartz diorite.

PIPE JACKING TO AVOID CONTAMINATED GROUNDWATER CONDITIONS

Marco D. Boscardin[1], M. ASCE, R. Lee Wooten[1], M. ASCE, and James M. Taylor[2]

ABSTRACT

A 430-meter-long segment of sanitary sewer force main was constructed using an earth pressure balance (EPB) tunnel boring machine (TBM) and pipe jacking to avoid contaminated soils and groundwater, and to minimize the potential for the construction to cause additional migration of contaminants. The casing, 2045-mm-outside-diameter reinforced concrete cylinder pipe (RCCP), was jacked through alluvial sands and gravels below the ground water level. The carrier pipe, 1220-mm-diameter ductile iron pipe, was installed within the casing. The construction method was chosen to eliminate the need for dewatering that could cause groundwater flow from deeper, contaminated strata up across a silt aquitard to shallower strata.

INTRODUCTION

A 430-meter-long segment of a 1220-mm-diameter sewer force main was constructed through an area where, during the investigation phase of the project, ground water contamination had been detected. The field studies indicated that the upper sands and gravels likely to be excavated during the force main construction were not contaminated. However, ground water contamination was detected in the lower sands and gravels. A silt layer which appears to serve as an aquitard was observed between the upper and lower sands and gravel zones. To minimize the potential for the force main construction to cause a spread of the contamination from the lower to the upper sands and gravels, several alternative construction methods were considered and evaluated during design. Earth pressure balance pipe jacking through the upper sands and gravels at a depth of 3 to 4 meters was selected.

BACKGROUND

The 430-meter segment of force main is located under University Road in the northwest corner of the Town of Canton, Massachusetts. This segment of force main is part of the Massachusetts Water Resources Authority (MWRA) New Neponset Valley Force Main project which involves the construction of approximately 5650 meters of 1220-mm-diameter force main connecting a new pump station to the existing New Neponset Valley Sewer. The alignment of the 430-meter

---

[1] Sr. Project Manager, GEI Consultants, Inc., 1021 Main St., Winchester, MA 01890

[2] Project Engineer, Fay, Spofford & Thorndike, Inc., 5 Burlington Woods, Burlington, MA 01803

segment of force main is beneath University Road within a 9-meter-wide sewer easement that also contains a previously constructed 1400-mm by 1500-mm concrete box sewer comprising the existing New Neponset Valley Sewer, Fig. 1. The new force main is approximately 3 to 4.6 meters east of and about 1 to 1.2 meters deeper than the concrete box sewer.

After following University Road north for 430 meters, the force main comes to within about 20 meters of two sets of MBTA commuter rail tracks, makes a right angle turn to the east, leaves University Road and traverses under the railroad tracks. After passing beneath the railroad tracks the force main turns northeast and passes through the Dedham Street highway embankment. Although not the subject of this paper, these crossings, about 50 and 65 meters, respectively, were also made using pipe jacking methods.

Approximately 135 meters of the force main under University Road passes within the 122-meter-radius wellhead protection area for Canton Well No. 2. Massachusetts Department of Environmental Protection requirements for public water supply wellhead protection mandate that the carrier pipe be placed within a casing pipe in this area. In addition, the force main crosses 300 mm below an active 254-mm-diameter natural gas transmission pipeline under a pressure of 4800 kPa. The gas pipeline remained in service during construction.

SUBSURFACE CONDITIONS

The project area is located within the Neponset River flood plain. The naturally deposited surficial soils of the Neponset river flood plain are mapped as fine-grained alluvial sand, silty sand, fine gravel and silt. Peat and organic silt were encountered in the upper natural strata during earlier construction in the area. Fill placed during the construction of the earlier concrete sewer and University Road comprises the upper soils at the project site.

Sixteen borings were performed along the University Road alignment during the subsurface investigations for the project. During the initial geologic investigations, three borings were drilled from 6.4 to 7.9 meters deep. Later, an additional 13 borings ranging from 5.2 to 18.1 meters deep were drilled during environmental site investigations. A subsurface profile based on the boring data is presented in Fig. 2 and described below.

Along University Road, the borings encountered fill materials from the ground surface to approximately El. 44.2 (meters). The ground surface ranges from about El. 46 along the southern unpaved 122 meters of University Road to El. 47.2 along the paved portion of University Road. The fill consisted of sand and gravel with cobbles and boulders. Below the fill, a 0.3-to-1-meter-thick layer of peat was encountered; however, at the north end of University Road, 1.2 to 1.5 meters of peat was encountered. The bottom of the peat layer was typically at El. 43.6 or higher. Below the peat, an upper sand layer consisting of medium dense to very dense, fine to medium grained sand often with a trace of fine gravel was encountered. The upper sand ranges from 1.5 to more than 3 meters thick. At most locations, a layer of medium dense to very dense, non-plastic to low plasticity silt to fine sandy silt was present below the upper sands. The silt layer ranged from 1 to more than 2.4 meters thick except for a 100-meter-long section of the alignment centered at the boundary of the paved and unpaved portions of University Road where the silt layer appeared to be absent. A lower sand layer consisting of medium dense to very dense, very fine to coarse sand with a trace of gravel was

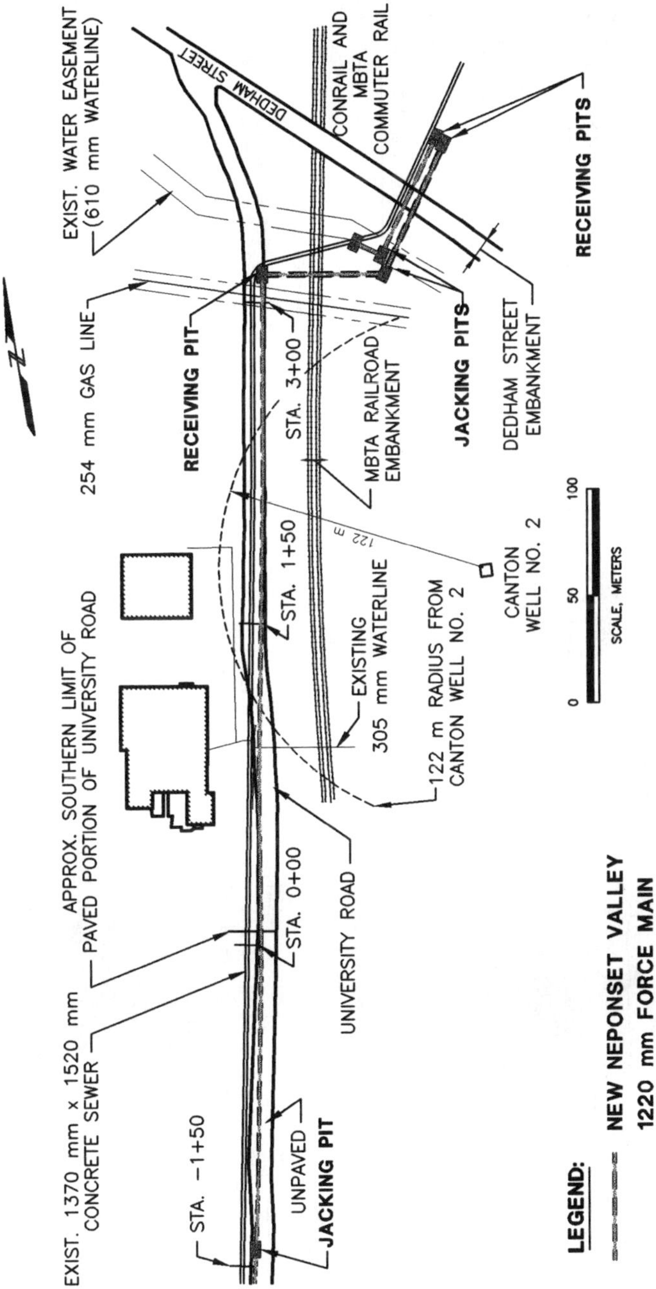

**FIGURE 1 - NEW NEPONSET VALLEY FORCE MAIN
SITE LOCATION PLAN**

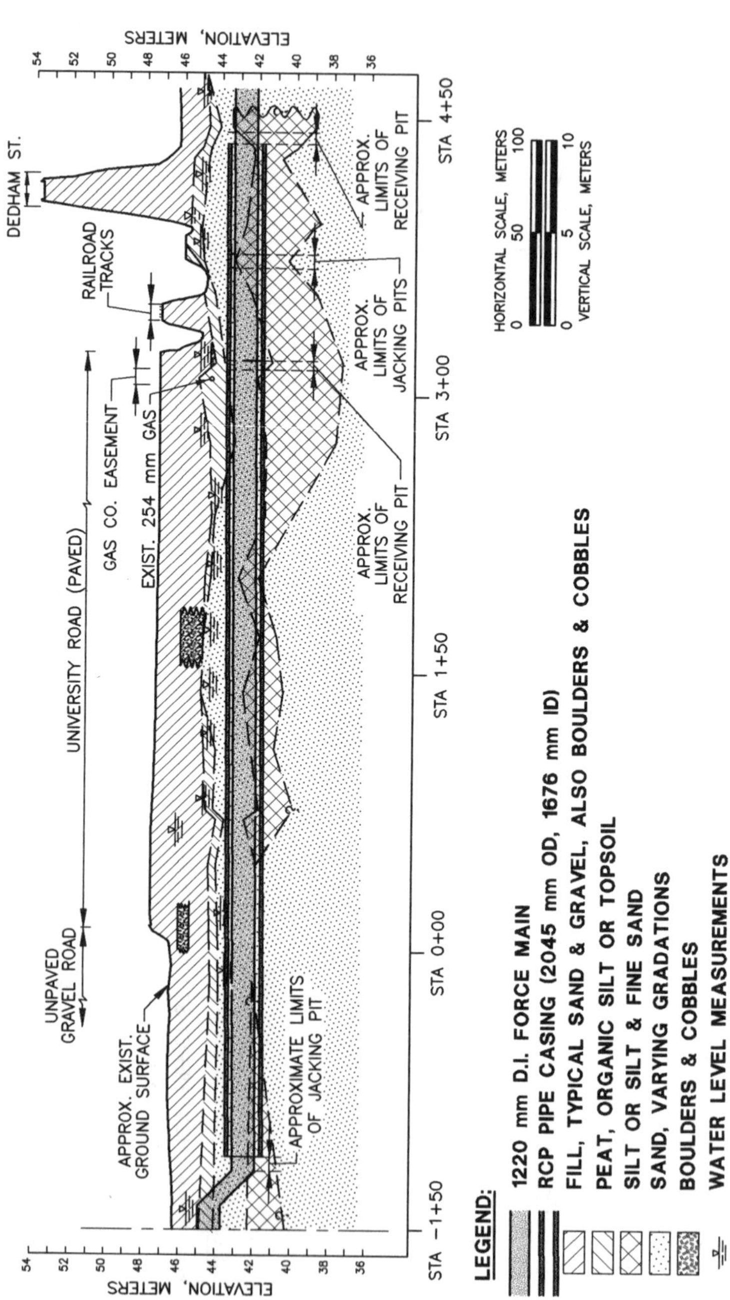

FIGURE 2 - NEW NEPONSET VALLEY FORCE MAIN - PROFILE

encountered below the silt layer . The upper portions of the lower sand were generally finer and become coarser with depth.

Ground water levels in the borings and monitoring wells ranged from El. 44.3 to 45.1. Prior to construction, it was evident that the invert of the force main or casing pipe would be about 1.8 to 3 meters below the ground water level and dewatering would be required if conventional open excavation with trench box support was used.

ALTERNATIVES FOR FORCE MAIN CONSTRUCTION

During design, alternatives for construction of the force main were examined including:

a) Conventional open excavation with trench box support;
b) Open excavation using a trench box and subaqueous construction;
c) Open excavation with sloped sides and subaqueous construction;
d) Open excavation with temporary sheeting;
e) Open excavation with permanent sheeting and tremie concrete base mat;
f) Conventional open face or auger methods of pipe jacking;
g) Slurry or earth pressure balance methods of pipe jacking; and
h) Open excavation using ground freezing for support and ground water control.

Initially, the technical merits of each alternative were evaluated relative to the amount of dewatering required, amount of excavation in the event that contaminated soils were encountered, impact on future construction, impact on use of the area, and constructibility. Those alternatives considered to be technically satisfactory were then compared on the basis of the estimated relative cost.

The method of construction selected had to meet the following technical criteria:

- Eliminate or minimize the need for dewatering and the potential for introducing hydraulic connections and ground water flow between the upper and lower aquifers;

- Minimize the potential volumes of contaminated soils and ground water to be handled and managed;

- Minimize the potential disruption of the use of and access to properties along University Road;

- Minimize the permanent impact on the hydrologic regime and future construction along the force main alignment; and

- Be constructible.

Based on the initial evaluation of the technical merits of the alternatives described, only slurry or earth pressure balance pipe jacking methods and open excavation using ground freezing were retained for further evaluation. The ground freezing approach was then eliminated because preliminary estimates were that an approach using ground freezing would cost several times that of an approach using a pressure balance TBM and pipe jacking. In addition, the surface disruption and the duration of construction were judged to be significantly greater with a ground freezing approach in this case.

Slurry or earth pressure balance methods of pipe jacking can be used to control the excavation face and ground water inflow. The pressure can be maintained sufficiently high to counterbalance the soil and water pressures without heaving or causing a blowout to the ground surface. This approach can also minimize the amount of soil excavated and managed, and the disturbance of the ground surface and to the traffic flow on University Road can be much less than open cut methods. The TBM with a pressurized face can provide better face stability and ground water inflow control in the sands than conventional pipe jacking. The ability of the equipment to handle obstructions such as cobbles and boulders was expected to be the main limitation with this approach.

DESIGN

Use of pressurized face pipe jacking required several changes in the design of the force main compared to the preliminary plans which were based on conventional open cut methods. When the crown of the jacked pipe is less than about two pipe diameters below the ground surface, surface settlement/heave and face stability tend to be difficult to control. Therefore, the depth of the force main was increased to improve face control, reduce ground surface settlements, and reduce the potential for the pressurized face causing a blowout or objectional heaving of the ground surface. The alignment of the force main was lowered so that the top of the jacked pipe was approximately El. 43.3. This placed the crown of the jacked pipe about 4 meters below surface of the paved portion of University Road and about 3 meters below the surface of the unpaved portion.

Lowering the force main alignment had the added benefit of placing the jacking operations below the level of the fill and thereby reducing the potential for interference by cobbles and boulders which were detected above El. 44.2. However, the TBM was required to be able to handle boulders up to 30 percent of the diameter of the TBM. If boulders larger than that size were encountered by the jacking, the pipe was shallow enough and in a relatively open area so a sheeted or trench box supported pit filled with water could have been excavated at the face of the TBM to remove the obstruction. The pit would be filled with a sand and gravel fill and jacking could then continue.

The entire jacked pipe length (430 meters) of the force main was designed as a 1220-mm-diameter ductile iron carrier pipe within a RCCP casing pipe to meet the requirements of the well head protection area. It was not considered feasible to jack two different sizes of pipe due to equipment mobilization costs. The jacking pit was located outside the area where contamination had been detected while the receiving pit was located at the north end of the run where the force main makes a right angle turn to pass under the railroad tracks. Contamination had been reported in the area of the receiving pit. The specifications did not permit dewatering outside the jacking and receiving pits and required provisions for grouting or sealing if the pits leaked and ground water levels outside the pits dropped.

A slurry pressure balance TBM was specified for the construction. The slurry method was selected because the cohesive nature of the bentonite slurry provides face support, permits control of the sands, and groundwater inflow by serving as a binder and preventing runs and flows. Control of the slurry pressure was required to avoid the potential for a blowout under the low cover conditions at this site. Intermediate jacking stations were specified for the jacking. The resultant settlement trough due to the construction was estimated to be about 6 meters wide with estimated centerline ground surface settlements in the 30 mm range.

CONSTRUCTION

The contractor requested that an earth pressure balance (EPB) TBM be permitted for the pipe jacking work. The TBM proposed was a 2045-mm-diameter Lovat Model M77 equipped with flood gates at the face and pressure relieving gates to control removal of muck from the muck chamber while maintaining pressure on the face. Provisions for injecting a binding medium, either bentonite or polymer based, into the muck chamber to assist control of the face, ground water infiltration or running sands were included. The contractor's reasons for the EPB TBM included less risk of a blowout through the shallow ground cover compared to a system actively pumping slurry under pressure, smaller muck handling and processing staging area requirements, and easier muck disposal because of the elimination of slurry mixed in with the muck.

The TBM included a bi-directional, high torque cutter head mounted in an articulated shield. The shield was composed of three sections with a total length of about 7 meters. Articulation of the 2-meter-long front section of the shield was controlled by hydraulic rams. Steering was monitored using a laser and twin target arrangement. The TBM could be thrust independent of the pipe string by using four, 750-kN hydraulic rams. The cutter head included gage cutters to create a 2108-mm-diameter opening. The pressure relieving gate doors of the TBM would permit passage of 360-mm-diameter boulders. The cutter head was equipped with carbide ripper teeth on the spokes and hardened picks at the perimeter to create an aggressive toothing arrangement so that larger boulders encountered might be broken up or cut through.

The main jacking station was equipped with 4,1800-kN capacity, 3-meter stroke hydraulic cylinders designed to jack up to 6 meters of 1676-mm-ID RCCP at a time using three strokes of the cylinders. The RCCP was supplied in 3-meter-long segments and had a rated thrust capacity of 8900 kN. The first intermediate jacking station (IJS) was located about 60 meters behind the TBM and at about 90 meter intervals thereafter. Each IJS included 10 hydraulic rams with a total rated capacity of 5600 kN and a 267 mm stroke.

Primary injection of bentonite lubricant was performed through injection ports directly behind the TBM tail shield. The volume injected was limited to 100 to 110 percent of the theoretical volume of the annulus between the outside of the pipe and the hole cut by the TBM. Volumetric control was used rather than pressure control to avoid heaving of the ground surface under the shallow cover conditions. Secondary injections along the pipe were used as needed to respond to jacking loads or ground surface settlements. The contractor proposed and used a lubrication/grout mix with a delayed setup time.

After the muck was discharged through the pressure relieving gates, it was removed from the heading area via a conveyor system and deposited in rail mounted muck cars. The conveyor system extended to about 10 meters behind the face. The contractor's estimate of the duration of the jacking was 35 working days. The jacking took 60 days to complete due to start up and mechanical delays.

GAS MAIN CROSSING

Several protective measures for the force main/gas main crossing were considered, including increasing the clearance between the TBM and the gas main, installing a split shell carrier pipe placed around the gas main prior to the crossing construction, depressurizing the gas main, and instrumenting the TBM approach and crossing with borros settlement points. Because the gas main could not be depressurized and the clearance could not be increased, the gas transmission company exposed the gas main crossing using a vacuum excavated pit to a depth just below the gas main. The pit allowed visual confirmation of the clearance between the gas main and the TBM and prevented confined heave pressures from affecting the gas main. The use of the earth pressure balance shield TBM instead of a slurry pressure balance TBM allowed the TBM to make the crossing with little to no cover.

GROUND SURFACE SETTLEMENTS

The borros point monitoring program along the path of the TBM indicated that at the time the TBM cross beneath the gas line, heave and settlements were typically less than 25 mm. However, more significant surface settlements and sinkholes occurred over about one half to one third of the pipe jacked force main alignment following rainfall events.

Much of the University Road paving between approximately Stations 1+50 (meters) and 3+15 had sustained damage in the form of road settlements, sinkholes, and pavement cracking. Numerous pavement settlements, sinkholes, and cracks existed on the east side of University Road. Settlements and sinkholes were located typically within about 1.2 meters of the pavement edge and directly over the force main alignment from about Station 1+50 to 3+15. Pavement drops up to 450 mm deep in the sinkholes were observed. Cracks in longitudinal and transverse directions were evident over much of the west side of University Road between approximately Stations 1+50 and 3+15. Spalling of the pavement topcoat was observed in two locations.

CONCLUSIONS

Use of an EPB TBM combined with pipe jacking permitted the successful construction of a segment of force main without dewatering that could have caused a spread of contamination from a lower sand stratum to an upper sand stratum. The method was also used successfully to pass under, within about 300 mm, of an active, 4800 kPa, natural gas transmission line without incident. Ground movements in shallow cover cases are highly dependent on the details of the TBM cutting profile, operation methods and control. Despite the ground surface settlements and pavement cracking that developed, EPB TBM pipe jacking was an appropriate method for this set of circumstances.

ACKNOWLEDGMENTS

The Massachusetts Water Resources Authority is the owner of the project. Fay Spofford and Thorndike, Inc. is the designer with GEI Consultants, Inc. as the geotechnical consultant. Caliacco Corporation is the contractor with Laxfield Corporation as the pipe jacking subcontractor.

SERVICE LATERAL REHABILITATION TECHNOLOGY OVERVIEW

Jeff P. Wells, M. ASCE[1]

Abstract

Service lateral rehabilitation technology is being given increasing consideration because of its ability to effectively reduce the infiltration/inflow (I/I) entering the sanitary sewer collection system. The numerous service lateral rehabilitation products available on the market today can be categorized as either chemical grouting, epoxy injection, linings, or pipe bursting techniques. This paper presents an overview of the different lateral rehabilitation technologies that are commercially available to the owners and designers today.

Introduction

Studies have shown that service laterals can contribute a significant portion of the sanitary sewer's extraneous infiltration/inflow (I/I). Owners and designers are now realizing the benefits of addressing all the sewer system piping components including mainline pipes, manholes, and service laterals. When only mainline pipe rehabilitation is performed, migration of groundwater to other sewer system entry points, such as manholes and service laterals, can occur. A comprehensive approach enabling contractors, owners, and designers to schedule evaluation and rehabilitation efforts more efficiently can result in a more cost-effective project.

Evaluation is the key first step to a successful rehabilitation program. The I/I sources must be identified as either infiltration or inflow. Smoke testing and dye water flooding are effective ways to locate direct connections such as roof drains. Internal television inspection is the best way to quantify and locate infiltration problems. Based on the evaluation report, the engineer and owner can decide on the most appropriate rehabilitation method.

---

[1]Senior Product Engineer, Insituform Technologies, Inc. 1770 Kirby Parkway, Suite 300, Memphis, TN 38138

Current Technology Overview

Service lateral rehabilitation technology is one of the fastest growing segments of the trenchless technology field. There are now an increasing number of options from which the owners and designers can choose for their service lateral rehabilitation needs. These technologies can be grouped into the following four categories:

- chemical grouting
- epoxy injection
- linings
- pipe bursting

Chemical Grouting

In this section, two different methods of applying chemical grout are discussed. The first is the more traditional method and one of the oldest remediation methods used in mainline pipes, and more recently in laterals. This method uses a number of different devices specifically engineered to address the lateral system. Some systems dispense grout from the mainline pipe up the lateral, while other systems have been designed to operate in the opposite direction. There are a number of different types of grout used today including acrylamide, acrylate, and polyurethane. While the techniques of sealing laterals with grout as well as the material used may differ, the process can be generally defined as the placement of a mechanical device that dispenses the grout between some type of forming bladder and the lateral itself. Once the grout has congealed, the forming bladder or packer is removed. The application of chemical grout is also used to terminate infiltration coming into the mainline pipe from around the hole cut in a liner so as to allow the lateral to function. Chemical grouting is not a structural rehabilitation method, i.e., the placement of grout is not intended to re-establish structural integrity of the pipe being repaired. Additionally, chemical grout has a rather low shear strength and is most effective if the grout can pass through the defect in the pipe and then into the surrounding soil matrix. The following is a sampling of the most widely known and used grouting systems that rely on some form of mechanical placement and injection system.

TRB Specialty Rehabilitation, Inc. helped develop and now has the rights to the CUES Lateral Sealing System (LSS). The LSS is a trenchless rehabilitation method developed to test and repair defective service connections. All entrance of equipment and access to the laterals is through existing manholes and sewer mainline pipe.

The lateral sealing packer typically performs the work in 200 mm (8-inch) and 250 mm (10-inch) diameter main line pipe. The lateral sealing inversion tubes accommodate two sizes of lateral openings: 100 mm (4-inch) and 150 mm (6-inch) diameters. The inversion tubes are of sufficient length to seal the lateral joints outside the main sewer pipe. The 100 mm (4-inch) diameter inversion tube sealing length is approximately 2 m (6 ft.). The 150 mm (6-inch) diameter inversion tube sealing length is a approximately 0.7 m (2 ft.). The circumference of each inversion tube is undersized to permit the passage of grout around the annulus. The last 0.7 m (2 ft.) of the inversion tube creates an air-tight seal with the lateral pipe so that no grout passes beyond the inversion tube.

Air testing house laterals is accomplished by isolating the area to be tested with the grouting packer and inversion tube, and applying positive pressure into the isolated void. A sensing unit is used for continuous monitoring of the void pressure. This sensing unit is located within the void and accurately transmits pressure read-out to the control panel.

House lateral sealing begins if the lateral does not pass the air test. The lateral packer remains in position, maintaining the isolated void. Chemical grout sealant is pressure injected through the lateral packer into the annular space between the inversion tube and the service lateral pipe. Under pressure, the grout material is forced out into the surrounding soil through leaking joints, cracks and other pipe defects.

Upon completion of the lateral sealing procedure, the house lateral is air tested a second time to verify the sealing of the connection. If the lateral fails the air test a second time, the grouting procedure is repeated until either the lateral is sealed or it is determined that the grout consumption is too high and may result in the blockage of the lateral pipe. The final determination to stop subsequent attempts to seal a lateral is typically made jointly between the Inspector and the grouting operator.

American Logiball's Lateral Injection Packer works internally from the main sewer to grout the service lateral pipe up to 6 m (20 ft.) from the main pipe. A sewer camera is used to position the Lateral Injection Packer at the service connection. A D.C. motor rotates the packer to align it with the connection. Air pressure is applied to inflate and extend the grouting plug from the packer housing into the lateral pipe. The end bladders are inflated and expanded against the mainline pipe wall, thereby isolating the lateral connection. After testing or grouting, the plug is retracted by vacuum and the two end bladders are deflated. The packer is then pulled to the next lateral connection, where the process is repeated.

The other type of chemical grout application has come to be known as flood grouting. This procedure uses no specific mechanical delivery system to place the grout and is quite different because it is intended to treat the whole line, whether that be the mainline pipe and the lateral system or just the lateral. Flood grouting is being commercialized here in the United States by Sanipor.

The Sanipor sewer renovation process is based on the seepage of a two-component material into the soil surrounding leaks in sewage pipes. The two components combine outside the pipe and harden in the void spaces in the soil or backfill surrounding the pipe. The procedure consist of carrying out a survey inspection of a section of suspected leaking pipe, whereupon it is cleaned and isolated. The section and manholes are pumped full of water and the height loss in the manhole is recorded as a function over time. This provides information on the expected material consumption. The section of sewer and manholes is then filled with the first fluid, Sanipor Solution No. 1. The fluid fills the entire section and rises up the manhole shafts. This vertical column of fluid in the manholes creates enough pressure for the fluid to exfiltrate or seep out of the sewer at the point of previous infiltration. After this has continued for a set time, the fluid is pumped out of the sewer and second fluid, Sanipor Solution No. 2 is pumped in. This fluid changeover is done as rapidly as possible. After the system is filled with Solution No. 2, the fluid exfiltrates where Solution No. 1 exfiltrated. Once the level of Solution No. 2 stabilizes or stops dropping in the manhole, the section and manholes are sealed. Solution No. 2 is then pumped out, any residual material is removed and the section is televised. Outside the manholes and pipe the two components react with each other and harden to form a rigid mass, which makes the soil locally impenetrable so that leakage can no longer occur.

Epoxy Injection

One of the primary differences between chemical grout and the use of epoxy is that epoxy has the ability to re-establish structural integrity of the pipe being rehabilitated. Although for decades epoxies have been used very effectively in man-entry pipes, it has only been in the last 4 to 5 years that epoxy-dispensing robots have been commercially available here in the United States. These type of robotic systems have operated throughout Europe for more than 20 years with good results.

Another primary difference between chemical grout and epoxy is that epoxy is much more viscous and generally can not be pumped long distances. With the exception where epoxy is used as a lateral lining material, as opposed to other resins, epoxy is only used at the connection of the lateral and the mainline pipe, and up the lateral approximately 150 to 250 mm (6 to 10 inches). As a result, the use of robotically-placed epoxies are generally limited to:

1. the structural reconstruction of recessed laterals, where the lateral is once again reconnected to the mainline pipe, free from infiltration at the connecting joint,

2. filling in the area around a hammer tap where sections of the mainline pipe have been broken away and the sealing of the lateral to the mainline pipe,

3. sealing of the lateral where cured-in-place or PVC-based fold and form liners are used. In this case, the epoxy is used to permanently seal the area around where the lateral opening has been made in the liner. A permanent and structural closure, joining the mainline pipe, the lateral and the liner, can be achieved depending on the way in which the robotic system prepares the contact surface areas.

There are two companies primarily involved in the commercialization of robotic-based, epoxy-delivery systems. They are:

- KA-TE Systems, and
- American Robotics, formally known as SIKA.

Both of these systems utilize specially designed grinding tools to remove contaminated and unwanted material, and the subsequent placement of epoxy in the defect area. Each of these two systems have different methods to prepare the defect area, as well as the placement of the epoxy, however, they both have a number of similar steps and recommendations associated with the preparation of the host pipe by grinding away contaminated material so as to ensure the bonding of the epoxy to the substrata. The injection of a specially formulated epoxy, capable of adhering to wet surfaces and able to harden underwater and in the absence of air, is completed using different procedures and mechanisms, all of which are designed to hold the epoxy in place until it hardens. Once the process is complete, the area can be ground smooth with no obstruction to the flow, either in the lateral or the mainline pipe. Additional information regarding the unique features, capabilities and restrictions applicable to each of these robotic systems can be secured by contacting the companies directly.

Linings

Insituform Technologies offers two lining methods for the rehabilitation of service laterals. The snap ring method utilizes a flexible ring embedded into the downstream end of the liner which physically matches the configuration of the service connection. This method requires access to the service lateral from the private property side. After first vacuum impregnating the liner with a thermoset resin, an inflation bladder is inverted through the length of the liner. The bladder is then deflated, the liner and bladder are folded and the snap ring end of the liner is attached to a rope pre-strung through the lateral pipe from the cleanout or access pit to the adjacent downstream manhole. With a camera positioned in the main pipe upstream from the service connection, the liner and bladder are pulled through the lateral pipe until the snap ring extends out into the main pipe. The bladder is inflated to circumferentially expand the folded liner and snap ring, and the liner is pulled back from the cleanout side until the flexible snap ring seats against the service connection. The ropes are removed from the liner and the curing cycle begins. Hot water may be used to accelerate the cure of the thermoset resin. Once the cure is completed, the inflation bladder is diverted from the lateral pipe leaving the hardened liner in place. The snap ring method has been used on 100 mm (4-inch) and 150 mm (6-inch) diameter laterals for lengths up to 26 m (85 ft.).

The top hat method allows for the remote lining of the service lateral from the main pipe without access to the lateral pipe from the private property side. The downstream end of the liner is constructed with a flexible felt flange or top hat that is physically similar to the configuration of the service connection. The liner is vacuum impregnated with a liquid thermoset resin and loaded inside a hose assembly. The assembly is attached to a robotic-positioning device and the unit is winched through the main pipe to the service connection from an adjacent manhole. The robotic device is controlled remotely to position the leading end of the liner at the service connection. Air pressure is then supplied to the hose assembly causing the liner's top hat to seat against the connection and inverting the liner and inversion hose into the lateral pipe. Pressure is maintained throughout the cure cycle to hold the top hat and liner tight against the pipe wall. Heat in the form of hot air, hot water, or steam may be supplied to the hose assembly to displace the air and accelerate the cure of the thermoset resin system. Once the cure cycle is completed and the liner is a hardened pipe fitted tightly against the host pipe, the inversion hose is diverted into the assembly and the installation unit is winched back to the adjacent manhole to be loaded with the next liner.

LMK Enterprises offer the Performance Liner lateral lining process. The Performance Liner consists of a non-woven fabric liner with a translucent inversion bladder. The felt liner is impregnated with a thermoset resin and cured at ambient temperatures.

The first step in this process is to make a small excavation pit and approximately 1 meter (3 feet) of the lateral pipe is removed. The pipe is cleaned and an internal inspection of the pipe is done. Next the liner is vacuum impregnated with a hybrid polyester resin and the liner is loaded into a launching device. Air pressure is then applied causing the liner/bladder assembly to invert forward out of the launching device. The liner protrudes into the main pipe approximately 75 mm (3 inches). The next step is the curing process which takes approximately 90 minutes at ambient temperatures. Once the liner is cured the bladder is diverted back into the launching device and removed from the pipeline. The liner protrusion is cut flush utilizing a robotic cutter with a diamond full-circle blade. A second internal inspection of the pipe is done. A cleanout is installed and the site is left clean. The result is a new smooth jointless pipe within the existing pipe.

The Performance Liner is available in diameters of 100 mm to 150 mm (4 to 6 inches) and up to 30 m (100 ft.) in length. The liner will negotiate and conform to 45° bends and sweep 90° bends. The end of the liner is open so only one access point is required. The liner can be installed from an access pit down the mainline pipe and back up to the house all from one location.

CSR Pipelining System's IntraLaminar Heat Cure (IHC) utilizes a system where a conductive fabric layer is located within the liner composite to generate heat for the thermoset resin. The first step in this service lateral lining process is to locate and excavate the service lateral on the private property side. The excavated pit need only be large enough to accommodate one person. After the pipe is exposed, a short section must be removed to allow for the insertion of the resin impregnated lining material. The excavated pit is then fitted with a launch tube through which the liner is fed.

Once the material is viewed entering the main pipe, air inflation begins at a pre-determined pressure. After the liner is inflated, the composite fabric is resistively heated to achieve cure. With the cure complete, the service can then be reinstated using a remotely operated robotic cutter. The flexible design of the composite material forms a flared seal at the service entrance into the main pipe and can be bonded to a spot repair material.

Pipe Bursting

Pipe bursting is a technique used to replace an existing pipe with a new pipe, either of the same size or upsized depending on the requirements of the line being rehabilitated and the depth of the line. For some time, pipe bursting has been used in the gas industry to replace small diameter cast iron lines that were either undersized or in the process of deterioration. The use of pipe bursting in the utility industry, particularly in laterals, is a newer application of this technology. An entrance pit is generally provided for the installation of the new pipe, and a second pit is generally required from where the replacement pipe is pulled in. In some instances, these pits can be avoided if manholes or other structures are used to introduce the new pipe into the old line and/or from which the pipe is pulled. The material normally used in pipe bursting is high density polyethylene pipe, although other piping materials are used.

Conclusion

The rehabilitation of sewer service laterals should be considered, particularly in those areas where high ground water tables contribute significantly to infiltration. There are differing opinions as to how much of the lateral should be rehabilitated, at what cost/benefit ratio and, at whose expense. Regardless, there are numerous trenchless methods commercially available that suit the individual requirements of today's decision makers. This paper merely outlines the different types of service lateral rehabilitation that are available. It is the intention of the author not to make product endorsements but to provide a guide from which these rehabilitation methods can be evaluated.

Cementitious Liner Design for In Situ Manhole Rehabilitation

Phil Forst[1] and John M. Pitt[2]

Abstract

This paper is an evaluation of design methods for thin shells of brittle materials to be used as rehabilitative liners for in situ manhole repair. Several combinations of elastic and plastic theory are available to determine loads due to water, soil, and traffic. Liner capacity is considered as a free standing shell capable of buckling under symmetric load and bending under non-symmetric loads. The result of this evaluation is that various methodologies produce similar results.

Introduction

Deterioration of manholes has contributed to overloading of wastewater treatment plants from infiltration. The cost of replacement is usually prohibitive and frequently causes unacceptable disruption of traffic. One common rehabilitation method is to line partially deteriorated and leaking manholes with a thin shell of mortar. These thin liners are best if centrifugally cast, but can be trowelled or sprayed. This paper summarizes and provides an evaluation of theories from the mechanics of soils and solids for rational design of thin cementitious shells applied to existing manholes. The effects of hydrostatic, soil, and traffic loads are considered. A comparison of solutions from elastic and plastic theory are applied to determining lateral pressures on manhole liners.

Liner Design

Hydrostatic Loads

The impetus behind rehabilitation in many sewer systems is not based primarily on loss of structural integrity, but on a need to reduce flow into treatment facilities overloaded by infiltration of ground water. One method of providing this reduction is with a liner system. In this situation, the thickness of a liner system is controlled by the hydrostatic pressure of ground water. Buckling stress is the main factor to consider in the design of a liner to withstand a uniform external pressure. Experiments at Iowa State University demonstrate that certain cementitious liners cast on existing manholes do not bond with the host structure, but shrink away and produce an independent internal shell (Pitt 1995). External hydraulic pressure is transmitted through the old structure and produces axisymmetric stresses on the new liner.

Buckling results from the inability of a material to withstand a given load as a result of exceeding tensile strength caused by internal moments and is the controlling mechanism for thin shells made of a material with low tensile strength. A thin shell is defined as structure with a radius

[1] Research Assistant, Spangler Geotechnical Laboratory, Iowa State University, Ames, Iowa, Student Member ASCE

[2] Associate Professor of Civil and Construction Engineering, Spangler Geotechnical Laboratory, Department of Civil and Construction Engineering, Iowa State University, Ames, Iowa, Member ASCE

to shell thickness ratio greater than ten. Manhole liners typically meet the criteria defining a thin shell. Buckling stress is the value used herein to determine the required liner thickness.

Assuming the water table is below the base of the cobel, the critical buckling pressure is based on a uniform external pressure acting on a circular ring. (Symbols are defined in list of variables.)

$$\sigma_{buckling} = \frac{3E_l I}{R^3} \text{ N/mm}^2 \text{ (Moser)} \quad (1)$$

$$I = \frac{t^3}{12} * 1mm \quad \text{mm}^4 \quad (2)$$

The value of the buckling stress is replaced by the value of the hydrostatic pressure at a given depth. The moment of inertia as defined by (2) is substituted in (1) and the required thickness for a hydrostatic load is determined by solving (1) for t.

$$t = \sqrt[3]{\frac{4R^3\sigma}{E_l}} \quad \text{mm} \quad (2a)$$

Soil Loads

Lateral symmetric soil pressures should also be considered in a design. These pressures can be estimated from classical soil mechanics if something is known about deformation at the soil-structure interface. As structures requiring rehabilitation have been in place for many years, it can be presumed that the soil-structure system is in equilibrium and that movement into or away from the structure is zero. This state is called the at-rest condition and stresses can be estimated from the earth pressure coefficient, K_o (Spangler and Handy 1982). K_o is that ratio of lateral pressure on the structure to the vertical stress from soil stacked upon itself.

For cohesionless soils, the at-rest earth pressure coefficient can be determined from Jaky's equation as follows:

$$K_o = 1\text{-}\sin\phi = \sigma_1 / \sigma_2 \quad (3)$$

When soil has cohesion it tends to hold itself up and causes no lateral pressure. The depth to which there is no lateral pressure can be estimated from a relationship used to determine the depth to which a cut will stand unbraced (Das 1990). The drained cohesion and unit weight of the soil are all that is needed to determine the critical depth.

$$d_{cr} = c_u / \gamma_s \quad (4)$$

The drained cohesion of the soil can be estimated from simple in situ tests using a pocket penetrometer or vane shear tests on soil in the vicinity of the manhole.

Once the critical depth is determined, the lateral earth pressure acting on the manhole structure can be determined by

$$\sigma_s = 0 \quad \text{if } z \leq d_{cr}$$
$$\sigma_s = K_o[z - d_{cr}]\gamma_s \quad \text{if } z \geq d_{cr} \qquad (5)$$

The total thickness of liner required for hydrostatic and soil pressures can be computed by replacing σ with the sum of the appropriate pressures in Equation 2a.

Traffic Loads

Unlike soil and hydrostatic loads, traffic loads are cyclic, non-axisymmetric, and produce impact. Also, a traffic load may result in a vertical and lateral load at the same time. The magnitude of stress on a manhole that results from traffic passing near such structures can be estimated by adaptation of elastic stress distributions or relationships determined by previous experiments.

The Boussinesq solution for distribution of stress on a manhole assumes a point load at the surface of a semi-infinite medium as defined in Equation 6. The Boussinesq solution

$$\sigma_h = \frac{x}{r}\left[\frac{P}{2\pi}\left[\frac{3r^2 z}{R_1^5} - \frac{1-2\nu_s}{R_1[R_1 + z]}\right]\right] \text{ N/mm}^2 \qquad (6)$$

assumes that the soil is an elastic, homogeneous, and isotropic material.

Spangler developed an empirical relationship; for distribution of lateral stress of a point load of P as this load is distributed on a vertical plane. This is equation 7.

$$h_c = \frac{Px^2 z}{R_1^5} \text{ N/mm}^2 \qquad (7)$$

Burmister determined the traffic induced vertical soil stress by compensating for the change in stiffness between the pavement and subgrade layers. Burmister developed a nomogram of two-layer stress influence curves based on a load distributed over a circular area and the ratio of the pavement modulus of elasticity to the subgrade modulus of elasticity as shown in Figure 1.

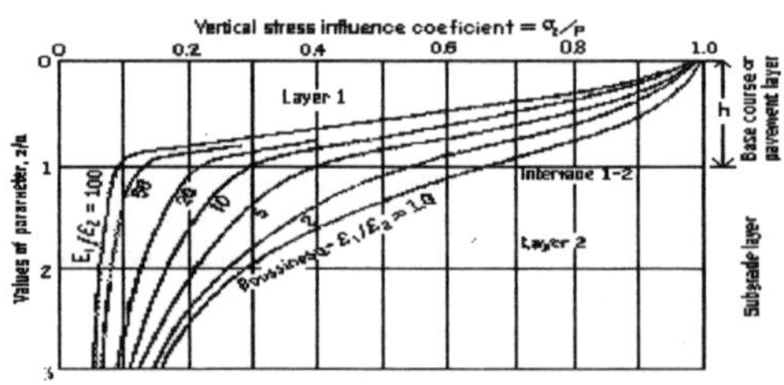

Figure 1: Burmister's Vertical Stress Influence Curves(from HRB 177)

Stiffness ratios of 100, 50, 10, 1 correspond to portland cement concrete, asphalt, gravel, and an unpaved surface, respectively. As in the Boussinesq solution, Burmister's work assumes that the soil is a semi-infinite medium. Unlike the Boussinesq solution, however, Burmister's solution assumes that the depth of one layer is finite. As shown in Figure 1, the vertical stress can be determined by multiplying the appropriate vertical stress influence coefficient by the applied wheel pressure. The vertical stress at the interface is defined in Equation 8 and simplified in Equation 9.

$$\sigma_z = \frac{cP}{A} \text{ N/mm}^2 \quad (8)$$

$$\sigma_z = cp \text{ N/mm}^2 \quad (9)$$

The vertical stress influence coefficient for determining the vertical stress at the pavement-soil interface is determined by substituting h for z in the z/a ratio and selecting the appropriate curve based on the pavement material.

Assuming the pavement is an elastic medium, the load spreads over an area greater than the original "footprint" at the tire-pavement contact. According to the American Concrete Pipe Association, the load spreads over an area defined at the pavement-soil interface as defined in Equation 10.

$$A_2 = AL1 * AL2 \quad (10)$$

where: L = length of contact area, mm
W = width of contact area, mm
AL1 = L + 1.75H, mm
AL2 = W + 1.75H, mm

Unlike Burmister's solution, Spangler's solution as defined in Equation 7 and the Boussinesq solution as defined in Equation 6 are not based on an area load. The Boussinesq equation can provide a solution for an area load by breaking the area into a discrete number of elements and summing the effect of all the elements at a given point. Spangler's solution, Equation 7, can be integrated in the x and y directions as shown in Equation 11(Spangler & Handy 1982).

$$h_a = \frac{p_a}{\pi}\left[\tan^{-1}\left(\frac{x_2 y_1}{zR_{x2}}\right) - \frac{x_2 y_1 z}{\left(x_2^2 + z^2\right)R_{x2}} - \tan^{-1}\left(\frac{x_1 y_1}{zR_{x1}}\right) + \frac{x_1 y_1 z}{\left(x_1^2 + z^2\right)R_{x1}}\right] \text{N/mm}^2 \quad (11)$$

With the exception of Burmister, none of the previously mentioned solutions account for ability of the pavement to attenuate or affect the stress distributions acting on the manhole.

Resistance to Non-Symmetric Loads

Research at Iowa State University (Pitt 1995) has shown that cementitious liners usually fail with an arc of 60 degrees This failure results from the inability of the liner to withstand internal moments. Although the most realistic depiction of the traffic load is shown in Figure 2a, a radial load over a 60^0 arc(Figure 2b) is used as an approximation for ease of calculation. The arch is assumed to have fixed ends. Assuming that lateral stress from traffic loads act over this 60 degrees, the required thickness can be determined by analyzing the section as a circular arch(Figure 2c) with a uniform radial loading over the entire arch.

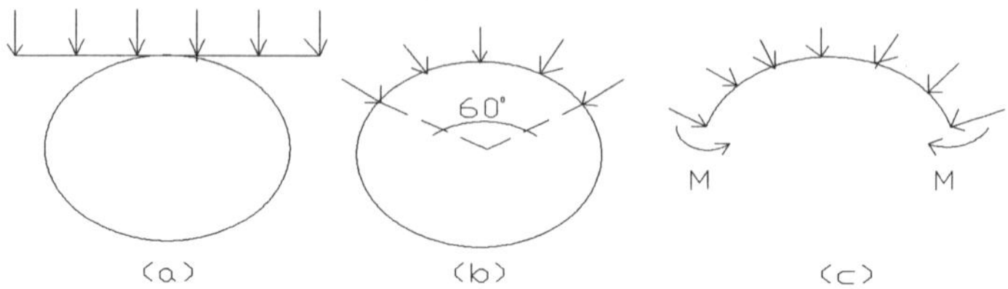

Figure 2: Distribution of Traffic Load on Manhole Structure

The critical load for a uniformly loaded circular arch, as defined in Equation 12,

$$w_t = \frac{E_L I}{R^3}\left(k^2 - 1\right) \text{ N/mm} \qquad \text{(Young)} \quad (12)$$

can be solved for t_t to determine the required thickness to prevent buckling.

$$t_t = \sqrt[3]{\frac{w_t R^3}{6.10 E_l (1mm)}} \text{ mm} \qquad (13)$$

The thickness required to withstand a wheel load that is directly applied to the vertical axis of the manhole can be determined with Equation 14.

$$t_a = \sqrt{\frac{\text{P}}{1.74 \text{E}_l \cos^2 \alpha}} \text{ mm} \qquad \text{(Young)} \quad (14)$$

Safety and Impact Factors

None of the methodologies discussed thus far include safety or impact factors. Work with reinforced concrete pipe suggests an appropriate safety factor ranges from 1.3 to 1.5 (Moser 1990). These factors should be applied to anticipated hydrostatic pressure, soil pressures, and traffic loads.

Impact factors account for the dynamic nature of traffic loads and need to be applied only to non-symmetric loads acting on the manhole. Impact factors in Table 1 were developed for horizontal pipe beneath a road but should be suitable for this application.

Table 1: Impact Factors for Cementitious Liner Design

Depth of Cover (mm) H	Uni-Bell Handbook (Moser 1990) I_f
0 < H < 304.8	1.50
304.8 < H < 609.6	1.35
609.6 <H < 914.4	1.15
914.4 < H	1.0

Comparative Evaluation

Figure 3 is a comparison of lateral stresses estimated by the methods previously described. The example is for a 86,539 N wheel load over a 254mm x 508mm area located at the edge of a manhole. The pavement structure is 304mm of portland cement concrete over a sandy soil unit weight = 1.75 x 10^{-5} N/mm^3. Friction angle of soil = 35°. Impact factors varied with depth according to Table 1 and SF = 1.3 was used.

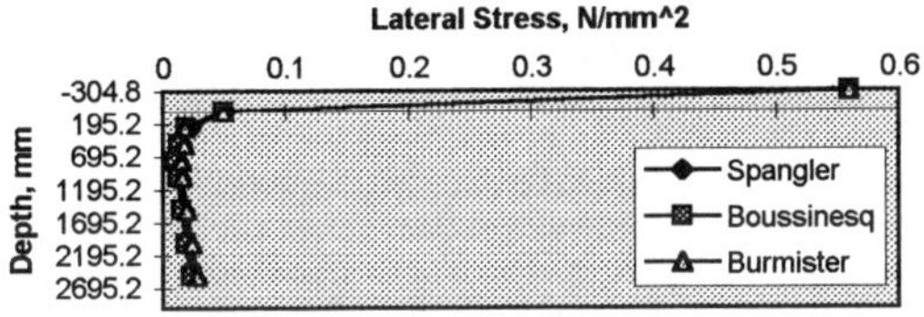

Figure 3: Lateral Stress from Soil and Traffic Loads

Conclusions

Results in Figure 3 show that for this example, the method of calculation makes little difference. Two solutions based on elastic theory produce results equivalent to that of classical and experimental soil mechanics.

List of Variables

a = Radius of circle with area equal to that of tire contact area, mm
A = Contact area of wheel used for traffic loading, mm^2
A_2 = Area of wheel load at pavement-soil interface, mm^2
c= Vertical stress influence coefficient
c_u = drained soil cohesion, N/mm^2
d_{cr} − critical depth of an unbraced cut, mm
E_l = Young's Modulus of the liner material, N/mm^2
h = Thickness of pavement layer, mm
h_a = Lateral unit pressure at a given point, N/mm^2
h_c = Lateral pressure, N/mm^2
H = Depth of cover, mm
I = Moment of inertia of structure about axis perpendicular to plane of buckling
I_f = Impact factor
k = constant = 8.62
K_o = Rankine at-rest lateral stress coefficient
p = Pressure applied to pavement by wheel, N/mm^2
p_a = Pressure at pavement-soil interface, N/mm^2
P = Wheel load, N
R = Radius of pipe or liner, mm
R_1 = Radial distance from point load to point of interest, mm
R_{x1} = Radial distance from point on manhole to near side of load, mm
R_{x2} = Radial distance from point on manhole to far side of load, mm
r = Horizontal radial distance in xy-plane from wheel load to given point, mm
t = Thickness of pipe or liner, mm

t_a = Thickness required to withstand axial loading, mm
t_t = Required liner thickness for traffic load, mm
w_t = Critical load for a uniformly loaded circular arch, N/mm
x = Distance along x-axis from wheel load to given point, mm
x_1 = Horizontal distance from manhole to near side of load, mm
x_2 = Horizontal distance from manhole to far side of load, mm
y_1 = Length of loaded area, mm
z = Distance along z-axis from wheel load to given point, mm
α = Angle at intersection of manhole vertical axis and slope of the cobel
ϕ = Drained angle of internal friction, degrees
γ_s = Soil unit weight, N/mm^3
v_s = Poisson's Ratio of soil
σ = Stress, N/mm^2
σ_1 = Horizontal stress, N/mm^2
σ_2 = Vertical stress, N/mm^2
$\sigma_{buckling}$ = Buckling strength of a structure, N/mm^2
σ_h = Lateral stress as determined by the Boussinesq solution, N/mm^2
σ_s = Lateral soil stress, N/mm^2
σ_z= Vertical stress, N/mm^2

Bibliography

American Concrete Pipe Association(ACPA). Concrete Pipe Technology Handbook. Vienna, VA: American Concrete Pipe Association, 1993.
American Concrete Pipe Association(ACPA). Concrete Pipe Handbook. Vienna, VA: American Concrete Pipe Association, 1988.
Burmister, D.M., "Evaluation of Pavement Systems of the WASHO Road Test by Layered Systems Methods," Highway Research Board Bulletin 177, 1958.
Das, Braja M., Principles of Foundation Engineering. Boston: PWS Publishing Company, 1990.
Moser, A.P., Buried Pipe Design. New York:McGraw-Hill, Inc., 1990.
Pitt, John, "Centrifugal Cast Mortar for In Situ Manhole Rehabilitation," No-Dig 1995, North American Society for Trenchless Technology, 1995.
Spangler, M.G., and Handy, R.L., Soil Engineering. Philadelphia: Harper & Row, 1982.
Young, Warren C., Roark's Formulas for Stress & Strain. New York: McGraw-Hill, Inc., 1989.

Acknowledgment

The authors wish to thank the Center for Advanced Technological Development(CATD) at Iowa State University and the NSF Small Business Innovation Research Program for their support and funding.

Innovative Methods for Rehabilitating Sanitary Sewer Service Laterals

Gregory K. Anderson[1]
Daniel A. Riley[2]

Abstract

This paper discusses two trenchless installation methods for rehabilitating gravity sanitary sewer service laterals. The methods discussed include pipe bursting and the cured-in-place liner installation processes. The paper identifies the procedures necessary for completing the rehabilitation and benefits associated with its implementation.

Introduction

The City of Norfolk, Virginia has one of the most aggressive sanitary sewer rehabilitation programs in the nation. The City's Department of Utilities manage the projects and annually allocates in excess of 15 million dollars to remediate wastewater collection system deficiencies and eliminate extraneous volumes of groundwater infiltration and stormwater inflow. While several ongoing projects have been mandated by the State of Virginia's Department of Environmental Quality, due to illegal discharges (overflows) of wastewater into state waters, the majority of the projects have been scheduled as part of a preventive maintenance program. The areas of the City in which the rehabilitation projects are scheduled are prioritized based on the age and condition of the gravity sewers. Of the 840 miles of sewer main within the City, the Department of Utilities has rehabilitated, or is in the process of rehabilitating, approximately 95 miles of piping. While their program has been under way for only 6 years the results achieved to date have had a great impact on the overall system operations.

[1]Project Manager, URS Greiner, Inc., 5606 Virginia Beach Boulevard, Virginia Beach, VA 23462
[2]Project Manager, Norfolk Department of Utilities, 400 Granby Street, Norfolk, VA 23501

Because of the City's aggressive attitude concerning rehabilitation of its wastewater infrastructure many state-of-the-art techniques for performing sewer rehabilitation have been initiated on their projects. This paper discusses a recent project, West Ghent Sanitary Sewer Rehabilitation, in which several innovative methods for rehabilitating service lateral piping has been implemented.

Project History

The West Ghent Sanitary Sewer Rehabilitation project first began as an Inflow/Infiltration (I/I) Analysis of the wastewater collection systems within the neighborhood and then developed into a Sewer System Evaluation Survey (SSES). Based on the identification of excessive I/I and structural deficiencies with the 70 year old collection system infrastructure a rehabilitation program for the entire gravity piping systems within the neighborhood was established.

Because the City of Norfolk, Department of Utilities had recently completed the entire renovation of the neighborhood's water distribution system it was determined that any proposed rehabilitation to the sanitary sewer system would encompass all piping within the City's right-of-way. These improvements entailed rehabilitation, or replacement, of the sewer mains, property line cleanouts and all service laterals located on City property.

The West Ghent neighborhood was constructed in the 1920's and much of the original wastewater conveyance system was installed at that time. The collection system piping consisted primarily of vitrified clay and terra cotta materials, and the manholes are of brick and mortar construction. The gravity piping ranges in size from 6-inch to 15-inch diameter and conveys the wastewater discharged from approximately 785 single family houses, several duplexes and apartment buildings, an elementary school, churches, and a few small businesses. The neighborhood was developed around the original Norfolk and Western (Norfolk Southern) Railroad terminal building and coal loading piers and many of the original inhabitants worked for the railroad. The majority of the buildings in the neighborhood have basements with plumbing connections for laundry discharge making the service lateral fairly deep below existing ground elevations. This condition compiled with the use of granite curbing which is difficult to remove, and size and age of trees within the City's right-of-way make replacement of the service laterals within the right-of-way difficult.

The construction bid documents for the project were assembled in a manner which allowed the contractor to bid either open cut excavation and replacement of the laterals or rehabilitation using a pre-approved 'No-Dig' lining process. The low bid contractor for the project originally bid the public lateral portion of the project as open cut excavation, but then provided the City an option of trenchless rehabilitation. The methods presented to the City for review included a cured-in-place lateral lining system and a replacement system utilizing the pipe bursting technology. Table 1 describes the various components and magnitude of the project and highlights the extent of service lateral rehabilitation.

Sewer Main Replacement		
Size(Inches)	**Length(Feet)**	**Cost**
6,8,10,12&15	16,575	$764,685
Sewer Main Rehabilitation		
Size(Inches)	**Length(Feet)**	**Cost**
8,10,12&15	18,859	$863,697
Sewer Lateral Replacement/Rehabilitation		
Size(Inches)	***Length(Feet)***	***Cost***
4&6	25,545	$1,190,000

Table 1.

The contractor's recommendation as to which rehabilitation system would be used on individual laterals was based primarily on the new piping material of the sewer main. If the sewer main was part of the replacement activities, and a new PVC sewer had been installed, the lateral rehabilitation was recommended to be a cured-in-place liner. If the sewer main was rehabilitated using a cured-in-place liner then the lateral was recommended to be rehabilitated in the same manner. Pipe bursting of the existing laterals was only recommended when the sewer main was rehabilitated using a deformed/reformed polyethylene liner.

Prior to the actual insertion of the cured-in-place liner the contractor was required to clean and TV all laterals to ensure that the process could achieve the specified rehabilitation results. If the lateral had joint offsets, root intrusions, or other miscellaneous defects, which would result in a significantly reduced cross sectional area after lining, the lateral rehabilitation would be defaulted to pipe bursting. The only recorded limitation of the pipe bursting activity was the inability to burst through cast iron 90 degree fittings. The following paragraphs describe in detail the preparation required for each rehabilitation technology and the actual installation process.

Pipebursting Laterals

Preparatory work included marking the underground utility locations adjacent to the sewer main and the property line cleanout. Once all utilities were identified excavation at the main and at the property line were made. The pipe bursting equipment, pipe and pipe fittings were then mobilized to the excavations. The contractor then disconnected the lateral at both the property line and at the main and the bursting cable was pulled through the lateral to be bursted. Any necessary dewatering equipment was then placed in service and the contractor began the pull. Figure 1 shows the polyethylene lateral being pulled to the main through the access pit at the property line.

Figure 1. Polyethylene Lateral

When the lateral is pulled to the main the bursting head is removed from the lateral and the contractor would then breakout the host pipe to expose the polyethylene liner. The new polyethylene lateral was then joined to the main utilizing an electrofusion polyethylene saddle and necessary electrofusion fittings. Figure 2 shows the final connection to the main with the saddle already connected and the lateral piping being joined to the saddle utilizing an electrofusion fitting.

Figure 2. Final Connection

Final connection at the property line entailed the use of a PVC wye and dresser couplings and installation of a new property line cleanout.

The piping/fitting material specifications are listed in Tables 2 and 3:
Pipe: Plexco PE 3408 HDPE (SDR 17)

Property	**Unit**	**Test**	**Typical Value**
Material Designation	---	PPI/ASTM	PE 3408
Material Classification	---	ASTM D-1248	III C 5 P34
Cell Classification	---	ASTM D-3350	345434C
Density [3]	g/cm^3	ASTM D-1505	0.955
Melt Index [4]	g/10 minutes	ASTM D-1238	0.1
Flexural Modulus [5]	MPa	ASTM D-790	917
Tensile Strength [4]	MPa	ASTM D-638	22
ESCR [3]	hours	ASTM D-1693	> 5000*
HDB [4]	MPa	ASTM D-2837	11
UV Stabilizer [C]	% Carbon Black	ASTM D-1603	2 to 3
Elastic Modulus	MPa	ASTM D-638	758

* Test discontinued after 5000 hours non-failure.

Table 2.

ElectroFusion Fittings:Central Plastics PE 3408

Property	**ASTM Test Method**	**NominalValue**
Density, gm/cc	D-1505	0.955
Melt Index, gm/10 min	D-128-E	0.12
High Load Melt Index, gm/10 min	D-1238-F	15.0
Tensile Strength @ Break, MPa @ Yield, MPa	D-638	 31 12
Elongation, %	D-638	>800
Flexural Modulus, MPa	D-790	827
Environmental Stress Cracking	D-1693	
Shore D Hardness	D-2240	68
Hydrostatic Design Basis @60°C, MPa @23°C, MPa	D-2837	 4 11
Notched Iod Impact (ft.-Lb/in)	D-256	6.0

Table 3.

As the contractor established a system for bursting the laterals, production generally varied from 4 to 8 per day, and each lateral service was usually reinstated within 2 hours of the initial disconnection from the main. The property line cleanout was installed during this time, and the site restoration was completed shortly thereafter. The excavation at the main was then backfilled and compacted and a temporary pavement patch placed to match the existing pavement elevation. If the contractor was able to expose more than one lateral at the street excavation production increased dramatically as the majority of time and effort was associated with exposing the connections at the main. In many cases connecting laterals for houses on either side of the street were in the same vicinity, and contractor was almost able to double his lateral installation production.

At the beginning of the project the contractor developed an elaborate winch system for pulling the bursting tool, but by the end of the project the bursting was being accomplished primarily by backhoes and the winch system was used only for the bursting of the laterals which were constructed of cast iron.

Cured-In-Place Laterals

Preparatory conditions were very similar as the pipe bursting method except access was only required at the property line. The existing utilities in the area were marked, and the excavation pit was dug. The existing lateral is then accessed and cleaned of all debris, roots, and other materials which would block the inspection of the piping. Portable CCTV equipment was then used to ensure in fact that the lateral could be lined. The resin impregnated ambient cure liner bag is kept at the site in a refrigerated truck where it remains until it is ready to be inserted. A CCTV inspection truck is located over an adjacent manhole, and the camera unit is inserted into the sewer main and is focused where the liner will enter the main. A compressor, pressurized blower and rubber inflatable bladder is mobilized next to the excavation pit.

At the excavation pit, a two foot section of the existing lateral is cut out and removed where the cured-in-place liner will be inserted. If the existing cleanout is to be replaced, the cuts are made on both sides of the cleanout where it connects to the lateral.

Depending on the depth of the lateral and the elevation of the water table, a dewatering pump maybe used. The liner is then impregnated with resin at the site within the refrigerated truck in what is described as a wet-out procedure. At this time the resin is injected into the center of the liner and then rolled throughout the liner to ensure complete saturation. The wet-out procedure is generally completed just prior to the installation.

A flexible fiberglass rod is then inserted into the lateral at the excavation pit and pushed towards the main line connection. The camera operator in the TV truck signals when the rod tip reaches the main. This method is used to measure the exact length of the lateral so the liner bag can be cut to the proper length. The next few steps must be completed very quickly as the impregnated liner begins to cure once it is exposed to sunlight. The resin impregnated liner is unloaded from the refrigerated truck and cut to the proper length. The impregnated liner is either pushed with rods, or pulled by cable, into place and is positioned so that the "cuffed" resin rich end is flush with the inside of the main. The camera operator signals when it has reached the main line.

A calibration hose (bladder) is inverted by air into the liner by the blower expanding the bladder and molding the liner to the host lateral piping. Ambient air is provided by the blower at approximately 10 psi, and the liner is cured for approximately 30-45 minutes. After the lateral is cured, the bladder is deflated and removed from the lateral. Figure 3 shows the blower inverting the bladder into the liner.

Figure 3. Inverting Bladder

After the curing process is completed, the next step is to reinstate the lateral connections. At the excavation pit PVC pipe and fittings along with Dresser Couplings are installed to replace the previously removed section of existing lateral. If the cleanout was removed, a PVC wye is installed to accommodate the new cleanout. Inside the sewer main the protruding liner is cut off flush using a remote control robotic cutting device. After these two connections have been made, the lateral is in service. Excluding any unforeseen delays, the total amount of time a resident will be without sewer service is approximately 2 hours.

When the lateral is placed back into service, back of the excavation pit is backfilled. The backfill material is carefully tamped in around the lateral and cleanout in 12 inch layers, to reduce settlement and stabilize the cleanout, until it is brought up to existing grade. Any areas where concrete sidewalks or driveways were disturbed, new concrete was poured and finished to match the existing conditions.

The previously removed turf, for grass areas, was placed back over the pit area and around the new cleanout box. Grass seed was also sown where needed.

The final step of this process is to perform an interior inspection of the lined lateral. This is done with a portable CCTV camera to assure that the liner was installed properly and without any defects.

The liner/resin material specifications are listed in Tables 4 and 5:

Cured Liner: Applied Felts, LTD.

Property	Test Method	Min Value
Flexural Strength	ASTM D-790	31 MPa
Flexural Modulus	ASTM D-790	1,723 MPa
Tensile Strength	ASTM D-68	20,684 MPa

Table 4.

Resin: Reichhold Polylite 33420

Property	Test Method	Min Value
Barcol Hardness	ASTM D-2583	45
Heat Distortion	ASTM D-648	122^{0}C
Water Absorption	ASTM D-570	*0.49%

* (2 hour @ 100^{0}C)

Table 5.

As with the pipe bursting process as the contractor became more experienced with the insertion of the cured-in-place liner his production rates increased dramatically. On an average 4 laterals were lined daily.

<u>Summary</u>

It is important to note that many of the same preparatory steps necessary for open cut excavation and rehabilitation of utilities were utilized during this project.

Prior to any digging, all utilities in the area of the existing piping were located and marked with paint. This helped to prevent the contractor from damaging other utilities around the excavation pits and along the lateral if problems were encountered while installing the new lateral, and it was necessary to dig down to the lateral. The rehabilitation plans were prepared in a manner such that if open cut excavation was required, adequate information was provided to determine piping depths and existing conditions.

Daily contact with the residents within the project limits, was critical to the success of the project. Homeowners were contacted when their service was to be interrupted. The installation process was described to them and assurance was given that disturbance to their property would be minimized.

The major benefits from using a trenchless rehabilitation solution for the lateral piping on this project include the following:

- Limited Disruption of Service to Residents
- Minimized Disturbance to Established Landscaping
- Less Conflicts with Other Utilities
- Reduced Safety Hazards to Pedestrians and Vehicles
- Earlier Completion of the Project

While the cost of providing trenchless rehabilitation of service laterals generally exceeds the cost of open cut excavation and replacement, proponents of the project believe the extra cost is warranted. From the homeowners perspective when the contractor completes the work at your house and all of your neighbors houses within one day, and the work areas are barely identifiable, it is difficult to say that the soft costs do not equal the additional expenditure.

Houston's Standard Wastewater Products Committee

Gary N. Oradat, P. E.
Todd L. Papes, P. E.

Abstract

Houston is currently the fourth largest city in the United States. It has over 9,662 kilometers (6,000 miles) of sewers, 100,000 manholes, and operates 52 wastewater treatment plants and 320 lift/pump stations. Much of the system is over 40 years old and the City has been under administrative orders filed by the Environmental Protection Agency (EPA) and Texas Natural Resource Conservation Commission (TNRCC) to control sanitary sewer overflows (SSO's). To that end, the City has committed to expenditures in excess of $1 billion to expand and rehabilitate its wastewater system.

In 1992 the City implemented the use of a Standard Wastewater Products Committee (SWPC) to assure that the utility users were afforded the highest value in the products used to construct, operate, and maintain its wastewater system. The Committee serves as a vehicle to review, evaluate, and approve wastewater products for use on City projects.

Based on experience gained here in Houston, the use of a Standard Products Committee, whether for wastewater or other disciplines, provides any municipality with a vehicle to control the quality of their capital investments, offer the rate payers assurance of funds well spent, and provide businesses and contractors with more opportunity at a reduced level of risk. Here in Houston, this process has also resulted in clear benefits to the City in terms "standardization" of products, reduced engineering effort, and ensured fairness in a competitive marketplace.

Introduction

It's a familiar occurrence in most municipalities across this country. A hard working design project manager is busy trying to make that last revision to the project specifications in order for the project to advertise on schedule, when there is a knock on the door. In walks a highly motivated and well-trained product representative whose only request is that the specifications be revised to include their product. The project manager is assured that the product will result in a lower construction cost, increased competition and a better overall project. What's the project manager to do?

Gary N. Oradat, P.E., Assistant Director with the City of Houston Department of Public Works and Engineering, Utility Maintenance Division.

Todd L. Papes, P.E., Principal with Montgomery Watson Americas, Inc.

Or maybe that same project manager is busy trying to get that last addendum issued before the project bids. It's just about ready for the appropriate approval signatures when, there's a knock on the door. This time, it's a different product representative. However, the request is the same.

Or it could be a construction project manager who is in the process of reviewing a "substitution" submittal for a significant element of the project. The contractor offers the change "at no additional cost to the City." Is it good for the City? More importantly, is it fair?

Pressures of the Marketplace Create the Need for a Standardized Approach

In an era of evolving technology, the pressure to stay "on the cutting edge" or to be "the leader in the industry" is significant for all parties involved. Rate payers demand more for their money. CEO's and stockholders want increased profits. Sales representatives need to get their product in "that job."

The City of Houston created its Standard Wastewater Products Committee (SWPC) in 1992 to assure that the utility users were afforded the highest value in the products used to construct, operate, and maintain its wastewater system. The Committee serves as a vehicle to review, evaluate, and approve wastewater products for use on City projects.

Houston's eleven member SWPC is an appointed committee composed of nine City and two non-City representatives. The City representatives consist of design, construction and maintenance personnel. The non-City personnel represent both the local contracting and engineering communities and provide valuable insight regarding the impact of the Committee's decisions. However, only the City personnel have "voting" status.

SWPC Focuses on the City's Wastewater Collection System

The SWPC has focused its review on those products involved in the construction or rehabilitation of the City's wastewater collection system. These products consist of new pipe materials, rehabilitation techniques, and sewer, manhole, and wet well liners and coatings. An application for approval of a product or technology may be submitted for any of the following categories:

- Sanitary sewer pipe materials for direct bury and auguring installations
- Sanitary force main pipe materials
- Trenchless sewer rehabilitation techniques, including:
    - Sliplining and sliplining grout materials
    - Cured-in-place pipe
    - Fold and form processes (PVC and PE)
    - Pipe bursting/pipe crushing
- Protective coatings and liners for sewers, manholes and wet wells for structural repair and corrosion protection
- Concrete additives to decrease permeability
- "Specialty" pipe products such as filled PVC, oriented PVC, and polymer concrete
- Alternative pipe embedment materials such as "flowable fill"

SWPC Reviews Products and Technologies with Clearly Defined Goals

The SWPC must have clearly defined goals in order to be effective in its reviews of products and technologies. The goals of Houston's SWPC are:

- Ensure the use of cost-effective, high quality proven products and technologies in the City's sanitary sewer construction and rehabilitation program;
- Encourage the development of new wastewater products and technologies;
- Develop a standardized process for the application, evaluation, and approval of wastewater products and technologies into City wastewater standard specifications; and
- Promote a fair and equitable bidding environment for contractors, manufacturers and vendors.

The SWPC Provides a Consistent and Fair Process

The SWPC provides a consistent and fair process to review, evaluate, and approve products and technologies for use in the City's wastewater collection system. All products are subjected to the same type of review; although the mechanisms to complete that review may differ depending on the product category. A flowchart delineating the entire review and approval process is shown in Exhibit 1. The following notes the steps of the process:

- Formal applications are completed by each vendor or manufacturer utilizing a standard application packet which is submitted directly to the SWPC. A great deal of latitude is given the manufacturer regarding the type of information which can be submitted on the product or sewer construction/rehabilitation methodology or technique. However, all applications must include product data, history of the product, previous testing results, and references. References are critical. Each application is initially reviewed for completeness.
- If complete, a determination of whether the product is eligible for consideration is made. To be eligible the product submitted must fall within the bounds of the established standard wastewater products categories.
- If eligible, a SWPC member is assigned responsibility for reviewing the application in detail to check the references, product data, history of previous installations, and previous testing results, if any.
- The next step may vary depending on the product itself; however, most products are given a "Demonstration Project" and may be required to participate in independent testing performed by the University of Houston. Following completion of either, or both, of the above, an independent consultant under contract with the City provides an evaluation report of the product or technology.
- The SWPC then reviews the report and, with a majority vote will forward a recommendation to either approve or reject the product or technology to the City Engineer.
- The City Engineer must concur with the recommendation or the process is repeated.
- Once a product or technology is approved it is incorporated into the City's Standard Specifications.

Exhibit 1
Standard Wastewater Products Committee
Application to Approval Process

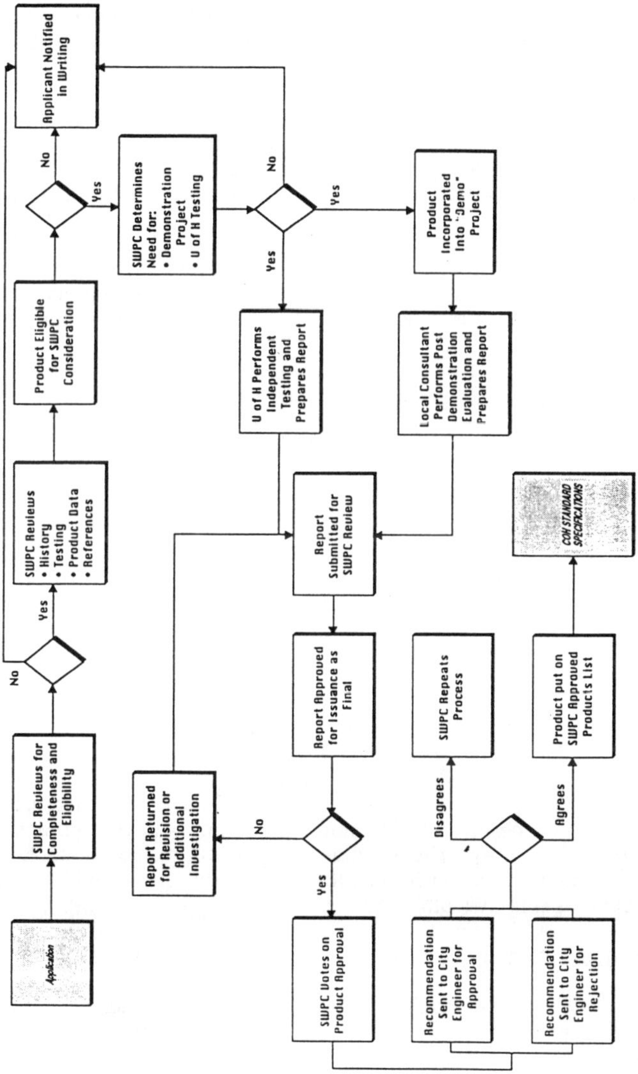

The SWPC Reviews All Applications for Eligibility First

To be eligible, the product or technology submitted must fall within the bounds of the established SWPC categories (i.e., sanitary sewer and force main materials, sewer rehabilitation techniques, protective coatings or liners, or other specialty items).

The SWPC does not accept applications for electrical or mechanical components. The City operates 52 wastewater treatment plants and 320 lift/pump stations which vary in needs and specifications. For this reason, it was deemed impractical to pursue a standardized approval process for such equipment.

Due to the proliferation of "trenchless" techniques for sanitary sewer rehabilitation, the Committee established more stringent requirements for these applicants. Before an application for a trenchless sanitary sewer rehabilitation method or product may be submitted, the following criteria must be met:

PRODUCT DIAMETER	LF INSTALLED IN US	YEARS IN SERVICE
0.2 - 0.53 m (8 - 21 In.)	10,668 m (35,000 Ft)	2 years
0.61 - 1.22 m (24 - 48 In.)	6,096 m (20,000 Ft)	2 years
1.37 m (54 In.) or greater	3,657 m (12,000 Ft)	2 years

Adherence to this criterion prevents the City of Houston from being exposed to undue risk associated with unproven or misapplied technologies. This criteria recognizes that it is more difficult to rehabilitate a 48-inch sewer than it would be for a 10-inch sewer. Thus, a range of diameter categories has been created. It also eliminates applications from products or technologies with a history of only a few hundred or thousand feet installed.

The SWPC does not engage itself in the qualification of contractors. Although the capabilities of the installer play a role in the performance of the product, the SWPC recognizes that licenses can be sold, and that companies can be acquired or merged. The Committee did not believe it prudent to tie a product's future with any specific installer. As a result, once a product, or methodology, is approved, the product is not subject to a subsequent review if the license for that product is sold to another installer.

Demonstration Projects Allow Evaluation Under Actual Field Conditions

The Demonstration Project provides an opportunity for the City to evaluate a product under actual field conditions. This provides an excellent source of data for the vendors and manufacturers, as well as the SWPC during its evaluation. The new product could be incorporated into a new design project or an ongoing construction project. Specification of the "demo" product on a new project could be sole-source, or if other similar products have been submitted, competition may be allowed. Inclusion into a new project also ensures that the project design engineer has worked with the vendor or manufacturer and understands the needs of the product related to preparation and installation.

Incorporation into an ongoing project is accomplished via a construction "Change Order". This would be done where engineering coordination is not necessarily required. Upon completion of the Demonstration Project, the evaluation period is initiated.

Evaluation by an Independent Third Party Ensures Objectivity

The Post Demonstration Project evaluation is the crucial factor in determining the ultimate fate of an applicant. Thus, objective evaluation is mandatory. The City of Houston has successfully used local consulting engineering firms for that evaluation. The evaluation may include a myriad of techniques (i.e., TV inspection, deflection measurements, physical inspection, etc.). The criterion is established by the SWPC with input from the applicant. All such criteria must conform to the appropriate ASTM or other accepted practices. The time allotted to conduct this evaluation is typically a minimum of six months.

Once the evaluation is completed, the third party issues a draft report for internal review by the SWPC. The Committee discusses the preliminary findings with the investigative firm, and may ask for either additional information or investigation or request that the report be issued as a final. Once the final report is issued, the Committee votes on whether or not to recommend approval of the product to the City Engineer. If the product is approved, the applicant is so informed, in writing, by the City Engineer. The same procedure applies for product rejections. Once approval is obtained, the product may be utilized on any project which has yet to receive bids. The product may not be used on projects which have already bid.

University of Houston Provides Independent Testing of Coating and Liners

In addition to the Demonstration Project, the SWPC may elect to perform independent testing by the University of Houston. This is a cooperative program to assist the SWPC in the testing of specific products related to protective coatings and liners. Although in its infancy, the program is yielding favorable results. To date, Houston's City Council has appropriated almost $300,000 to fund research efforts in support of, or in coordination with, the SWPC. In addition, the City has committed approximately $500,000 to UH for the "Intelligent Renewal of Urban Wastewater Systems" project; an endeavor for which UH was awarded a $1,000,000 grant from the National Science Foundation. These efforts both directly and indirectly test and evaluate products related to sanitary sewer collection systems.

Typically, each vendor or manufacturer is required to pay a nominal fee for inclusion into the testing program. The fee accomplishes two purposes. First, it offsets any expenses incurred by UH and secondly, it ensures that only "serious" parties are provided the opportunity to have their product evaluated. Currently, UH is testing products categorized as "liners" or "coatings" which would be utilized in the protection of manholes or wet wells, or the rehabilitation of existing sewers. Any vendor or manufacturer seeking approval of such a liner or coating must participate in the UH testing program. Refusal to do so results in rejection of the product. The testing program is designed to evaluate:

- Chemical Resistance. The product is tested (with and without pinholes) by immersion in 3% and 30% sulfuric acid solution and deionized water. The impact of pinholes, and whether or not the presence of a pinhole in a coating results in a significant deterioration of the cementitious surface behind the coating is also evaluated. The samples to be tested are coated or lined by the manufacturer.

- Hydrostatic Pressure Resistance. This is especially important in Houston where the water table can be within a few feet of the ground surface. A product's ability to survive in a corrosive environment is immaterial if the bond between the cementitious surface and the coating is unable to overcome the hydrostatic pressure in a given location. Products are applied to concrete media (wet and/or dry) and tested to 15 psi.

Standard Products Committee Requires Commitment of Time and Money

The concept of a "committee review" format for various products is not new. Houston had availed itself of this vehicle in the past. However, those instances were limited to a specific project which had a significant construction cost. The implementation of a dedicated committee for ongoing review was unchartered territory. The experiences of Houston's SWPC can benefit other municipalities desiring to implement their own Standard Products Committee.

First of all, give serious consideration to the time involved. As economics within a city government mandate that more is to be accomplished with reduced resources, an individual's involvement in a Standard Products Committee must be balanced against the daily commitments of one's position. Economics may not lend itself to utilizing employees whose only jobs are the Standard Products Committee.

The Committee should consist of the most qualified individuals. In most organizations, these individuals will already occupy positions of influence in middle to top management. The Standard Products Committee cannot be viewed as a body more suited for "staff" personnel. Decision makers need to be involved and supported. If a member has to "check with the boss" before voting the process is unnecessarily prolonged.

The time commitment to the Committee cannot be overemphasized. Members will need time for checking references, reviewing prior installations (if possible), completing follow-up action items, and an establishing ongoing constructive dialogue with the participants in the municipal marketplace. A Standard Products Committee must not only be an open forum for the established product representatives and manufacturers, but for the newest entry as well. Committee members must always strive to "make time," even when it doesn't seem possible to do so.

Of course, every applicant wants their product evaluated immediately so that they can be allowed on projects as soon as possible. And yet, because Committee involvement is "part time," only so many products can be evaluated between meetings and only so many items can be covered at any one meeting. Factor in meetings being canceled for one reason or another, and the potential for not being able to make "all of the people happy, all of the time" becomes overwhelming.

The other important factor to be considered is support. In order for a Standard Products Committee to function as it should, it must be able to adequately investigate prior installations and stay abreast of the latest developments in the marketplace. This translates to a monetary investment by the municipality to allow Committee members to actively participate in organizations, visit trade shows, and travel to various parts of the country to personally evaluate a product's past performance.

Support also takes the form of upholding the Committee's actions and methodologies. If the impression is ever given that the Committee can be influenced by "outside" pressure, its ability to objectively evaluate products has been severely compromised.

Most challenges can be met if everyone understands that the Committee members are doing their best to fairly and equitably evaluate as many products as possible.

The SWPC Has Yielded Many Benefits

Houston's experience with a Standard Products Committee has been most favorable. The benefits which have been derived from the SWPC have been many. The more tangible benefits are:

- Reduces contractor's risk associated with pre-bid speculation regarding approval of substitutions after contract award. This promotes better pricing.
- Ensures specification of quality, proven, cost-effective products and technologies for the City's municipal projects.
- Reduces Engineer's risk associated with specifying a product which appears to be an "equal," but is unproven.
- Reduction in the complexity of submittals during construction thereby allowing quicker turnarounds and subsequent issuance of material purchase orders.
- Enhancement of the City's operation and maintenance functions through the standardization of products utilized.

In conclusion, the use of a Standard Products Committee, whether for wastewater or any other design discipline, allows a municipality to control the quality of its capital investment. It also promotes an atmosphere of fairness for bidders and product suppliers. In Houston, the Standard Wastewater Products Committee's "Mission Statement" accurately reflects the value of such a program:

> "To promote competition, to fairly and equitably evaluate those products or construction methodologies requesting entry into the Houston municipal marketplace, to create opportunity for new and better products, to protect and enhance the City's infrastructure and to provide our rate payers with the most cost-effective collection and treatment system anywhere in the nation."

TRENCHLESS TECHNOLOGY IN THE YEAR 2000 AND BEYOND

Steven R. Kramer, Jason Consultants International, Inc., Washington, DC[1]
James C. Thomson, Jason Consultants SA, Geneva, Switzerland[2]

ABSTRACT

Trenchless technology has grown from a number of techniques used to install crossings under roads and rail tracks, to a design concept and a well practiced methodology in the design and construction industry. The growth was driven by a decaying underground infrastructure and increased demand for utility services throughout the world. Cities and towns are now regular proponents of trenchless technology. It is a cost effective way for the owner to install, replace and renovate utilities with minimum disturbance and cost to the public and surrounding communities. This paper analyzes the trends and factors that will influence trenchless technology in the next decade and what new ideas are being explored.

INTRODUCTION

Trenchless technology has matured from a concept to equipment and techniques that are regularly used in almost all forms of underground utility construction. The advancements have been evolutionary in technology, tools and equipment. Trenchless technology has been used to install telecommunication and electric cables, and gas, water and sewer pipes. As the underground infrastructure throughout the world continues to age and deteriorate, the applications for trenchless tools have expanded. Today, the world of trenchless technology includes many conventional methods developed decades ago and new equipment developed only in the last few years. All of the following technologies are used in trenchless work: guided/horizontal drilling, piercing tools, impact ramming, auger boring, pipe jacking, microtunnelling, pipe bursting, cured-in-place liners, compressed-fit and deformed liners, sliplining, cement-mortar lining to name some of the major methods used.

In this paper, we will analyze how trenchless technology is being used today and what are the key domestic and international trends that will influence the trenchless business in the next millennium. We will be bold and attempt to make some predictions on how trenchless technology will be used in the Year 2000 and beyond.

----

[1] Member ASCE; Jason Consultants International, Inc., 2000 Massachusetts Ave., NW, Washington, DC 20036
[2] Member ASCE; Jason Consultants SA, International Centre Cointrin, CP 1289, CH-1215 Geneva 15, Switzerland

Our predictions will likely not be all correct, but they will help you think about the years ahead. Since we are engineers, we have sought the assistance of several leading books and publications to help define the future directions. We hope that this paper will provoke others in the trenchless and related industries to think about how to better utilize these technologies to solve our most pressing underground infrastructure issues.

It is most interesting to note how the application of trenchless technology has quickly spread across the world. We are no longer many national economies instead we are now a world economy. Unfortunately, the problems with older cities and growing cities are not unique. Public officials, engineers and the general public prefer to improve our cities with minimum disruption and inconvenience to commerce and our daily activities. The economies of the world are growing and many are growing rapidly. This growth has pressed the utility systems to capacity in many cities. Cities and towns are regular proponents of trenchless technology as an effective way to install, replace and renovate utilities with minimum disturbance and cost to the public and surrounding communities.

MAJOR USES OF TRENCHLESS TECHNOLOGY

We offer the following perspectives on how trenchless technology is being used in various parts of the world. It is important that we understand how far we have come in the relatively short period of 10 years. The most spectacular recent success story has to be the development of guided drilling in North America for the installation of pressure pipe and cable. Guided drilling is a steerable system for the trenchless installation of pipes, conduits and cables using a surface launched rig. A pilot bore is drilled using a rotating drill string an is then enlarged by a backreamer. Guided/directional drilling has grown to an everyday technology that is used in almost every state. In many cities, you will pass by a guided drilling rig in a typical drive to and from work. We have gone from the first machines being introduced in 1985 to over 3,000 rigs in use today. Although the growth has been less spectacular outside North America, the technology has become established throughout the world. Guided drilling is more of a specialist technique in Europe and is just recently experiencing high growth. For a variety of reasons, it has not been able to find the widespread acceptance and application as in the US. The European country where guided drilling finds greatest use is Germany where tough environmental laws have created a demand

The use of microtunnelling and mechanized pipejacking continues to gain ground in Europe, Japan and other parts of the world. Microtunnelling is a miniaturized version of pipe jacking for installing pipes to a predetermined line and level by remote control. Germany alone has several hundred pressure balance machines and the total in Japan runs into thousands. In North America, we have seen solid progress in establishing the technology. However the market has not developed as many of its proponents had hoped. On the basis of population size, we might expect to see around a 1,000 machines in US when compared to utilization in Germany. However in reality there are probably still only about 50 machines in operation. Pipejacking and microtunnelling are specialists systems that have become popular in portions of the country such as Houston, Dallas, Boston, San Francisco, and parts of Florida. However, a great deal of work is still the special crossing of a roadway or rail track rather than line installation.

Rehabilitation technology has also been an area where we have witnessed an explosion of new technology. Rehabilitation includes a variety of methods including sliplining, modified sliplining, cured-in-place liners, coatings, on-line replacement, segmental linings and point repair techniques. The rehabilitation market is large and being served by many players. In 1995, it was estimated that the value of rehabilitation services was about $1.1 billion in the U.S. and about half this size in Europe. Many of the rehabilitation products are now mature and highly accepted, especially in the wastewater industry. This is a very competitive business with decreasing prices in several markets.

Ten years ago we had a number of established methods like Insituform and Paltem from Japan. Many of the rehabilitation technologies found their first homes in Europe and are now proven tools of the engineering trade. In Japan, the CIP (cured-in-place) technology found use for gas mains but in the rest of the world and particularly in North America and Europe the rehabilitation of sewers was the dominant market for people like Insituform. In the mid-1980's, we were beginning to see the emergence of technologies like pipe bursting and modified sliplining methods such as Swagelining and Rolldown in the UK. The primary use then was for gas pipe. Today, we have literally hundreds of competing systems for use in sewers, gas and water applications.

What has also developed alongside these main technologies has been many support techniques such as improved cleaning, survey and location equipment. The robotic restoration of connections and robotic repair tools that can operate from within the sewer have all come to the market in the last 10 years.

Trenchless technology has also found a home in several parts of the Middle East and South America. Many developing countries and emerging economies have used trenchless methods to install new pipes and cables in existing cities to minimize disruption to commerce and tourism.

KEY DOMESTIC AND INTERNATIONAL TRENDS INFLUENCING TRENCHLESS TECHNOLOGY

With the vast amount of information that comes our way each day, it is becoming increasingly difficult to capture what is useful versus what is useless. We are drowning in information but starved for knowledge. We need some frame of reference in which to work from. We will provide this point of reference for the underground utility infrastructure.

The following is a discussion of some of the key trends that will impact the operation, installation and maintenance of the underground utility networks.

- **Aging Underground Infrastructure Throughout the World**

The developed countries and so called economic powers have utility systems that may be 50 to 100 years old or older. These systems are national assets of the countries of origin and are fundamental to a good quality of life. The value of these national assets is billions of dollars and can not be easily replaced or repaired.

- **Doing More with Less**

City managers and engineers are continually being asked to maintain and improve the utility systems with minimal budget. It has become a difficult balancing act of trying to satisfy increasing needs with decreasing resources.

- **Protecting the Environment**

Throughout the globe, it is no longer acceptable to blatantly harm the environment without evaluating the economic consequences of your actions. Wastes can not simply be deposited in our rivers or oceans. Harmful chemicals can not be used in manufacturing and energy systems can not spew their by-products into the atmosphere. However, there is a balance between reasonable environmental protection and using the best available technology at any cost.

- **Increasing Congestion in Urban and Suburban Centers**

People continue to congregate in the urban and suburban cores of our major cities. The suburban rings around the cities are becoming larger with traffic congestion spreading from downtown to the outer suburbs. The use of automobiles and miles traveled has far out paced population growth in most of the developed countries. People are more dependent on their automobiles, especially in the U.S.

▸ **Faster Rate of Transfer of Knowledge and Information**

The common technologies of information and service are unifying North America, Europe and the Pacific Rim. Methods being used in other regions are quickly transferred throughout the world. All users and owners benefit from this ease of access to new data and knowledge. Almost overnight, the new information is communicated from one country to another. It is as easy to deal with a colleague in a foreign country as one located five miles away.

▸ **Privatizing Utility Companies**

Privatization of utility companies owned by the government or state has been happening in many parts of the world. In some countries the operation is privatized along with the capital assets. In US, the gas, power and telecommunication companies are generally in the private sector, but sewer and water operations for the most part are still operated as public services. Competition and the inability to subsidize inefficient operations is changing the way that prior public companies are operated. Whether public operations are maintained, they will be leaner and more cost-effective.

▸ **Growing Needs for Utility Services**

Our existing utility networks are forced to operate at their maximum output. The water, sewer, gas, electric and telephone systems have not been expanded to meet increasing demands. In addition, maintenance funds have been frequently diverted to other areas. The end result is over used and over worked systems which have a higher frequency of breakdown and failure. Maintenance of the utility network is dull and mundane business which makes it an easy target for slashing by politicians or company executives. It is difficult to win votes for repairing water or sewer systems.

▸ **Contractual Barriers**

The basis of contract for governmental, state and municipal contracts for sewer and water can discourage innovation and alternative technologies. In the gas and cable markets the contract situation is less restrictive and is probably a key reason why trenchless installation has gained greater market share.

Undoubtedly the ability of contractors in Europe and other parts of the world to exploit new technology and expertise has been a major factor in the establishment of techniques like microtunnelling and remote control pipejacking.

Each of the these eight trends represents opportunities and helps to define the market for using trenchless technology.

TEN PREDICTIONS FOR THE YEAR 2000

Let's look at the above-trends and see how they will influence the trenchless business in the years ahead. When we think of the 21st century, we imagine high technology such as exotic robots to perform work, automated vehicles traveling our highways and breakthroughs in medicine through biotechnology. However, it is likely that the next century is more complex than just new technology. The 20th century was the period of a manufacturing revolution and the late 1900's became the period the of an information revolution which will likely continue into the 21st century.

The previously cited trends serve as our benchmarks to define how trenchless technology will advance and hopefully prosper in the coming years. As is frequently the case, there will be winners and losers and some in between.

Based on our analysis of the current world-wide events and trends, we make the following predictions.

Overall Trenchless Market

The issue of an aging underground infrastructure defines the opportunity in the industrialized nations. In contrast, the emerging economies have increasing needs for basic services. Both of these trends lends optimism for the use of trenchless products in the industrialized regions of the United States, Canada, Europe and Japan and in the emerging economies in Asia and Latin America. The overall needs for replacement, renovation and installation of the underground utility networks will remain high. However, growth will be limited by other items of higher priority and the lack of financial resources for infrastructure work.

1) The world-wide construction market for trenchless technology will be over $10 billion in the year 2000 and increase to $15 billion by 2005. Future growth regions for trenchless technology include Latin America such as Mexico, Brazil, Chile and Argentina and the Far East such as China, India, and South Korea.

Guided/Directional Drilling

Environmental issues will continue to expand the applications for guided drilling. The ability to minimize disruption to pristine and protected areas will continue to be drivers. Expanding telecommunications and electricity needs will present more business opportunities. Automation and the integration of advanced electronics on the machinery will expand the envelope of utilization and simplify operations.

2) A consolidation of the manufacturing industry will occur and two to three manufacturers will control over 70% of the new rigs sold. Growth for new rigs will continue to be strong with total operating rigs approaching 7,000 to 8,000 by the year 2000.

3) Guided drilling will expand to impact the Pacific Rim in a similar manner to North America by the year 2000. However, the density of rigs will be unlikely to match the U.S. because of varying needs.

4) New mechanical and instrumentation systems will be introduced that permit sewer pipes to be installed on precise grade and line. Information systems will be integrated onto the machines to provide system feedback and to improve the installation process such as obstacle identification and avoidance.

Pipe Rehabilitation

The rehabilitation of the existing networks will dominate the overall trenchless market. The demands to maintain the network will not diminish. However much the politicians and accountants would like to defer spending on the underground infrastructure, the problem does not go away, but gets worse with neglect. Increasing privatization or contracted-out operations will create a change of climate. More parts of the underground network are likely to be operated by outside private organizations and contractors. The privatized owner or contractor will be required to maintain certain standards of performance. He will in turn be seeking sub-contractors or services which allows him to meet his serviceability criteria at least cost over an extended time period. Competition will keep prices low and allow entry of more systems. Design options will become more simplified and permit operators to easily select the best alternative.

5) Sewer rehabilitation by on-line replacement and renovation will continue as the most important sector. Pressure pipe renovation will grow and technology will improve. We will see the development of technology which will permit the reconnection of service connections onto pressure lines from within the pipe.

   However the pressure pipe market will not develop to the extent that some of the more optimistic forecasts suggest. The comparatively high cost of rehabilitation against the cost of a new installation for pressure pipe being the key factor. Combined market value for sewer and pressure pipe rehabilitation will be at $5 billion world-wide by 2000.

6) Continued competition from new rehabilitation systems will cause further turmoil in the market place and a commoditization of the products. Many of the existing players will not be around in 10 years time The present licensee

system will tend to fall out of favor. It is apparent that the interests and goals of licensor and licensee are not always consistent. Quality control by a licensor of his system to maintain his standards can be difficult. Increasing competition will make installers reluctant to pay any more than they have to. Many products and systems will become more user friendly for installation. We will see the growth of systems as commodity items which will be available for a contractor to buy off the shelf of his local stockist. This is already beginning to emerge in Europe.

7) There will be increasing emphasis on evaluating the network condition to allow the most cost effective maintenance and renovation strategies to be employed. The demand for improved inspection devices of all kinds will be strong. There will be a continuing trend towards localized repair methods to deal with specific problems. The use of robotic methods for repairs and renovations will grow. With competition and greater utilization, the cost of these operations will decrease making them more attractive to clients.

Pipejacking/Microtunnelling

Similar to other parts of the trenchless business, the market for pipejacking and microtunnelling was influenced by the need to replace and expand existing networks. It was expected that our crowded urban cores would create high demand for this service. However due to the lack of recognition of the "true costs" of construction, open-cut methods continued to be used in situations where pipejacking offered advantages. Based on the existing trends and markets, our predictions are:

8) The 1996 market in North America for microtunnelling installation is probably valued at $75 million. This market will grow over the next 10 years to at least $250 million. The unit price of installation will drop in real terms by at least 20% over current prices. There will be increasing use of the technology for line installation of sewers rather than the crossing work which is the main market today in the U.S. Other markets will emerge such as water transmission lines and cable tunnels. These will require purpose designed pipes for jacking with a pressure joint which are available in some countries already.

9) A new generation of machines will be developed which will be designed to overcome some of the present limitations and inefficiencies . Machines and heads will be designed to deal with specific conditions rather than being multi-purpose. Heads will be available to deal with rock, mixed conditions including obstructions, and soft ground.

The range of diameters and machine variations required will favor rental with a skilled operator rather than purchase by contractors in the USA. In the rest of the world the present trend towards purchase will continue especially for the larger schemes such as in Bangkok.

We will see the development of much better methods for safely and quickly sinking drive and reception shafts economically. This should realize substantial cost savings.

10) We are already seeing the introduction of new forms of pipe specifically designed for microtunnelling and pipe jacking. This trend will continue and there will be a greater demand in all forms of sewer installation for a higher quality pipe product which will be resistant to degradation and have a longer life. Pipe production methods with mobile plants will allow high quality pipes to be manufactured on site for larger jobs. This will be another factor in reducing costs.

CONCLUDING THOUGHTS

A consolidation of the trenchless business is beginning to occur and will continue for the next several years. Trenchless will lose its role as a pioneer in the construction industry. It will be placed on the table as an option at the beginning of a project instead of as an afterthought. This is a result of a significant education process throughout the world to contractors, engineers and operators who know understand the technical merits and cost benefits of trenchless technology. The missing piece for the optimal application and exploitation remains the lack of adequate funding for infrastructure works in all utility sectors.

BISCAYNE BOULEVARD INTERCEPTOR- TRENCHLESS SEWER REHABILITATION UNDER HEAVY TRAFFIC

Luis Aguiar[1], Thomas G. Scheller[2], James T. Cowgill[3] and Iqbal Noor[4]

Abstract

The Miami-Dade Water and Sewer Department (MDWASD), maintains the largest sewer collection system in the Southeastern United States. The Biscayne Boulevard Interceptor is an integral component of MDWASD's sewage collection system. The Interceptor is 45 years old, 7,514 meters (24,653 feet) long and contains 96 manholes and inspection chambers. It is constructed of reinforced concrete pipe with sizes ranging from 91.4 centimeters (36 inches) to 182.9 centimeters (72 inches). A closed circuit television inspection of the Interceptor was performed and extensive corrosion with exposed reinforcing steel was detected in most of the line segments. Based on a hydraulic study, both sliplining and cured-in-place liners were chosen to repair the Interceptor.

The 691 meters (2,266 feet) long, 91.4 centimeters (36 inch) diameter portion of the Interceptor was repaired using a cured-in-place liner. As the hydraulic capacity of this portion of the Interceptor had to be optimized, cured-in-place lining was chosen. To facilitate the installation of the cured-in-place liner, special bypass pumping arrangements were necessary. The larger diameter portions of the Interceptor were repaired by sliplining with a fiberglass liner. As these portions of the Interceptor had excess hydraulic capacity, the reduction in diameter caused by sliplining did not adversely affect flows in the pipe. The annular space between the liner and the host pipe was grouted to prevent liner movement. Although the sliplining process did not require extensive bypass pumping as was the case with the cured-in-place liner, insertion pits were required at bends and at other locations along the pipe. Deteriorated manholes were also rehabilitated.

This paper presents a detailed description of all of the phases of the rehabilitation of the Biscayne Boulevard Interceptor. Constructability concerns and other issues raised in the Value Engineering Study are discussed. Steps taken to minimize the impact of the construction work on traffic and commerce on the busy Biscayne Boulevard (downtown Miami) corridor are also discussed.

[1]Assistant Director, MDWASD, 4200 Salzedo Street, Coral Gables, FL 33146.
[2]Program Manager, Rust Environment and Infrastructure, 2600 Douglas Rd., Coral Gables, FL 33134.
[3]Vice-President, Hazen and Sawyer P.C., 2600 Douglas Rd., Coral Gables, FL 33134.
[4]Project Engineer, Hazen and Sawyer P.C., 730 NW 107th Ave.,Miami, FL 33172.

Introduction

The Miami-Dade Water and Sewer Department (MDWASD), the regional wastewater utility of Metropolitan Dade County, is currently undertaking one of the country's largest Infiltration/Exfiltration/Inflow (I/E/I) Reduction Programs. MDWASD's gravity sewer collection system consists of 57,211 manholes and 3,996 kilometers (2,483 miles) of gravity sewers, which make up 889 collection basins. The Biscayne Boulevard Interceptor is a key component of MDWASD's gravity collection system. Located in downtown Miami, the Biscayne Boulevard Interceptor sewer runs along the busy Biscayne Boulevard Corridor and is located next to beautiful Biscayne Bay.

The Interceptor, which is constructed of reinforced concrete pipe, was inspected with closed circuit television inspection cameras. The inspection revealed that extensive corrosion of the concrete was present throughout the Interceptor. In several places the corrosion was so severe that the reinforcing steel was exposed. The results of the close circuit television inspection prompted MDWASD together with its team of consulting engineers to formulate a strategy for the rehabilitation of the Interceptor.

The rehabilitation of the Interceptor was performed in stages. The first stage involved the identification and evaluation of the various rehabilitation alternatives. In the next phase, contract documents were prepared and contracts for the rehabilitation were awarded. Once the rehabilitation work commenced, efforts concentrated on ensuring that costs were within budget and that project milestones were being met. This task was complicated by the high groundwater table and pressures from local businesses and government agencies whose activities were affected by the construction activities.

Physical Characteristics of The Interceptor

The Interceptor is 45 years old and has 96 manholes and inspection chambers. Its total length is 7,514 meters (24,653 feet), comprising of 691 meters (2,266 feet) of 91.4 centimeter (36 inch) diameter pipe, 3,273 meters (10,738 feet) of 121.9 centimeter (48 inch) diameter pipe, 904 meters (2,967 feet) of 152.4 centimeter (60 inch) pipe and 2,646 meters (8,682 feet) of 182.9 centimeter (72 inch) pipe. A schematic layout of the Interceptor is shown in Figure 1.

The Interceptor collects flows from a total of ten collection basins and a shopping center. The Interceptor transmits collected flows via gravity to Regional Pump Station 2. Flows from Pump Station 2 are then routed via force main to the Central District Wastewater Treatment Plant at Virginia Key.

The Biscayne Boulevard Interceptor serves a population of about 100,000. The northern third of the Interceptor serves single family homes and light commercial

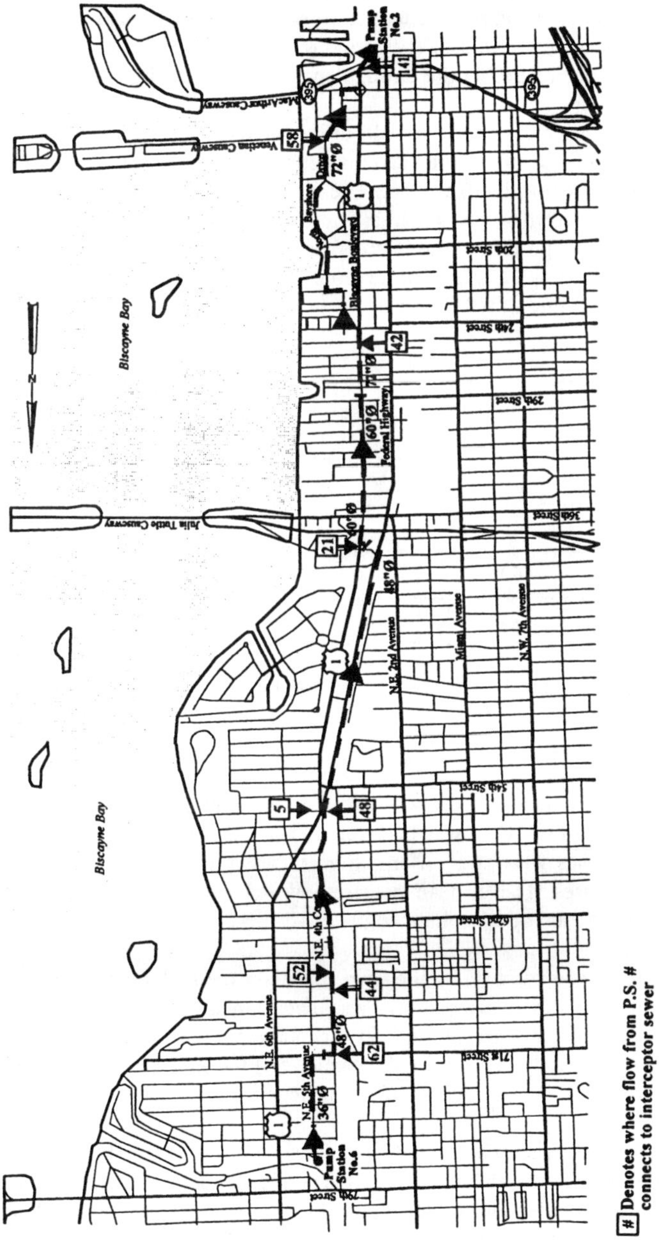

Figure 1: Route of the Biscayne Boulevard Interceptor

areas. One quarter of the Interceptor is located along a railroad right-of-way adjacent to office buildings and factories. The remainder of the route is through the downtown Miami area between the Julia Tuttle (I-95) and MacArthur (I-395) causeways, except for a run of about one half of a mile through another single family housing area.

The extent of the corrosion in the walls of the Interceptor is summarized in Table 1. In the most severe cases about 7.62 centimeters (3 inches) of the walls were eroded in the 121.9 centimeter (48 inch) and 152.4 centimeter (60 inch) diameter portions of the Interceptor. The corrosion was neither confined to any part of the circumference of the pipe, nor, to any portion of the Interceptor. In some sections of the Interceptor, the inner cage was completely eroded near the spring line of the pipe. Based on the results of the closed circuit television inspection, it was decided to rehabilitate the entire Interceptor.

Diameter (cm)	Diameter (inch)	Average Depth Of Corrosion (cm)	Average % of Wall Thickness Eroded
91.4	36	3.9	38
121.9	48	3.8	30
152.4	60	5.1	33
182.9	72	2.9	17

Table 1: Extent of Corrosion In Interceptor

The Value Engineering (VE) Study

It order to ensure that the most cost effective rehabilitation option was chosen, a Value Engineering (VE) study was performed. The VE study was broken into five phases. These phases were the information, creativity, evaluation, proposal and implementation phases. The VE team consisted of three independent representatives from MDWASD's team of consultant engineers.

The VE team was guided by the need to satisfy the hydraulic capacity of the various portions of the Interceptor. It was determined that the portions of the Interceptor with diameters greater than 91.4 centimeters (36 inch) in diameter had excess hydraulic capacity. The hydraulic capacity of the 91.4 centimeters (36 inch) diameter portion of the Interceptor was just sufficient.

The forgoing criteria together with cost considerations, assisted the VE team in evaluating the wide variety of repair options. Traditional dig and replace repairs were ruled out due to the potential problems posed by the high water table, high surface restoration costs and the associated traffic disruptions. Trenchless repair alternatives were evaluated by the VE team as they had the potential of satisfying the repair criteria.

After a process of elimination among the various trenchless repair options, the Value Engineering team concurred with the design engineers and a cured-in-place liner was chosen to repair the 91.4 centimeter (36 inch) portion of the Interceptor and sliplining was chosen to repair the other portions. This decision was in conformance with the hydraulic capacity restrictions. Compared to sliplining, the cured-in-place liner will not significantly reduce the hydraulic capacity of the Interceptor. The sliplining method of repair could be performed with the Interceptor sewer remaining in service thus eliminating the need for expensive bypass pumping operations. The decision to use two different trenchless repair techniques to repair the Interceptor, resulted in the preparation of two construction contracts.

The Construction Contracts

Based on the repair methods chosen, two separate construction contracts were prepared. The contract for lining the approximately 691 meters (2,266 feet) of the 91.4 centimeters (36 inch) diameter portion of the Interceptor was awarded to Lanzo Lining Systems Incorporated. This cured-in-place lining contract was valued at $636,000. Work commenced in June 1995 and was completed in August 1995 for a total cost of $537,041. The sliplining contract, on the other hand, was awarded to the Kimmins Contracting Corporation. This contract was valued at $10,158,700 and involved sliplining approximately 3,273 meters (10,738 feet) of 121.9 centimeter (48 inch) diameter pipe, 904 meters (2,967 feet) of 152.4 centimeter (60 inch) diameter pipe and 2,646 meters (8,682 feet) of 182.8 centimeter (72 inch) diameter pipe. Work on the sliplining contract commenced in August 1995 and was completed in August, 1996 for a sum of $9,792,083.

The Sliplining Process

In order to facilitate the sliplining process, insertion pits were required at bends along the length of the Interceptor. The approximate plan dimensions of a typical insertion pit were 3.7 meters (12 feet) by 7.6 meters (25 feet). In compliance with the State of Florida Trench Safety Act, the sides of the insertion pits were shored once they were excavated. This measure also helped to minimize any possible infiltration of groundwater into the Interceptor and/or exfiltration of sewage from the Interceptor to the surrounding ground water. The sliplining material was Reinforced Plastic Mortar Pipe (RPMP) and was supplied in standard 6.1 meter (20 foot) sections. In situations where sudden changes in alignment existed, shorter lengths of pipe were used and insertions pits were dug on either side of the section.

Once the pits were dug, the top half of the existing pipe was then removed and the liner sections were then positioned in the host pipe and pushed into place using hydraulic force. Once the liner was installed, the annular space between the host pipe and the liner was grouted. The contractor had the option of using a grout mix which would allow the liner pipe to be kept down on the invert of the host pipe or be floated up to the crown of the host pipe. The contractor used a lightweight grout which allowed the liner pipe to be kept down on the invert of the host pipe. This option allowed the grout to fill the voids in the crown of the host pipe which were created by the heavy corrosion. After the annular space between the liner and the host pipe was grouted, the trench was then backfilled and surface restoration performed. A schematic layout of the sliplining process is shown in Figure 2.

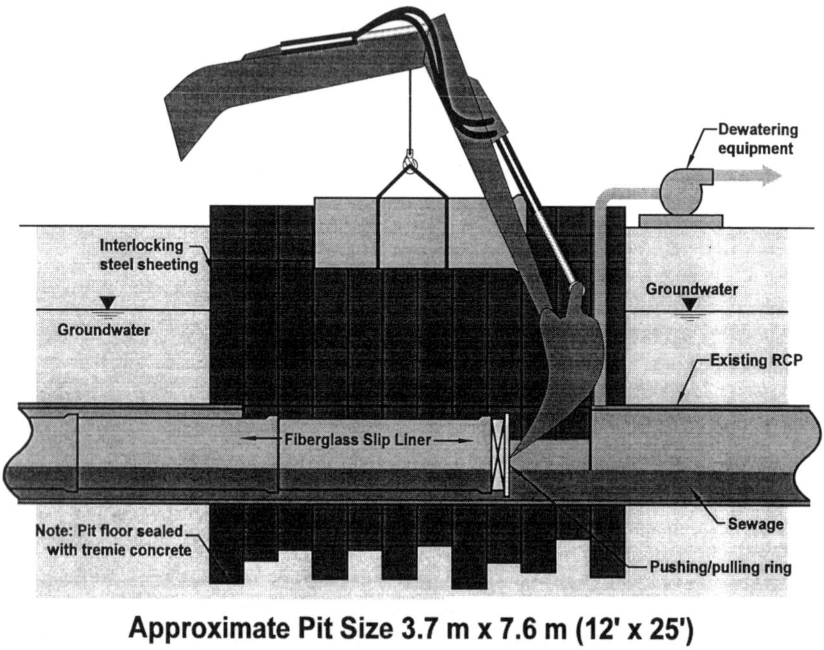

Figure 2: Sliplining of the Interceptor

Cured-In-Place Lining Process

In the cured-in-place liner installation process, a resin impregnated felt tube was inserted in the manhole to manhole line segment to be repaired. The point of insertion was either one of the two manholes in the line segment. Once the impregnated felt tube was placed in the line segment, it was inflated against the host pipe using hydrostatic pressure. The liner was then cured by circulating hot water at a specified temperature through the tube for a predetermined amount of time to ensure that the liner was properly cured.

On some occasions, the felt tube was pulled through two line segments. When these "double pulls" were used, the cured liner was cut at the intermediate manhole and the connections to the manhole were reinstated. The use of "double pulls" helped to expedite the lining process. Since the insertion of the felt tube into the line segment disrupted flow in the Interceptor, bypass pumping was required. To facilitate the installation of the cured-in-place liners, flows were bypass pumped from a manhole upstream of the start of the 91.4 centimeter (36 inch) diameter portion of the Interceptor to a manhole located downstream of that portion.

Public Involvement/Participation

Approximately 150 businesses were identified along the Interceptor. These businesses were comprised of restaurants, hotels, service companies and the main office of a local Miami newspaper. Prior to the commencement of construction work, a concerted effort was made by MDWASD and the project management team to minimize the disruption to local businesses. Residents and local businesses were mailed information on the proposed construction work and were invited to a pre-construction meeting. Included in the information mailed to the local businesses and residents was a tentative construction schedule. The tentative construction schedule allowed the local businesses and residents to implement steps to minimize the impact of the necessary construction work.

The intent of the pre-construction meeting was to address any concerns that the local businesses and residents may have. The meeting was well attended with about 50 of the businesses and a group of local residents participating. The major concerns of the participants were the potential loss of business, traffic congestion and the impact of the construction work on the general aesthetics of the neighborhood. These concerns were directly addressed by the representatives of the project management team.

In order to address the traffic concerns of the meeting participants, assurances were given that police officers would be available for traffic control, on a 24 hour basis throughout the construction period. In addition, the project management team were sensitive to the occurrence of community events such as the Caribbean Carnival. When the Caribbean Carnival took place, construction work was shifted from the area.

Aesthetic concerns related primarily to the stockpiling of construction debris and to the presence of unrestored surfaces in excavated areas. The project management team gave assurances that there would be a regular clean up of construction areas and that all surface restoration would be performed expeditiously.

Effectiveness of Trenchless Repairs

By using trenchless methods to repair the Biscayne Boulevard Interceptor, repairs were able to be completed within the time period specified in the construction contracts. Moreover, the costs of the repairs where within the contract budget. The sliplining contract was completed for approximately $366,000 less than the contract value and the cured-in-place contract was completed for approximately $100,000 less than its contract value.

The first order objective of providing a long term repair solution to the Interceptor while minimizing the inconveniences caused by construction related traffic delays was met. The disruption to local businesses was kept to a minimum. In essence, the project provided an example of how state-of-the-art trenchless repair techniques, together with effective public relations can be used to provide a comprehensive and cost effective repair solution.

The New St. James Avenue Interceptor
Boston, MA

Howard Goldberg, P.E.[1] & Irene McSweeney Woodfall, P.E.[2]

Abstract
The New St. James Avenue Interceptor Project with a construction cost of $17.1 million utilized a combination of Trenchless Technologies for the installation or replacement of the sanitary, storm drain and combined sewer systems located within thc St. James Avenue area in Boston. This paper will provide an overview of the innovative planning and design parameters which resulted in a trenchless approach for the project and the implementation of these techniques during construction.

History and Purpose of Project: Summary of Planning Studies
After more than sixty years of continuous and repetitive maintenance problems with the St. James Avenue Sewer System, the Boston Water and Sewer Commission (Commission) and Bryant Associates, Inc. prepared a Design Study Report for the St. James Avenue Sewer "System" ***(See Project Area Map)***. The System is a complex network of interconnected sanitary sewers, storm drains, combined sewers and underdrains between Copley and Park Squares. The Study was to review the condition of the System and to identify an economically, environmentally and socially-acceptable solution for the rehabilitation of the System.

The project area was a former tidal estuary that was subsequently filled with sandy gravel soils without proper compaction between 1836 and 1871. Most of the sanitary and storm systems were installed in the early 1900's to improve an original system of wooden box conduits. The existing sanitary sewer carries on average is twenty feet below street grade. The project area includes the John Hancock Insurance buildings, Trinity Church, the first Boston Public Library, numerous hotels, office developments and retail establishments.

The System was evaluated by television inspection where feasible and from previous television and physical investigations. Overall, the System was found to be in a deteriorated condition including non-uniform settlement of the pipe, sags and structural deficiencies, surcharging in the System and extensive groundwater infiltration.

---

[1] Vice President and Project Manager at Bryant Associates [2] Deputy Director of Construction at Boston Water and Sewer Commission

An extensive groundwater study was performed to determine areas of groundwater depletion and to understand the relationship of groundwater flows and groundwater depletion rates. Certain buildings within the area were constructed with wood pile foundations. The tops of most of the piles were cut off at or driven below Elevation 5.0± Boston City Base Datum (BCB). To ensure that the system did not cause groundwater to fall below elevation, 5.0 (BCB), the City of Boston installed a dam to maintain the flow within the system at or above elevation 5.0. However, the dam, while alleviating infiltration problems and groundwater issues, aggravated sediment and organic grease build-up in the system.

Plans were prepared identifying all surface and known subsurface features including the sizes and types of all utilities, present and abandoned, as well as all roadway, railway and buildings in the project area. The System's hydraulic needs were analyzed and a base hydraulic flow model prepared. Based upon a review of this information with the Commission the project scope was increased and evolved beyond the initial project area of St. James Avenue only.

With all the constraints identified and the project limits defined, methods for Rehabilitation or Replacement of the System had to be selected. This process needed to limit the impact of construction on the surface activities, abutters and underground utilities. The methods included conventional open cut construction, jacking, a non-gravity installation, trenchless techniques, or a combination of these methods.

Each method was evaluated to decide which techniques were practical and cost effective for the constraints imposed. Further review and analysis was performed comparing several trenchless technologies based upon the existing hydraulic condition and types of pipes used in the original construction. The direction of the project was to incorporate several methods emphasizing the use of trenchless technologies for the replacement or rehabilitation of the system and to select the specific techniques in conjunction with the system characteristics and the capabilities of the selected methods.

Existing Flow and System Characteristics

The project area has an average dry weather flow rate of 1.75 million gallons per day. The System has a combination of sanitary sewers, storm drains and combined sewers which are interconnected to relieve the system during peak rainfall events. The project goals were to separate the system where feasible, provide a sound hydraulic system and remove all groundwater infiltration to reduce inflow into the system. The artificial surcharge of the system at the downstream end would be eliminated by removing the dam once the infiltration was controlled.

In order to evaluate the construction methods, the materials, sizes, shapes, and design capacities of the conduits were compiled by field investigation and record documents. The investigation determined that there were numerous types, shapes, and sizes of conduits, including vitrified clay pipe, cast- in-place concrete, formed brick, cast iron, concrete encased clay and ductile iron with several portions of the conduits being pile supported. With all of these varying pipe materials and the need to increase the capacity of some of the existing system, using one specific construction method was not practical.

In order to limit the range of proposed pipe sizes, one size of pipe was primarily selected for some of the methods to achieve economy of installation.

Trenchless Methods

Numerous methods were evaluated against the design parameters and it was determined that four main trenchless methods would satisfy the project objectives.

***MICROTUNNELLING*:** Microtunnelling was utilized for the majority of the mainline pipe installation. This method was selected for pipes that required replacement where the physical condition for rehabilitation was suspect but could not be verified due to the field conditions or where extensive settlement was known to have occurred. During the planning stage it was discovered that in the early 1900's the project area consisted of a major train terminal where the present roadway system exists. Within this area the potential to encounter abandoned wood piles and granite block foundations was high. The concept to replace the existing system in its existing (in line) location was specified to minimize the potential for encountering the abandoned foundations and wood piles. In line replacement appeared to be the only feasible construction method even though it had not been attempted in the United States before. Microtunnelling was also selected where clay pipes existed. The System where encased in concrete was not conducive to be microtunnelled. The pipe size was standardized at 36 inches even though a smaller diameter could have been utilized in some areas.

CURED IN PLACE: Cured in place was utilized for the pipes that had cracked allowing groundwater infiltration, but had not totally lost their structural integrity which could be verified by television inspection. Irregularly shaped cast in place or brick pipes were also rehabilitated. The System was cleaned and active services were confirmed. The existing or "host" pipe was then lined with a resin impregnated structural liner and the services reopened by remote controlled cutters. While other types of liner systems such as fold and form rehabilitation methods were evaluated, the shapes, sizes and ovality of the conduits did not permit these methods to be utilized.

PIPE BURSTING: For the clay pipes that had lost their shape and however had maintained a positive gradient, but could not be cleaned and televised, pipe bursting was the method chosen for replacement. The major benefit of pipe bursting is the ability to increase the hydraulic capacity of the system by increasing the pipe size where necessary. The actual type of pipe bursting equipment utilized was selected by the Contractor, which resulted in the lengths of runs being maximized.

SHOTCRETE LINING: In the largest conduits where physical personnel entry was feasible, the combined sewers were shotcrete lined and a polyvinyl chloride half pipe invert was installed. This was to provide a hydraulically efficient means for both low (sanitary only) and wet weather (sanitary and storm runoff) flows.

Open Cut Excavation

The pipe bursting and microtunnelling methods required thirty-two access shafts to be excavated in order to install the new pipes. The shafts were utilized to set up the necessary equipment to install the new pipes by either microtunnelling or pipe bursting. These access shaft excavations are generally small, with a surface area of forty to forty-five square yards each. The Contractor was required to pre-excavate each of the shafts to locate utilities for use in designing earth support systems for each of the shafts. The shafts were constructed sequentially and decked with wood beams and paved over until needed. The jacking and pipe bursting insertion shafts were located to limit impacts to vehicular traffic movement, while providing building access, and the work areas were defined in order to reduce the impact on other surface activities. The shafts were primarily constructed of wood sheeting and soldier piles with lagging due to the required configuration of the shafts. Many of the shafts had unique shapes in order to support utilities or circumvent the relocation of utilities.

There were two types of service connections that required additional excavations to be made. The first was the excavation and reconnection of over 100 existing service connections that were replaced in their existing locations by either microtunnelling or pipe bursting. The second included the extension and reconnection of over thirty existing service connections that needed to be reconnected to the new microtunnelled pipes that were in new locations from the existing pipes.

In certain areas conventional open cut excavation had to be employed to replace the pipe because of the need to connect to existing manholes or replace pipes that could not be replaced by one of the trenchless methods. Open cut was also specified where the depth of installation was shallow. The Contractor utilized wood sheeting with wood or steel internal bracing, trench boxes, steel (vertical) roadway plates, wood lagging, or a combination to provide the earth support system for all open cut work.

Existing Utility Networks and Utility Relocations

As part of the design, a complete utility inventory was assembled. All surface features were horizontally located and plotted on the base survey plans. Rims and invert elevations were obtained by a combination of field survey, record information, and actual manhole investigations. Record plans of all utilities were obtained and added into the base plans, including, sanitary sewers, storm drains, combined sewers, water mains, gas lines, telephone and other telecommunication systems, cable television, steam, subway traction power, electric, street lighting, and traffic signals. Private utilities between commonly owned buildings were also present in the project area. In addition to the "active" utilities, there were numerous abandoned utilities identified.
The locations for excavations were determined based upon the design requirements of the system and an attempt to minimize utility relocations. Due to the cost and time associated with relocating electric, communications and steam systems, a concerted effort was made to not produce a design which impacted these utilities. Numerous utilities were supported in place by the Contractor within the limits of excavation.

Traffic Maintenance and Coordination with Abutters

Approximately 20,000 vehicles per day traverse each of the streets in the project area in a grid-like pattern. Other surface activities included an open air farmers market during the spring, summer and fall, cultural activities such as musical groups playing for the public in Copley Square Park, and many tourists. A method of construction was needed which would limit its impact on surface traffic and other surface activities.

Roadway occupancy locations for construction activities had to be strategically located to limit the project's impact on surface traffic. The different trenchless construction methods required different work zone strategies and the work zones were optimized based on several criteria including methods of control for pedestrian and vehicular traffic movement, building access areas and other impacts. The locations of an abutters' loading dock, as well as its entrances to parking areas and public access points to the building, had to be evaluated. The disruption of businesses could cause economic loss not only to that business but also to the general public, by the loss of tax revenue for local and state governments. As a result of the extensive amount of work being performed by Boston's Central Artery Project, it was prudent to minimize the traffic impact in the project area. One of the major benefits of trenchless construction is a reduction in the surface area required to perform the work.

Numerous project meetings and individual meetings were held with building owners and abutters in the project area. Each had expressed concerns with maintaining public and commercial access to their buildings. Most understood the need for the project, and were very cooperative during the planning, design and construction phases. A monthly project update was mailed to all abutters, impacted agencies and individuals to provide them with the projected work for the following month and the progress to date. Individual telephone conversations, letters and meetings were held where required to provide additional information such that the construction schedule could be reviewed and modified to incorporate the abutter's concerns and needs. Loading dock deliveries and individual tenant and building matters were coordinated separately.

The Contract Documents also required the Contractor to provide street sweeping, a rodent control program, and to assist the City of Boston with snow removal within the project area to ensure that the abutters were not negatively impacted by these issues. The finish line for the Boston Marathon was adjacent to the project area and many of the hotels played important roles in the event. The 1996 marathon was the 100th running of the event and attracted nearly 50,000 participants and one million spectators. Boston's First Night Celebration held numerous activities, including ice sculptures. A Farmer's Market is set up during the week to sell produce and goods. The impact on these activities was minimized through continuous coordination.

Groundwater Issues

The direction of groundwater flow in the project area was evaluated during the design phase. The groundwater levels were determined with the dam in place and with the dam removed. Areas of the greatest and most rapid groundwater lowering were along St. James Avenue between Arlington and Berkeley Streets, west of Clarendon Street and the area beyond St. James Avenue along Blagdon Street towards Exeter Street. The results of this exercise was to confirm that the groundwater depletion in the study area was directly correlated to the poor condition of the existing conduits. Based upon the results of the groundwater study and the location and configuration of the wood piled supported building foundations, the Contract Documents established the minimum groundwater level that the Contractor needed to maintain during construction. Different groundwater levels were established for the Boston Public Library and the Trinity Church, two of the more historic and groundwater sensitive structures within the project.

During construction, the groundwater levels were monitored and recharge wells were utilized when the levels dropped below those specified. Once the project is completed and the existing system is abandoned, the groundwater will be monitored and the data compiled and evaluated. In addition, the Boston Public Library and Trinity Church continuously monitor the groundwater levels with their respective monitoring systems. This information was made available to the Project Team and utilized throughout the project.

Maintaining Existing Flows

The Contract Documents provided the Contractor with all existing building and catch basin service connections and an overall system map which indicated flow directions and approximate invert elevations. The project area has a significant amount of sewers, drains or combined sewers that allowed the Contractor to by-pass flows by a combination of pumping, temporary pipes and sequencing the construction to take advantage of the "new system" that in some areas could be installed without any flow diversion areas.

Obstructions

The one word that frightens people whenever microtunnelling is mentioned is the word "OBSTRUCTION". During the design phase various methods for locating and identifying potential obstructions were reviewed. These included borings, probes, test pits, historical data research, review of inspector's daily notes from the installation of previous utilities and buildings, ground penetration radar and directional drilling of a pilot bore along the alignments.

The information gathered from the methods utilized were identified on the contract drawings and helped establish the work zones. When an obstruction was known to exist a shaft or test pit was indicated on the Contract Documents in an attempt to remove the obstruction without delaying the microtunnelling process.

The Contract Documents defined an obstruction as objects or portions of objects consisting of (a.) granite blocks and boulders with maximum lateral dimension greater than one foot, (b.) metal, (c.) reinforced concrete and (d.) clusters of timber (wood) piles with less than two feet clear space between individual piles, located within the cross sectional area to be excavated by the microtunnelling boring machine (MTBM) and stopped the forward progress of the MTBM. In addition, the contract documents indicated that operating and standby equipment rates and labor rates needed to be furnished by the contractor and agreed to prior to commencing any work. If an obstruction was encountered as previously defined, the contractor was directed to excavate and remove the obstruction to allow for the microtunnelling to proceed. The value of the change order would be the actual time and materials utilized to remove the obstruction and was compensated at the rates previously established. This process allowed for a quick turn around time for compensating the contractor and keeping the microtunnelling moving.

We commonly say that there is theory, practice and reality. The reality of the obstruction issue on this project was that everyone's fear of the isolated piece of granite stopping the machine only occurred twice, once within twenty feet of the start of the first drive and then again thirty feet further. Even these two instances may have been related to an existing steam manholes' foundation, or remnants of granite blocks from the construction of the Boston Public Library.

There were fourteen drives performed on the project and the machine was stopped eleven times due to obstructions. In addition to the two obstructions mentioned above the machine encountered the remains of granite block walls/foundations eight times. The most costly and frustrating obstruction encountered was a steel strand tieback from a temporary building subsurface earth support system which was abandoned in place. As a result of the tieback being located in one of the major intersections on the project, the contractor had to limp the MTBM to a location where an access shaft could be installed. While the MTBM was able to cut the steel strands, the remnants produces clogged the slurry lines and forced the contractor to continually backflush the system in an attempt to clear the blockages. A mini television camera was inserted into the slurry lines to determine the extent of the problem. It was determined that the tieback had been pulled into the MTBM and sheared into hundreds of twisted pieces. Once the MTBM was beyond the maze of utilities, a shaft was constructed and the steel remnants removed.

All obstructions were removed by the excavation of access pits. During the design it was assumed that some of these obstructions would be able to be removed by installing a caisson. The reality was that the head of the microtunneling machine needed to be viewed and the size of the obstructions needed to be verified and removed prior to continuing with the microtunnelling which would have made this method impractical.

The Project Team spent considerable time reviewing the procedures used to determine the locations of potential obstructions even during the Construction phase. One example probably sums up the conclusions that were reached. A jacking shaft was installed at the location where an abandoned granite block railroad abutment was believed to have existed. The abutment was found and the Contractor was asked to probe beyond the shaft sheeting with a six foot long drill to confirm if all of the granite had been removed. A six inch horizontal and vertical pattern of offset holes were drilled with no granite blocks encountered. The Contractor also pushed a twelve foot long reinforcing bar at a few locations through the sheeting without encountering any granite blocks. Prior to launching the microtunnelling machine the Contractor cut through the wood sheeting and to everyone's surprise found a five inch high, eighteen inch wide and two foot long granite block that was neatly positioned between the probe holes! Luckily it was easily removed and it proved to be the only piece of granite encountered on that drive.

Excavated Materials and Sediment

During the preparation of the Contract Documents a review of the Department of Environmental Protection's (DEP) records was made to identify any potentially contaminated soil areas. In addition, the soil borings were also utilized to classify the characteristics of the existing materials that would be removed by the various types of construction. To date, the Contractor has excavated, tested and disposed of over 36,000 tons of material from the various items of work. This material was classified into one of eight different bid items, ranging from clean non-reusable fill to hazardous waste. Once classified this matter was disposed of at the various landfills and processing facilities accordingly.

In addition to the excavated materials, portions of the project contained contaminated groundwater which required that a portable treatment unit be brought to the site to treat the groundwater.

Another area that required disposal of material was the removal and testing of over 1,400 tons of conduit, manhole and catch basin sediments by the Contractor. During the design phase, representative samples of sediment was obtained, tested and the results sent to DEP for their review and classification. This sediment was classified into three different types and disposed of accordingly. All testing and disposal requirements for excavated material and pipe sediment were indicated in the Contract Documents.

Project Benefits

"The New St. James Avenue Interceptor Project" provided a cost-effective and sensitive solution to a problem that existed for over sixty years with the following conclusions:

1. The Commission was finally able to rehabilitate The New St. James Avenue Interceptor Project area by taking into account all aspects and impacts of the project with Trenchless Technology. Prior to trenchless technology, the Commission had not undertaken performing the work because of the negative impacts open cut excavation would have on customers, tourists, utilities, workers and groundwater.
2. Trenchless technology was utilized to reduce the construction duration and impacts to abutters and their businesses. Trenchless technology is reducing the financial impacts to other utilities including electric, gas, telephone and steam by avoiding the need to extensively relocate or support in place these utilities as required by open cut utility installation.
3. The project removed a significant amount of groundwater infiltration that previously discharged into the sewer. The elimination of this infiltration translates into a reduction in the cost for treatment at the Deer Island Wastewater Treatment Plant. By the separation of sanitary and storm flows, combined sewer overflows into the Charles River will also be decreased.

In summary, the Project Team successfully replaced or rehabilitated over three miles of sewers and drains utilizing a variety of trenchless technologies to solve infrastructure issues in a congested urban environment. The knowledge obtained from this project will not only benefit the Commission and its rate payers, but the industry in general because of the complexity of the processes used.

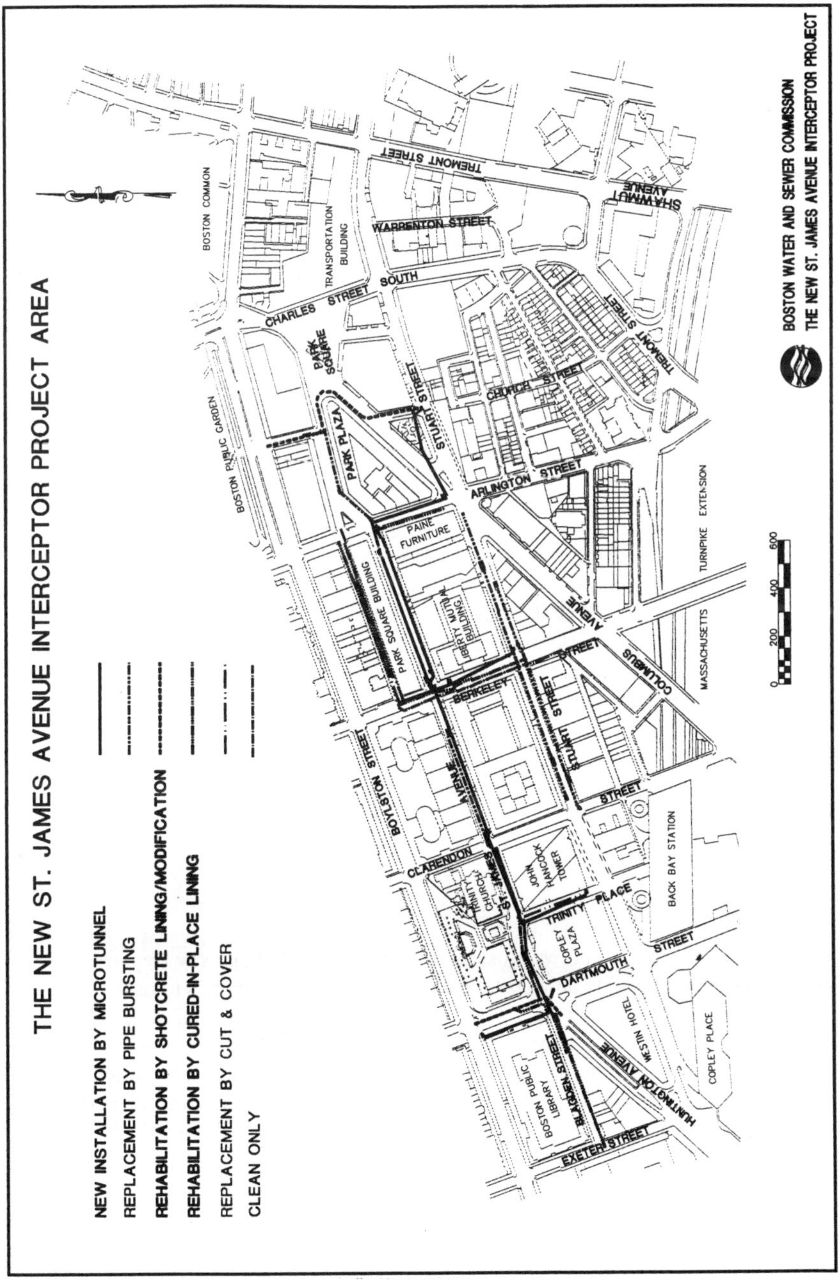
THE NEW ST. JAMES AVENUE INTERCEPTOR PROJECT AREA
NEW INSTALLATION BY MICROTUNNEL
REPLACEMENT BY PIPE BURSTING
REHABILITATION BY SHOTCRETE LINING/MODIFICATION
REHABILITATION BY CURED-IN-PLACE LINING
REPLACEMENT BY CUT & COVER
CLEAN ONLY
BOSTON COMMON
BOSTON PUBLIC GARDEN
TREMONT STREET
SHAWMUT AVENUE
WARRENTON STREET
TRANSPORTATION BUILDING
CHARLES STREET SOUTH
PARK SQUARE
PARK PLAZA
STUART STREET
CHURCH STREET
ARLINGTON STREET
PAINE FURNITURE
PARK SQUARE BUILDING
LIBERTY MUTUAL BUILDING
BERKELEY STREET
COLUMBUS AVENUE
TURNPIKE EXTENSION
MASSACHUSETTS
BOYLSTON STREET
ST. JAMES AVENUE
CLARENDON STREET
TRINITY CHURCH
JOHN HANCOCK TOWER
TRINITY PLACE
BACK BAY STATION
COPLEY PLAZA
DARTMOUTH STREET
WESTIN HOTEL
COPLEY PLACE
HUNTINGTON AVENUE
BOSTON PUBLIC LIBRARY
BLAGDEN STREET
EXETER STREET
0 200 400 600
BOSTON WATER AND SEWER COMMISSION
THE NEW ST. JAMES AVENUE INTERCEPTOR PROJECT

Installation of an Reinforced Concrete Sewer Through the Basement of an Existing Six-Story Building

Kevin R. Krawiec, P.E., Member ASCE[1]
Russel J. Ross, P.E., Member ASCE[2]

Abstract

In fulfillment of a major element of "corridorizing" existing utilities in downtown Boston as part of the Central/Artery Tunnel (CA/T) Project, a 72 inch reinforced concrete sewer was permanently relocated beneath the basement of a six-story brick building at 200 High Street. This relocation required the underpinning of several of the building's columns using transfer beams supported by drilled mini-piles. The granite block/masonry foundation walls were also to be underpinned utilizing steel needle beams supported on hand-excavated, concrete backfilled pits. The design of the underpinning system was revised twice, once to accommodate actual conditions encountered in the building foundations, and again as part of a contractor-proposed revision to use a posttensioned beam for underpinning the foundation walls. This paper will detail the design elements of the work and discuss the construction methods used in completing the installation.

Project Overview

In the area of downtown Boston, the Massachusetts Highway Department's CA/T Project will provide new northbound and southbound tunnels to replace the existing artery's elevated viaduct. Due to the close proximity of buildings on each side of the artery, the new tunnel alignment will be restricted to that of the existing viaduct. The area is abutted by the downtown area on the west, and the waterfront on the east. Construction of the tunnels will be done by cut-and-cover methods, utilizing soldier pile - tremie concrete (SPTC) walls, constructed in bentonite

[1]Senior Structural Engineer, Camp Dresser & McKee Inc., Ten Cambridge Center Cambridge, MA 02142
[2]Vice President, Camp Dresser & McKee Inc.

slurry-stabilized trenches. The walls will serve as lateral excavation support, vertical support for the underpinning of the viaduct during construction, and as the permanent walls of the completed tunnel. Unlike formed concrete walls, SPTC walls cannot be constructed around penetrating elements such as utility pipes and electrical ductbanks. Therefore, prior to mainline tunnel construction, utilities are being relocated out the alignment area, except at several "crossings," where the installation of utilities is being preceded by construction of those SPTC wall segments penetrated by them. The crossings connect to utility "corridors" on each side of the alignment, and in turn to the existing collection and distribution systems. The work described herein was required as a result of the relocation of the Boston Water and Sewer Commission's existing New East Side Interceptor (NESI) from the east side of the alignment to the new west corridor, which accommodates connections to the downtown area.

One of the closest approaches of the artery alignment to an existing building occurs at 200 High Street, with a separation of approximately 6 feet at the northeast building corner. Here, insufficient clearance is available for installation of the relocated NESI between the future west SPTC wall and the building foundation. As a result, the concept of routing the sewer beneath the basement at the building's northeast corner was selected during preliminary design. Since the building is soil supported, underpinning of its foundation was necessary for two principle reasons. During the sewer installation, it would protect the structure against potential loss of soil due to jacking, tunneling, or excavation operations. Upon completion, it would protect against potential foundation undermining due to any future leakage or failure of the pipe.

As of the writing of this paper, underpinning construction is completed and utility installation is in progress. The final design of the utilities and underpinning was done by Camp Dresser & McKee, Inc., as the Section Design Consultant (SDC), with geotechnical consultation by McPhail Associates, Inc. Preliminary design, resident engineering, and overall project management was done by the joint venture of Bechtel/Parsons Brinkerhoff as the Management Consultant (MC), and geotechnical investigation and recommendations provided by GEI Consultants, Inc., as the Area Geotechnical Consultant. The general contractor is Perini Corporation, and the underpinning subcontractor is Schnabel Foundation Company.

Existing Conditions

200 High Street is a six story office building immediately adjacent to the west side of the artery viaduct. The original building faces High Street on the south (Figure 1) and Broad Street on the north (Figure 2). A new addition designed in 1988 was built on the east, facing the viaduct and Purchase Street below. It has a curved exterior face to match the alignment of a previously demolished artery ramp. A plan of the building and sewer is shown in Figure 3.

Figure 1. Southeast Elevation

Figure 2. North Elevation

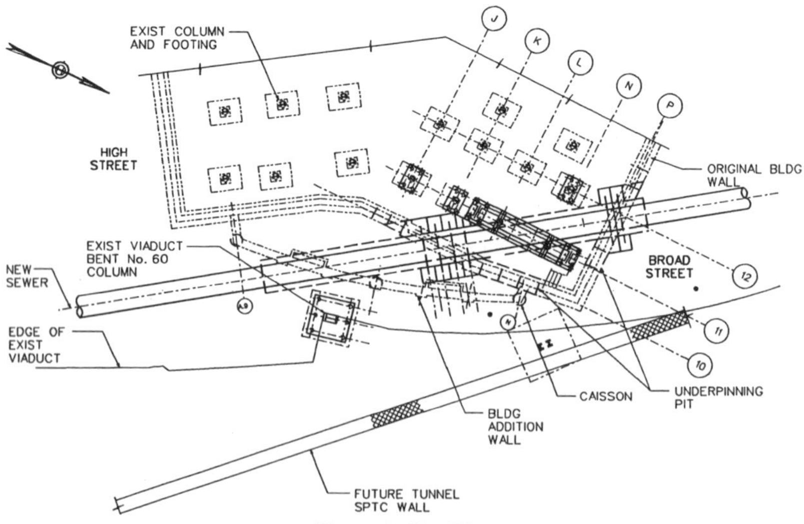

Figure 3. Site Plan

Construction documents of the original building were unavailable. An on-site inspection showed the superstructure to consist of exterior brick masonry walls and interior wood floors and framing. Beneath the original building is a one story basement with exterior brick/granite/concrete masonry walls and a concrete floor slab. The interior basement columns are built-up metal members, fabricated from riveted plate and angle shapes, and clad with a cylindrical metal shell with masonry infill. Test pits performed in 1988 at the exterior wall and Column N-11 indicated both to be founded on granite block footings bearing at approximately 7 feet below the basement floor, with a brick pier beneath the column from the floor to the footing.

The design drawings for the new east addition indicate it to be a steel framed structure with concrete floors on metal deck. The addition includes a mezzanine above the sixth floor, which is partially supported on the original building. The first floor is a concrete slab-on-grade with no basement. The building foundation design accounts for the future installation of the sewer by supporting the front wall on a reinforced concrete grade beam spanning to three 3 foot diameter concrete caissons, which would allow installation of the sewer beneath the grade beam without undermining the wall.

The artery viaduct is a steel framed deck, supported on steel rigid frame bents. Of several bents adjacent to the building, the closest to the proposed sewer alignment is Bent No. 60, the west column of which is approximately 13 feet from the sewer centerline at closest approach. The column is supported on an a 12 foot square concrete spread footing, founded approximately 10 feet below grade.

Test borings indicate the general subsurface profile in the area to consist of 10 to 14 feet of fill, overlying roughly 100 feet of a silty or clayey glacial till over bedrock. Standard Penetration Tests in the till yielded blow counts from 40 per foot to 100 per 2 inches.

Design Concept

The proposed sewer route crosses the northeast building corner obliquely. From a manhole in Broad Street, its alignment passes beneath the north wall footing, columns N-11 and L-11, the east wall footing of the original building, and the new addition grade beam to a manhole in High Street. In addition to these walls and columns, other foundation elements selected for underpinning included those located within the influence zone of the sewer, defined by 1 vertical to 1 horizontal lines extended up from the pipe invert. These included building columns J-11, K-11, and N-12, as well as the west column of Bent No. 60. The following describes the original design concept and work sequence contained in the contract documents.

The first elements to be underpinned were the building columns. Since columns L-11 and N-11 were located directly over the sewer, transfer of the column loads to new foundations was required. An elevation of the underpinning is shown in Figure 4. Groups of four mini-piles were installed to north of N-11 and south of L-11. These drilled and grouted, 8 inch diameter piles could be installed in relatively low headroom areas and were designed to develop a capacity of 60 tons in friction through the till. Two W36 steel transfer beams spanned to new concrete pile caps constructed on each pile group.

In order to transfer the load from the columns to the new foundations, temporary steel framing was installed, as shown in Figure 5. A number of the existing rivets were removed from the column flanges, and steel C15 brackets bolted to the flanges, perpendicular to the transfer beams. W18 needle beams were installed beneath the channels, with a space between them and the transfer beams for a hydraulic jack on each side of the column. The jacks were engaged to gradually transfer the column load to the transfer beams until the column's calculated design load was reached or the column deflected upward 1/32 inch, whichever came first. After jacking, a W12 post was installed on each side of the jack, and tightly shimmed. The existing brick pier was demolished, and a W21 needle beam installed and shimmed beneath each column. The temporary framing was removed, and the permanent transfer and needle beams encased in concrete for corrosion protection.

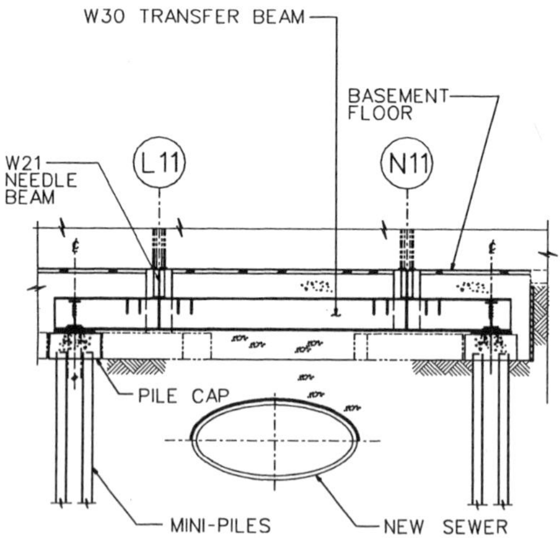

Figure 4. Column Underpinning

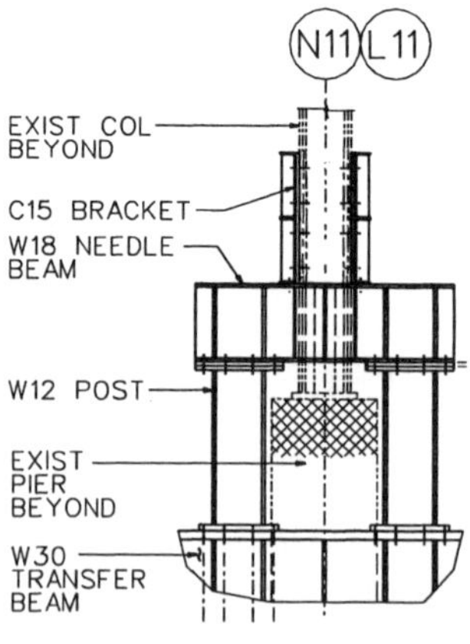

Figure 5. Temporary Column Support

Mini-piles were also used to stabilize the foundations of Columns J-11, K-11, and N-12, and were drilled directly through the existing granite block footings. An assemblage of brackets, needle beams and posts similar to that at L-11 and N-11 was used for temporary support, in this case bearing directly on the piles at the top of the footing. After support installation, the brick pier was demolished, and a new concrete pile cap installed over the existing footing (encasing the post bases), with its top 3 inches below the column baseplate. This 3 inch space was rammed full with a drypack grout (a mixture of sand and cement with just enough water for hydration), and the temporary support removed.

A similar approach was taken for the Bent No. 60 column (Figure 7). Here, the load transfer to the underpinning was made by a shear connection to the column. After drilling the piles through the existing concrete footing, automatically-welded steel headed studs were attached to the column flange and web. A new concrete pile cap and collar were constructed on the piles and around the column, engaging the studs for transfer of vertical load and moment.

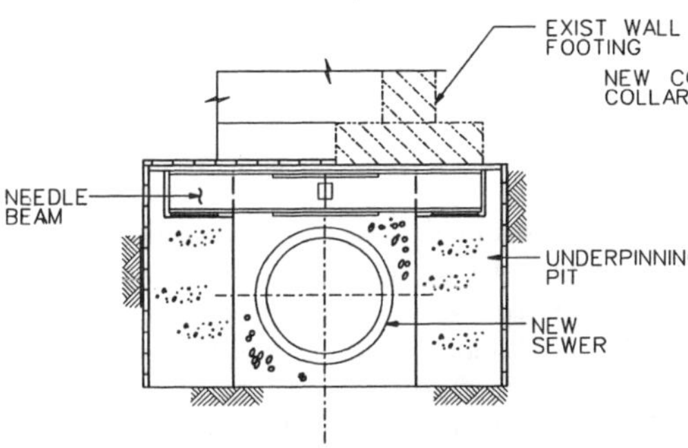

Figure 6. Wall Underpinning

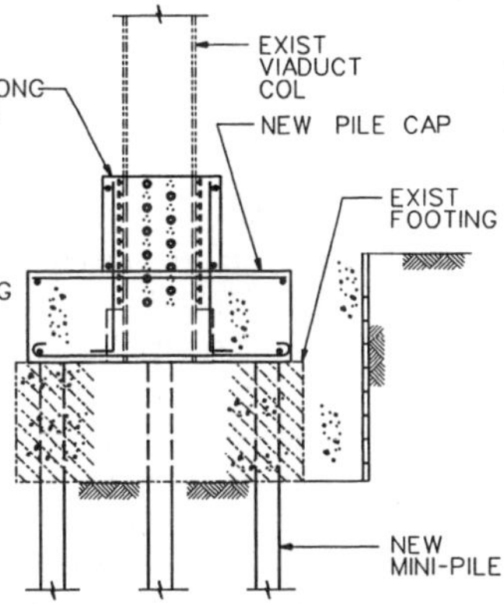

Figure 7. Viaduct Underpinning

Given the considerable bearing capacity and stability of the till on an excavation face, conventional pit underpinning was selected for the exterior foundation walls. A section through the sewer and foundation wall is shown in Figure 6. A 4 foot long by 4 foot deep approach pit was excavated in the basement at the footing edge. A 4 foot long hand-excavated, sheeted pit was dug beneath the full width of the footing, down to a new foundation bearing level below the sewer invert. Upon completion, the pit was concreted to within 3 inches of the footing bottom, and drypacked after concrete set. Similar adjacent pits were constructed until the required length of wall was underpinned. The pits immediately adjacent to the sewer were constructed with pockets formed at the top of the pit. Steel needle beams were set into the pockets, 3 inches below the footing bottom. The beams were preloaded by driving pairs of steel wedges into the space, the spaces drypacked, and the beams encased in concrete for corrosion protection. Since the approach of the sewer to the walls is oblique, the needle beams were oriented perpendicular to the axis of the sewer, rather than parallel to the wall, to significantly reduce the span and size of the beams. This required the supporting pits to be extended into and out of the basement area.

After completion of all underpinning work, the sewer was installed beneath the building. A jacking operation between manholes in High and Broad Streets was selected during preliminary design. Given the close tolerances required to avoid conflict with the piles and caissons, the contractor was also allowed the option of installation in an open excavation in the basement and a tunnel beneath the building addition.

Throughout construction, the project required continuous monitoring of the building for deformations and vibration. Limiting values were established at 0.02 feet for vertical deformation, 0.015 feet for horizontal deformation, and a peak particle velocity of 0.12 inches per second for vibration due to construction operations.

Construction Phase Design Changes

The first major design changes resulted from encountering a different foundation configuration than that expected based on the pre-design test pits. At columns N-12, L-11, and J-11, no brick footing pier was present. The columns were founded directly on stepped granite block footings, with the 5 foot square top block just below floor level. This block conflicted with the design spans of the brackets and needle beams, and spacing of the transfer beams. Also, at columns N-11 and K-11, the brick pier observed in the test pits was actually a brick encasement of the metal column, which extended down to the footing approximately 4 feet below the floor slab. This extension precluded use of the permanent W21 needle beam beneath the column.

Revisions were required for each element of the underpinning. Steel members were resized for increased spans. In place of the W21 needle beam, flange-mounted channel brackets were used, similar to the temporary support brackets, but framing into the W30 transfer beam web. The mini-piles were spaced farther apart to accommodate the size of the top footing block. The required increase in spacing was somewhat reduced by removal of the corners of the block, which was accomplished by line drilling and splitting of the granite to minimize vibrations to the structure. The pile caps were redesigned as well to accommodate the increased pile spacing and decreased thickness, now limited to that of the top footing block.

Since construction was underway, the redesign had significant ramifications for the project schedule. Through a cooperative effort between the SDC, MC, contractor, and subcontractor, a "fast-track" schedule for delivery of various components of the installation was established. This schedule was based on the specified sequencing requirements, various material fabrication times, and the contractor's sequencing. The resulting schedule required delivery of the revised design for pile locations at each column in a specific sequence, followed by the structural steel framing, and finally the concrete pile cap sizes and reinforcement.

At the north and east wall footings, significantly different conditions were encountered as well. Elevations of the granite blocks varied considerably along the walls. At the north wall, the footing was not continuous, but rather consisted of isolated footings beneath the superstructure masonry piers. The redesign in this area was done in conjunction with a revised concept patented and proposed by the underpinning contractor. The concept retained the underpinning pits along the walls on each side of the sewer. Rather than steel needle beams installed perpendicular to the pipe axis, the use of a posttensioned concrete beam parallel to the wall was proposed. The use of posttensioning accommodated the increased span for installation parallel to the wall with a member of sufficiently shallow depth to fit between the sewer and wall footing bottom.

The beam was constructed in 4 foot long segments similarly to the underpinning pits, and was 3 feet deep at the east wall and 2 foot 4 inches at the north. Corrugated steel ducts were installed at the base each segment in the direction of span, and shear keys were formed on the ends to promote shear transfer between segments. The end segments of the beam were supported by the full-depth underpinning pits immediately on each side of the sewer, while the pit beyond was left unconstructed. Once all segments were built, 1 3/8" high strength steel bars were inserted through the ducts and spliced with threaded couplers. Anchorage plates were provided at each end, and the bars posttensioned. The ducts were then grouted, and the remaining underpinning pits constructed. The design of the beams was performed by the underpinning contractor, with review and approval by the SDC and MC.

Figure 8. Installed Temporary Column Support

Figure 9. Posttensioned Beam Below North Wall Footing

Towngas Lantau Link
Beach Approach
Hong Kong & China Gas Co. Ltd.[1]

Dan Callnon[2]
Larry Bertolucci[3]

Abstract

Horizontal directional drilling (HDD) laid dual 12-in. OD, Grade API 5LX-52 natural gas pipelines beneath a critical sea wall on Lantau island, Hong Kong New Territories. This was part of a 30-mile gas pipeline crossing by prime contractor NKK Corp., Tokyo, under the South China Sea. The Project was part of the Towngas Lantau construction project for Hong Kong's new Chep Lap Kok International Airport on the island.

Between July 11 and August 2, 1995, Cherrington Corp. drilled and forward-reamed two, 20-in., 1,294-ft. holes to pull back the twin pipelines. The project was completed during typhoon weather, strong currents and logistical problems associated with operation in a remote uninhabited area.

The design profile necessary to clear the sea wall was the result of the relationship between the entry point, the exit point, sea wall barrier geometry, and pipe diameter. The successful installation of the twin gas lines was the result of high-quality surveying techniques. Innovative directional drilling methods enabled each pilot hole to clear the lower extent of the sea wall and to exit within 2 ft. of their planned target. Each hole exited approximately 90-ft. from the pre-installed product pipe in the sea floor trench. A 20-in. reamer with bull-nose and rear stabilizer was used to open both holes from 9-in. to 20-in. to accommodate the 12-in. OD product pipe.

Introduction

As part of the construction of Hong Kong's new Chek Lap Kok International Airport on Lantau Island, dual 12 inch natural gas pipelines were laid from Tai Lam Chung, near Brothers Point, across the South China Sea as part of the Towngas Lantau Link. The Towngas Lantau Link project was managed by NKK Corporation

---

[1] Copyright PennWell Publishing Co., April 8, 1996, reprinted with permission of Oil & Gas Journal
[2] V.P., Business Development, Cherrington Corp., 7398 San Joaquin St, Sacramento, CA 95833
[3] V.P., Estimating & Engineering, Cherrington Corp., 7398 San Joaquin St, Sacramento, CA 95833

of Japan, Prime Contractor (an engineering firm specializing in engineering, construction and ship building), for Hong Kong & China Gas Company Ltd. To avoid disturbing a newly installed seawall at the Ta Pang Po site, parallel beach approaches were sub-contracted to Cherrington Corporation, Sacramento, California. Cherrington Corporation drilled and forward reamed two, 20 inch holes to pull back the twin pipelines between July 11 and August 2, 1995. The project was completed during Typhoon weather, high seas, strong currents and logistical problems associated with operating in a remote uninhabited area.

Scope of Project

Twin 12" OD, Grade API 5LX-52 pipe, were laid in a sub-sea trench from the mainland to within 938 feet (286 meters) of the Ta Pang Po sea wall. HDD was employed to drill and forward ream two parallel beach approach holes, 1294 feet (394 meters), beneath the sea wall and into the sub sea trench. Each hole was designed to exit approximately 90 feet (30 meters) from the product pipe already laid in the sea floor trench to allow for connection by divers before being pulled into place (See Figure 1).

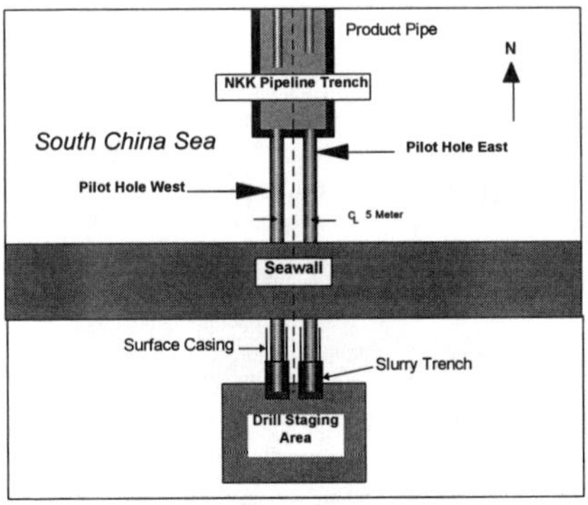

Figure 1. Plan view of beach approach site, Ta Pang Po.

The design curve to clear the sea wall was determined by the relationship of the entry point, the exit point, seawall barrier geometry and pipe diameter. An entry angle of approximately 15°, necessary to clear the lowest portion of the seawall from the entry point, and a radius of curvature of 1200 feet (365.67 meters) was maintained to achieve an exit angle of 90°, or horizontal, into the sub sea trench.

Entry points were 170.6 feet (52 meters) south of the seawall at Ta Pang Po Beach and 18.04 feet (5.5 meters) above mean sea level (See Figure 3). The exit point was approximately 1184.38 feet (361 meters) north of the seawall and 55.45 feet (16.9 meters) below mean sea level. Horizontal distance of each hole was 1294 feet (394 meters) from the entry point. Soil conditions through which the twin holes were drilled consisted of large diameter rock fill and marine deposits of soft silty clay with sand, angular rocks and gravels.

Figure 2. Rig was slid 5 meters to the west to drill parallel entry hole.

Nine inch (22.86 cm) diameter pilot holes were drilled sequentially from each site. Each was forward reamed with a 20 inch (50.8 cm) hole opener tipped with a bull nose and rear mounted stabilizer.

Installation of the dual gas pipeline beach approach was assisted by 5 professional divers supported by two surface vessels. Divers were critical to the success of the project facilitating the installation of the product pipe during pull back operations as well as locating the exit of each pilot hole in relationship to the existing trench.

Figure 3. Diving Crew.

Logistics required special consideration as Lantau Island had very few facilities or public infrastructure. All equipment, supplies and temporary facilities had to be mobilized to the island before work could start. Weather and rough seas became a major problem to logistically support the project, as crew and supplies had to be transported across the South China Sea, a distance of some 30 miles (48.28 km) each day. Strong currents, low under water visibility, rough seas and weather also complicated the efforts of the divers which were critical to the pull back operation.

Site Preparation

Prior to the installation of the HDD drilling equipment at Ta Pang Po Beach, three survey hubs (markers) were placed along the proposed path of the two pilot holes in line with each of the entry points. Each hub, or survey point, served as the reference for azimuth and inclination calculations during drilling operations. In addition to these reference hubs, buoys marked the center of each of the proposed exit points within the trench as well as the ends of the product pipes that were to be pulled into place from the sea floor.

The dual strings of product pipe were laid approximately 90 feet (30 meters) away from the planned exit points to allow room for alignment and connection of the drill pipe prior to pull back. Divers checked the actual exit point of each pilot hole to confirm its specific horizontal and vertical position.

Before installing the HDD rig, a deadman anchoring system was installed at the site to withstand 300,000 lbs (1,334,466.6 newtons) of pulling force. The deadman system consisted of "I" beams placed in a trench with a concrete cap. Because of the close proximity to a box culvert that traversed under rig location (See Fig. 4), a special anchoring system consisting of conventional "H" beams with two additional concrete blocks on each side of the box culvert was used for further stability. The rig was placed on a timber mat to protect the box culvert, otherwise, no special preparation was necessary for the rig staging area.

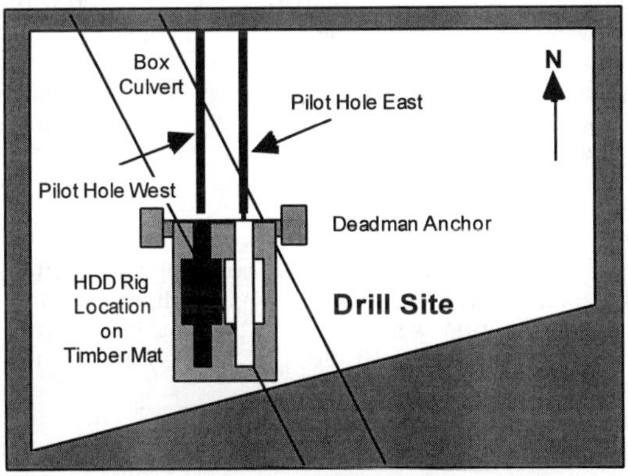

Figure 4. Location of Box Culvert and Rig anchoring system

Equipment

Rig - Cherrington rig #1 was used for the project. Rig #1 has 75,000 lbs (333,750 deca-newtons) of thrust, up to 300,000 lbs (1,334,500 deca-newtons) of pull back capacity and 30,000 ft-lbs (40,674 newton-meters) of rotary torque. The rig requires a crew of 5, including a surveyor and a superintendent.

Figure 5. HDD Rig used on project

<u>BHA Drilling Components</u>

Spud Jet - A spud jet was used to hydraulically penetrate softer formations like sands, silts and clays.

Directional Control - The successful installation of the twin gas lines at Ta Pang Po Beach was the result of high quality surveying techniques. The innovative methods employed not only enabled each pilot hole to clear the lower extent of the seawall (approximately 20 feet, 6 meters, below the surface and 170.6 feet, 52 meters, north of the entry point) but to exit within ± 2 feet (0.6 meters) from their planned targets. Furthermore, both pilot holes exited 16.4 feet (5 meters) apart and within 90 feet (27.4 meters) of the product pipe in the subsea trench(See Figure 7).

Wire Line Steering Tool - Directional heading information (azimuth and inclination) were updated every 8 seconds during drilling by a wireline steering tool. The steering tool's north referencing 3 axis magnetometer and a tri-axial accelerometer sensor package provided azimuth and inclination data to the directional driller while drilling each pilot hole.

The steering tool was placed within a non-magnetic bottom hole assembly (BHA). This was typically three, 30 foot sections of aluminum drill pipe just above the bit. Azimuth and inclination were monitored during the entire drilling operation with stationary surveys taken every 30 feet (~9 meters) and 15 feet (~5 meters), if required.

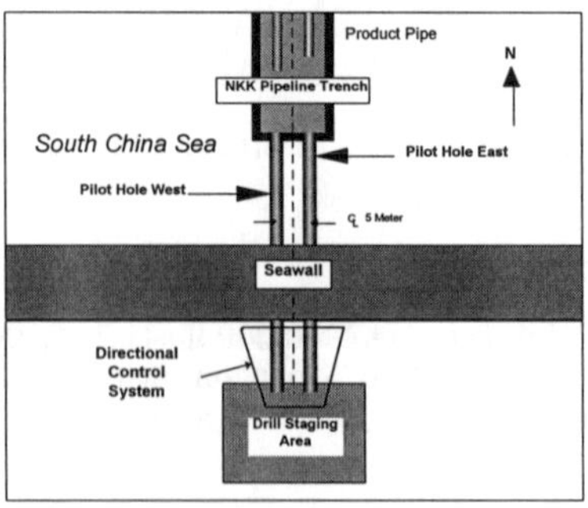

Figure 6. Guidance control system used for directionally controlling the path of each pilot hole

A guidance tracking system, in addition to the wireline steering tool, was used (See Fig. 6). The tracking grid or coil was laid out according to the depth of the proposed trajectory of each hole and was set-up on the land surface, adjacent to the seawall, to correspond with the proposed depth of the drilled path.

A magnetic field is set up from current flow through the coils. The downhole probe, or special steering tool, senses this current and triangulates the location of the drill head by using this known geometry.

Drilling Mud

Prior to drilling operations, soil samples were taken to anticipate mud properties required. Drilling Mud was used for jetting each pilot hole to reduce the friction and to support the borehole during pre-reaming, reaming and pull-back operations.

Drilling mud was a mixture of water and bentonite. Viscosity and density of the bentonite mud ranged from 60 sec/cc and 8.7-8.9 ppg (1.04-1.06 gm/cc) during drilling. Viscosity was increased to 100+ seconds to maintain hole integrity as drill pipe was withdrawn from each pilot hole prior to reaming . Reserve pits were constructed on location to store used mud. It is worth noting that a minimal amount of drilling mud escaped into the sea upon exiting the sub sea trench.

Project

After setting 9 5/8" (24.45 cm) OD casing through the slurry trench and fill zone (See Figure 7), two pilot holes, beginning with the easterly one, were drilled due north by jetting a 9" (22.86 cm) OD hole, 1294 feet (394 meters), into the sea floor trench. Upon exiting into the trench, air was pumped through the drill pipe so that divers could verify the exit point and mark them with buoys.

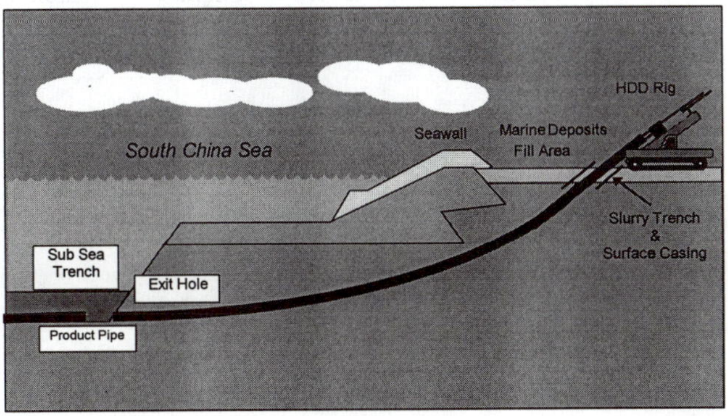

Figure 7. The object of the project was to circumvent seawall on Ta Pang Po beach

As the pipe was tripped out of the hole, high viscosity bentonite mud (viscosity = 100+ sec., density = 8.7-8.9 ppg, 1.04-1.06 gm/cc) was injected into the hole to maintain its integrity. The rig was then moved 16.4 feet (5 meters) to the west and a parallel pilot hole, identical to the first, was jetted into the sub sea trench within ± 2 feet (0.6 meters) of projected target. Air was once again pumped through the drill pipe while divers verified and marked its exit point. As in the case of the first pilot hole, high viscosity bentonite mud was injected into the hole as the pipe was tripped out to maintain hole integrity.

A 20 inch (50.8 cm) forward reamer, with bull-nose and rear stabilizer, was then used to open the westerly hole from 9 inches (22.86 cm) to 20 inches (50.8 cm) to accommodate the 12 inch (30.48 cm) OD product pipe. Bentonite mud was circulated during the reaming process to clean the hole and create a suitable medium in which to pull back the product pipe.

Once the 20 inch (50.8 cm) reamer had exited into the sub sea trench, divers removed the special bull nose and attached a swivel assembly to the drill pipe. The drill pipe was then pushed 90 feet (27.43 m) forward into the trench so divers could maneuver the reamer and swivel assemblies for attachment to the product pipe and pull-back.

Upon pulling in the product pipe into the westerly hole, the rig was again moved to the East Hole. Drill pipe was run back into the hole, with a specially designed bull nose, to insure that the hole had not lost its integrity. The 9 5/8" (22.86 cm) surface casing was removed and a 20 inch (50.8 cm) reamer with bull nose and rear stabilizer were then used to open the hole from 9 inches (22.86 cm) to 20 inches (50.8 cm). The same procedure used on the westerly hole was followed at the exit point. Air was pumped through the drill pipe for divers to locate and mark. The pulling assembly was affixed and the drill pipe extended 90 feet (27.43 meters) so it could be maneuvered into place for connection the product pipe and pull back.

After installing both product pipes, divers welded the final sections together within the sea floor trench. Likewise, surface connections were made and the slurry hole at the entry point of each hole was back filled to complete the installation.

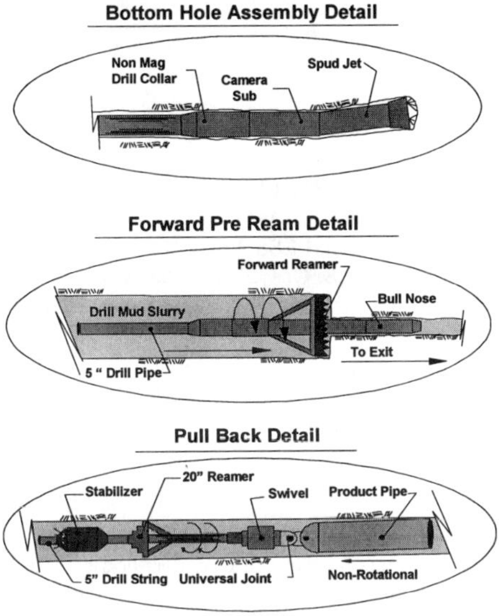

Figure 8. BHA to drill, forward ream and pull back product pipe

Summary & Conclusion

In spite of several problems, both pilot holes were placed into the subsea trench within ± 2 feet (±0.6 meters) of their projected targets and the product pipe was pulled into place ahead of schedule.

Weather conditions were less than optimum. Typhoon force winds, limited sub sea visibility (down to 3.28 foot or 1 meter) and six knot currents made work for the divers difficult at best. Product pipe laid in the trench in preparation of the placement beneath the seawall was covered with sediments on the sea floor by strong cross currents prior to its placement. This required special effort by the divers to uncover the product pipe so it could be moved into place for pull back. These adverse circumstances required special operating techniques and a BHA configuration to facilitate product pipe connections to minimize the time required for divers to work under water and to successfully complete the project. In retrospect, had the product pipe not been laid on the sea floor in advance, the project could have been an ideal candidate for a forward thrust pipe installation.

PROJECT PROFITABILITY
WITH
TRENCHLESS TECHNOLOGY

By: William W.S. Gray, P.E. - Past Chair NASTT
with Parsons Brinckerhoff Quade and Douglas, Boston, MA

Abstract

Project Profitability

Infrastructure installation or rehabilitation by trenchless technology is in its childhood. The profitability of this technology must be more fully understood by municipalities, engineers and contractors to allow for an increase in the application of the technology.

This paper will discuss the aspects of project profitability with regard to those three factors and offer some case histories. The following points will be included:

Municipalities - indicate advantages

- Lessened surface disturbance
- Savings in restoration
- Reduced traffic impacts
- Reduced Public disruption
- Improved facilities
- Reduced excavated material handling and disposal

Engineer - understand extent of work to be done to produce a trenchless contract:

- Knowledge of system requirements
- Facility design
- Material selection
- Subsurface investigation
- Procedure selection
- Standard plans and specs
- Construction cost and schedule estimates

Contractor

- Understand that more work to follow to justify capital expenditures
- Know that training available for his crews
- Realize profits available over long-term

Introduction

Profit is one of the many words in our language which can have a variety of connotations. The word comes from the Latin root of the prefix *pro* which means for or to the advantage of and *feco* the verb to make or achieve. The word today is used as meaning to our benefit.

Trenchless Technology is a procedure which is to our benefit and must, therefore, be a profitable procedure. There are a variety of definitions to the profit from trenchless, which include:

- Financial gain to the design engineer, to the equipment supplier, to the contractor
- Indirect financial advantage to the owner, whether a municipality, industry, utility or transportation authority
- Gain to the public because of reduction in time of construction disturbance and surface disruption
- Environmental advantage due to protection of ground cover, minimizing of dust and nuisance
- Gain to the Engineering profession in knowledge and expertise of new technology
- Opportunity to the construction industry with new procedures and methods
- Increase in construction safety
- Opportunities for construction workers other than just manual labor

The full benefits of trenchless technology must be understood by municipalities, industry, utilities, transportation authorities, engineers, suppliers and contractors to allow an increase in the application of the technology to occur.

What is Trenchless Technology?

Trenchless technology, or No-Dig as it is known in Europe, is the installation of new infrastructure, or the rehabilitation of existing infrastructure, by methods that are non-obtrusive to the surface. The technology is used in all infrastructure; water, sewer, gas, hydro, and telecommunications.

The following methods can be considered for installation of new facilities:

Microtunneling
Directional drilling
Augering
Jack boring
Small diameter directional drilling
- Water jetting
Pneumatic ramming
Pneumatic piercing

And the following methods might be considered for rehabilitation of existing facilities:

Relining with suitable pipe material
Slip lining
Swage lining
Roll-down
Thread-jointing
Spiral Winding
Mechanical spot repair
Grouting
Cured-in-place liners
Spray-on-liners
Pipe Bursting

The advantages of trenchless technology include the following:

- Reduction in traffic and surface disruption
- Reduction in disruption to other services
- Reduction in restoration costs
- Prevention of future settlement or compaction problems which can occur after open cut construction
- Prevention of loss of support of existing services
- Protection of the tree root systems
- Reduction in excavated material and costs to dispose of same
- Equipment and methodology is advanced and "state-of-the-art"
- Safety and productivity are improved

What Causes or Prevents the Use of No-Dig Procedures?

The comprehension of trenchless technology requires a re-definition of concepts. Some methods have been in use for some time and, as such, are not necessarily perceived to be trenchless. It is interesting to note that a water authority considers cleaning and lining of cast iron or older, small diameter ductile iron watermain as an acceptable practice, bid as a contract without being included in a project to excavate and replace the watermain, while on the other hand, a contract that could be installed by No-Dig generally is tendered as a regular open cut procedure where No-Dig will be considered if that method is the lowest price. This is often difficult to achieve if the trenchless contractor has to include equipment purchase in his price. Much of the apparent interest seems to be a desire or hope that trenchless methods are a panacea to all construction problems however, if not less costly than accepted practices, are generally not considered.

It is important, then, that all members of a project, owner, engineer, supplier and contractor understand the technology and consider its use for appropriate projects. If traditional methods are the correct construction procedures then that should be the design and construction method followed. On the other hand, if a trenchless procedure is appropriate then that is the way the design and bidding of the project should proceed.

Continuing this discussion then, the acceptance or non-acceptance of a methodology depends on the knowledge and understanding of that methodology. Cleaning and lining of watermains is an accepted practice in the water industry while swage-lining is an accepted procedure in the gas industry. Some twenty years ago sliplining of sewers was used for rehabilitation and was also a means used in trunk watermain repair with thick wall P.E. pipe. There was a learning period with all of these procedures before their acceptance.

It is quite possible that swage-lining could be used in the water business just as sliplining has some application. Hence the redefinition of concepts. Each of the trenchless technologies in new construction and in rehabilitation could well have more than one application. It is quite understandable that the project procedures be appropriate and successful. To be required to use open cut to correct a trenchless error or to have to repair a rehabilitated pipe is unacceptable.

Nevertheless, it is important that all continue to review trenchless procedures with an open mind and be willing to apply these methods where suitable.

Why is North America Behind?

It is interesting to note that many things operate in a sine curve. A country that developed earlier is often overtaken by one that develops later. Matters such as street widths, density of population, age of and need for infrastructure and public awareness dictate construction methods.

As Japan and Germany were installing much of their infrastructure as part of their rebuilding programs after the second World War, the value of trenchless technology became apparent. As a result, microtunneling machines, pipe materials and trenchless research has been quite advanced in these countries. In the same way, rehabilitation of aging infrastructure in countries such as the U.K. and France has seen considerable advancement in relining techniques such as sliplining, swagelining and CIPP procedures.

In certain areas of the U.K. and Germany, trenchless is the preferred procedure. This in turn leads to modified layouts such as radial alignment for sewer systems which deserve review as to applicability, at least in new systems.

North America is now catching up. The profitability of trenchless is becoming more widely understood. As more and more North American cities rehabilitate their infrastructure, so will No-Dig grow just as happened in Houston which has been involved in an extensive infrastructure rehabilitation program over recent years.

Who Has Seized on Trenchless Profitability So Far?

Of the four major members of a trenchless team, the materials supplier seems to have taken the greatest effort to establish the market. Consider the following examples in the new pipe industry:

- Fiber reinforced concrete jacking pipe has come from Europe
- Concrete pipe suppliers have re-tooled in North America to supply jacking pipe
- The clay pipe industry is making advances with jacking pipe
- New joints have been developed in steel pipe and ductile iron pipe
- The advantages of polyethylene pipe in internal frictional value, capacity for continuity through welded joints and flexibility are better understood

With the availability of proven products the microtunneling machine suppliers from Japan, Germany and U.K. are seeing a continually expanding business while rehabilitation methods such as pipe bursting are starting to make inroads.

The contracting industry is still wary because they cannot see a large market to justify great investment in equipment and training.

The engineering profession tends to face a reluctant client base where concern over construction costs especially in the leaner years of the nineties are a major concern. Clients generally wish to use only those procedures where costs and long-term results are known.

As already stated, it is important that all continue to review, learn and understand trenchless procedures which can only result in the desired cost savings for the clients and profits for the industry.

Work Involved in Developing, Designing and Constructing a Project by Trenchless Means

There seems to be a conception that there is no money to be saved or made if a project is to be trenchless. This is not the case. Consider the installation of a new sanitary sewer. The following work will be required to take such a project to completion. Consider that the need for the project and general design parameters of the work have been determined by the client.

Engineering Tasks
Check service area
Check design flows
Determine size of pipe
Layout limits of the contract and connecting pipes
Conduct a geotechnical investigation
Prepare detailed plans of the existing infrastructure
Check for any property requirements
Prepare plan and profile designs
Review alternative types of pipe and construction procedures
Review structural design requirements for the installed conduit
Finalize contract documents
Obtain permits and approvals
Provide a construction cost estimate

It can be said that the extent of engineering design work for a trenchless contract is similar to that for an open-cut contract.

Design engineers, then, should consider that the work to be done and the profitability available in designing a project for installation by trenchless methods is as much, if not more, than in an open cut project.

An important aspect of the engineering work is that provision of knowledge and understanding of the various No-Dig methods available and the selection of the procedure most suited for the project. Inclusion of this knowledge in the contract documents will allow those suppliers and contractors skilled in the procedure to reply to the tender in the most effective and positive manner since the project requirements will be clear.

In the same way, if the client understands the value of the No-Dig method and can foresee or provide additional projects requiring similar procedures, then local contractors will be able to determine their present and future profitability and decide whether or not to enter the trenchless market.

As has been shown there are a variety of procedures that are trenchless. A contractor then has to decide if microtunneling or directional drilling or slip lining are particular fields that are of interest.

When the City of Houston decided that microtunneling and directional drilling were suitable procedures for use in that City's large sewer installation and rehabilitation program, many of the oil drillers in the area turned from the vertical to the horizontal and entered the trenchless field. The local association - Gulf Coast Trenchless - was born. A manufacturer of a pipe specifically designed for pipe jacking, developed in Germany, opened up for business in Houston and now supplies pipe across North America. Due to the existence of the oil drilling industry, one of the most advanced suppliers of horizontal or directional drilling fluids is also in Houston and has seen a greater expansion of business.

Rewards of Trenchless

As the design engineer realizes that trenchless designs require engineering expertise and will provide profit by fee earnings both because of the provision of more regular engineering services as well as specialized expertise then engineering activity will increase through learned societies such as the North American Society for Trenchless Technology (NASTT), ASCE, WEF, AWWA, etc.

The development of this newer field will create new job opportunities, different fields of expertise and a general availability of professional advancement. Research and development will be a continuing area of growth as many aspects of the industry require a greater detail of understanding.

Consider, for example, pipe ovality. Pipe materials come in two definitions - flexible and rigid. In open-cut construction the type of bedding differs depending on the type of pipe. A rigid pipe will normally be selected to transfer the load of the backfill above the pipe to the bedding whereas a flexible pipe requires sidewall support to develop the necessary passive resistance to deflection of the conduit. It is reasonable to consider that a rigid pipe will be round whereas a flexible pipe will be slightly out of round when new. Older conduits, in need of rehabilitation, may be totally out of round. The effects of ovality within ovality needs to be more fully defined.

There are also rewards in the contracting business. Some thirty years ago open cut construction methods were those with the best safety record. At that time, tunnels were drilled by hand with poling plates or shields. In recent years, tunneling procedures have advanced to earth pressure balance shields with slurry systems or conveyor belts to remove excavate. Closed face tunnel boring machines (TBMs) are quite common. As a result, the safest place in construction today is at the tunnel face. As hand mining has decreased somewhat in application then so has the number of skilled workers decreased, while the need for skilled machine operators has increased.

In the case of trenchless, microtunneling machines require skilled crews to set up the equipment, operate it and maintain line and grade with careful pipe placement behind the machine. There is considerable profit in the increase in the microtunneling business as more and more skilled operators will be required. As an example of this need, one of the microtunneling companies in North America brings experienced professionals from U.K. to fill the voids in operations on this side of the Atlantic. This can occur since trenchless is much more advanced in Europe than in North America.

We must seize this profit of technical development of people.

The same applies to other aspects of trenchless such as directional drilling, pipe bursting and CIPP lining where the skill of the crew is the success of the job.

Profit to the Clients

There are a number of areas of advantage in the use of trenchless technology.

As has been noted, there is a difference in the support systems required for rigid or flexible pipes. Matters such as trench support of existing utilities, relocation of some facilities to allow installation of the proposed work, disposal of excavated material, especially if some is defined as hazardous, backfill and compaction and surface restoration are major aspects of open-cut construction. Other matters such as traffic control, site safety, working space, street closures, protection of trees and surface features must be considered.

Trenchless construction offers many advantages. Surface disruption is minimized. The amount of excavate to be disposed of is reduced. This can be most important in older cities where the intrusion of hazardous material into street subgrades creates disposal problems. Microtunneling or directional drilling is the installation of conduits in a compressive mode which should reduce the impact on existing infrastructure.

In many procedures the trenchless work is faster than open-cut with the longest time required for the installation of the mining, drilling or reception pits. In a major operation, all the pits and shafts can be dug and plated and then reopened as required when the trenchless work proceeds.

The profit to a municipal client then is multifaceted. By precluding the open-cut excavation the disturbance to the ground, the existing infrastructure, the surface, the surface features, environmental or man-made and the people, advantage is realized. These advantages when related to the surface are termed social costs. Unfortunately, social costs are not yet normally included in project estimates. Nevertheless, again, as more and more trenchless procedures occur the social benefits will be realized.

These profits have been referred to earlier in this paper as advantages of trenchless technology.

Conclusion

It can be considered that there is a place for the use of trenchless procedures due to the reasonably extensive use today in all facets of the infrastructure industry.

It can also be considered that such use should become more widespread to allow the full advantages and profit to be attained.

There is financial, social, technical, professional, and personal profit to be found in the use and advancement of Trenchless Technology.

Case Histories

The following case histories are offered as examples of the advantage or profit of Trenchless Technology:

Project	Description
MDC Hartford, CT (PB)	Connection to a deep tunnel with high flows and difficult access Alignment through two major intersections Microtunneled in rock from central traffic island beneath each intersection Completed in three weeks after shafts in place.
Landfill By-Pass Santa Cruz, CA (PB)	Required to prevent storm water from flowing down a valley through a landfill Creek by-passed the landfill by flowing in a pipe directionally drilled beneath the site.
Hydro Installation Toronto, ON (Toronto Hydro)	Hydro lines required to be buried Mini-horizontal directional drilling used under the sidewalks Shafts at 400 foot centers. Homeowners virtually undisturbed.
St. James Avenue BWSC Boston, MA	Sanitary and storm sewers required rehabilitation or replacement Located in Back Bay area of downtown Boston Work completed with minimum problem using microtunneling, pipe bursting, pipe replacement with a microtunnel and CIP lining.
Central Artery Boston, MA (PB)	Installation of utility systems below the active Central Artery and above the future Artery. These systems to be consolidated at selected crossing locations. In this one, combined and sanitary sewer to be installed by pipe jacking Other utilities to be installed in seven microtunneled sleeves.

Live Internal Sealing Cast Iron Gas Mains

Donald Eagleson P.Eng[1]
Robert Zlokovitz[2]

Abstract

Consumers Gas is a gas distribution utility servicing over 1 1/4 million residential and commercial customers in the province of Ontario. Consumers Gas has approximately 1,000 miles (1,600 kilometres) of cast iron main that has been in service for over 60 years. Consolidated Edison is a combination utility that supplies electricity, gas and steam to portions of New York City and Westchester County. Consolidated Edison serves about one million gas customers and has approximately 1,500 miles (2,500 kilometres) of cast iron, mostly located under existing roadways.

There are approximately 400 bell and spigot joints per mile. These joints that were sealed with jute and lead are prone to leakage. Many gas distribution utilities located on the eastern coast of the United States are also facing similar pipeline repairs.

Over the past 10 years, Consumers Gas has used anaerobic sealant to seal leaking cast iron joints. This anaerobic procedure is less costly and more effective than clamping or encapsulating leaking joints. Consumers Gas began and has since been joined by Consolidated Edison in developing a robotic tool to duplicate the external procedure from inside an operating gas main. The tool that has been developed is capable of repairing up to 24 joints in a live main from a single excavation. Insertion points can be conveniently located away from intersections. Disruption to both customers and traffic is eliminated or curtailed. Excavation and re-instatement costs are significantly reduced.

[1] Program Manager Consumers Gas Company P.O. Box 650
Scarborough, Ontario M1K 5E3 Phone 416-496-7173 Fax 416-496-7148
E-mail deagleson@consumersgas.com

[2] Manager Gas Research Consolidated Edison of New York 4 Irving Place
New York NY 10003-3598 Phone 212-460-2621 Fax 212-529-8511
E-mail zlokovir@coned.com

Live Internal Sealing Cast Iron Gas Mains

The major tooling components include: (a) robot drilling head, (b) video camera, (c) umbilical cable containing the various electrical, hydraulic and sealant lines to operate the robot, (d) stiff fibreglass rod to push the robot into the main, (e) pumping system, (f) turret wheel used to store the umbilical cable, (g) hydraulic pumps, (h) sealant storage tanks, (i) operators console and (j) television monitor.

The process entails: (a) drilling one or two 80 mm (3 inch) holes at 45° angles to the live main to insert the robot, (b) inserting the robot into the main, (c) using the television monitor, the operator locates the joints and raises the robot head for drilling a 4.76mm (0.187 inch) hole at each joint, (d) injecting a measured amount of anaerobic sealant through the hole under pressure into the porous jute packing of the joint, (e) moving the robot to the next joint.

Field trials of the robot system are underway at Consumers Gas in Toronto and Consolidated Edison in New York.

REQUIREMENTS FOR ANAEROBIC SEALING SYSTEM

Consumers initial introduction with internally sealing cast iron joints was a procedure developed to spray anaerobic sealant at each joint. This procedure did not meet our requirements but it did lead us to consider other methods of delivering anaerobic sealants to each joint.

When our development was first visualised, we specified several design criteria. The first robot developed would be for NPS (Nominal Pipe Size) 6 Cast Iron. The system must be inserted easily into the live cast iron piping system. The tap hole into the main must be 80 mm (3") in diameter. The robot head must be positioned precisely at the joint. The system must maintain an effective seal between the injection head and the cast iron main. Anaerobic sealant must not be spilled into the gas main. The robot head must be capable of being inserted into the main at least 45 metres (150 feet). The robot must be capable of being launched in either direction from a single excavation. The gas main will remain live during the repair procedure. The robot must be easily cleaned. The drill bit must drill a minimum of 15 joints before exchange. Wherever possible the tool is to be automated. The operators of the robot are intended to be field technicans, therefore ease of operation is very important.

The major components of the robot are:

<u>Sled Assembly</u>: The robot is mounted on a rigid sled assembly that allows the robot to be pushed or pulled easily inside the live main. The sled has two parallel runners

spaced to ensure the robot remains stable when the drill head is raised into drilling position. Mounted on the sled are the following components:

Support Arms: These expandable arms stabilise the robot when it is operated inside the main. These arms are controlled by the support arms actuator.

Tool Orientation Actuator: The tool orientation actuator folds or unfolds the drill head / sealant injection nozzle. The robot head must be unfolded for insertion and while moving in the main. The robot head is folded 90 degrees into an upright position to seal the joint.

Thrust Actuator: The thrust actuator raises or lowers the drill / injection nozzle.

Drill Head: The drill head contains a small low voltage high torque motor capable of powering the unit to drill into the cast iron joint. The drill motor selection is critical to ensure that it has sufficient torque to perform its intended task while being as small as possible. The combination 4.76mm (0.187 inch) drill bit / chamfering tool is made of carbide for longer life.

Sealant Injection Nozzle: The sealant injection nozzle is used to inject the sealant into the joint. It is non-metallic to ensure that the sealant does not cure prematurely. O rings are used to ensure a complete seal against the drill chamfer.

Tool Selection Actuator: The tool selection actuator is used to position either the drill head or the injection head into position.

Video Camera: The video camera with three halogen lamps and the television monitor allow the operator to observe the robot positioning and sealant injection. This camera gives the operator a full view of the joint. If sealant leakage should occur, the operator is able to observe this fact and stop the injection process.

Umbilical Cable: The total umbilical cable length is 54 metres (180 feet). This cable has a clear outer polyethylene hose. Inside this hose are the various hydraulic, electrical and video lines required to operate the tool.

The tool components that support the operation of the robot but remain outside the main are:

Umbilical Cable Storage Turret: The umbilical cable storage turret system is used to store the umbilical cable. The cable is stored in a ring shaped channel approximately 2 metres in diameter. The turret also stores the hydraulic components, pumping system and anaerobic sealant tanks.

Power Insertion and Power Storage Units: These power units are synchronised to ensure that the umbilical cable length between the insertion point and the cable turret is kept constant. The power insertion unit either pushes the umbilical cable into the main or pulls it out of the main. The power storage unit loads or unloads the umbilical cable.

Control Panel: The control panel allows the operator to control the entire system. From this panel he is able to: activate the lighting system, activate the tool delivery system in either forward or reverse, operate the support arm's actuator, fold or unfold the tool, switch position of the drill head and injection head, operate the drill, operate the injection pumps, raise or lower the robot head assembly.

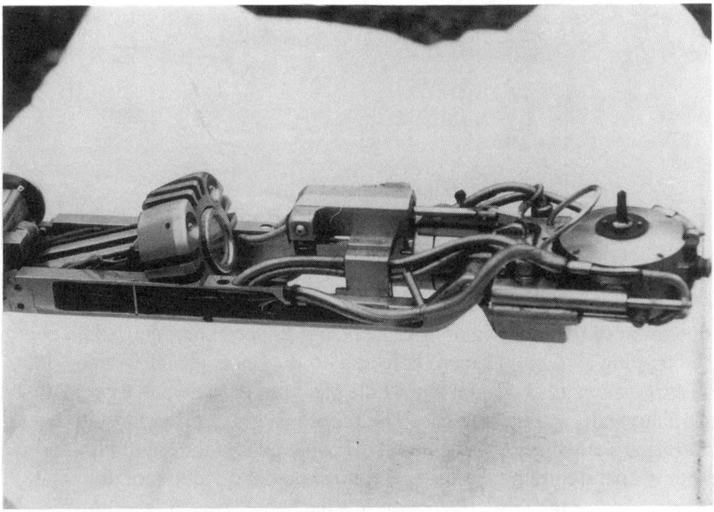

Picture # 1
Robot Head in Relaxed Position

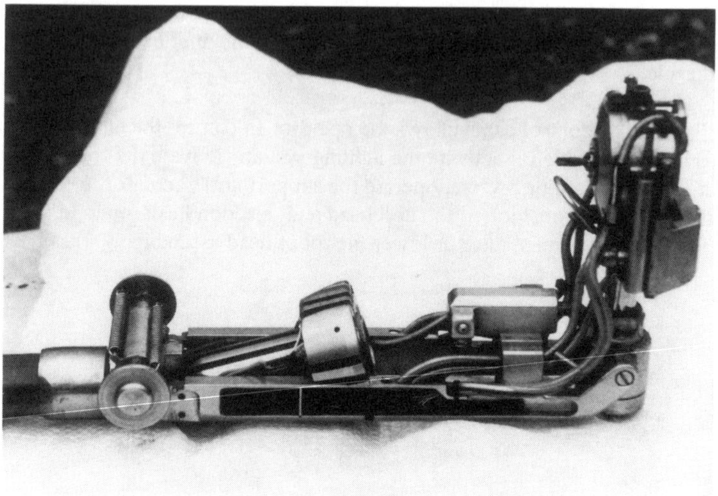

Picture # 2
Robot Head in Upright Injection Position

TOOL OPERATION

The basic operation of the tool begins with installing a special tapping saddle over the main then tapping a 76 mm (3 inch) hole at a 45 degree angle into a live NPS 6 (Diagram 1) cast iron main. This tapping saddle has been designed to work with a Mueller R Drilling and Tapping machine. The robot head is inserted through the tap, the support arms are extended and the umbilical cable pushes the robot into the main. The tool is operated remotely using a control box and video monitor. The operator visually locates the joint, then positions the robot toward the jute area of the joint and raises the robot head (diagrams 2a and 2b). The drill head is rotated into position for drilling (diagram 3). The drill bit has been designed to allow it to penetrate the spigot pipe wall, but it is not long enough to penetrate the bell pipe wall. The drill bit is shaped to prepare a chamfer above the hole. This chamfer provides a smooth clean surface for the sealant head gasket. The thrust actuator lowers the robot head but keeps the entire assembly firmly in place. The sealant head is then rotated into position and the thrust actuator is engaged to firmly press the sealant head against the main. A measured amount of sealant is injected into the joint. The thrust actuator is fully lowered and the robot head is rotated parallel to the main. The unit is then moved to the next joint. Once the robot has repaired all joints in one direction it is removed. If repairs are being completed in both directions from the tap hole, then the main is drilled to allow the robot to be inserted in the second direction. Typically, twenty-four joints can be repaired from a single excavation.

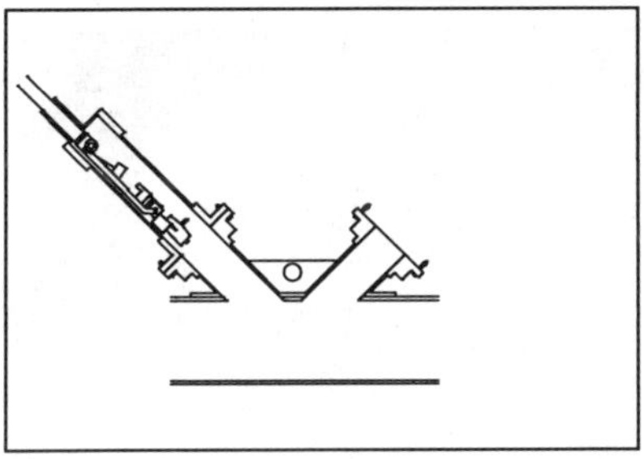

Diagram 1

The robot enters the main through a specially designed service saddle. A second hole must be drilled into the main to allow the robot to repair joints in both directions.

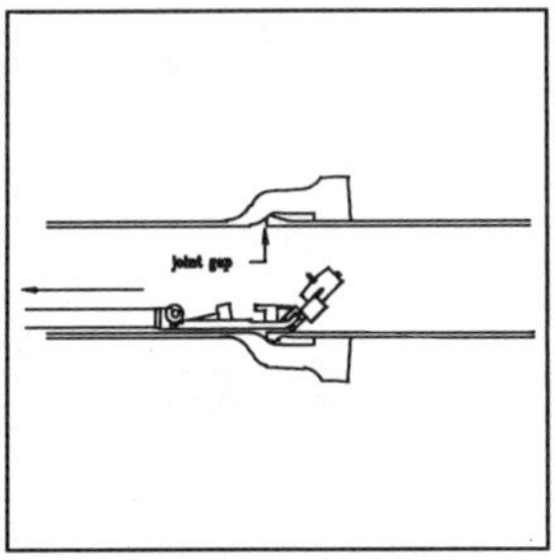

Diagram 2a

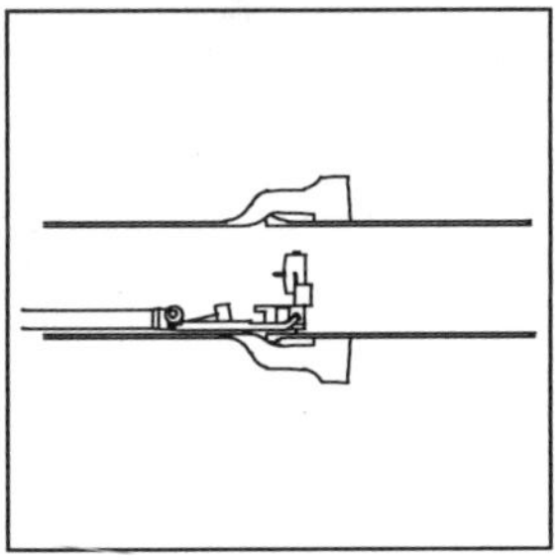

Diagram 2b

The robot head is raised. In diagram 2a the robot is approximately half raised and diagram 2b shows it in its full upright position

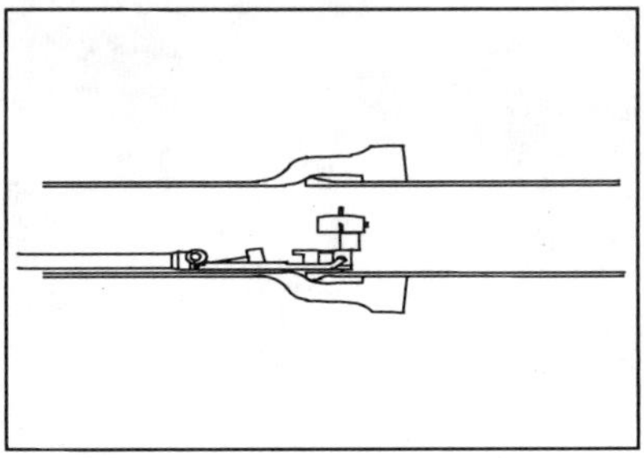

Diagram 3

The robot head is rotated to drilling position

Picture # 3
Turret Wheel

Picture # 4
Video Monitor, Control Panel and Electrical Connection Box

FIELD TRIAL RESULTS

1995

Several above ground trials and one field trial were completed during 1995. The cast iron main at the field trial site in Toronto was approximately 8 feet deep. The robot was inserted into the main approximately 120 feet before encountering an obstruction that prevented further insertion. A single exposed leaking joint was repaired using the robot. The joint was visually confirmed as no longer leaking when approximately 180 cc of regular viscosity sealant was injected into the joint. While this success was limited, we were able to use the results of this trial to determine areas for re-design. Pumping sealant was our major problem at this point and we have been successful in designing a much improved pumping system.

1996

Several field trials of the redesigned robot were held from August through October. The objectives for these trials were to confirm the viability of the current design, identify operational problems, train two gas technicians to operate the robot and to perform a field trial in Consolidated Edison's service territory.

1996 Successes

During our 1996 field trial all major components of the robot operated as designed. The two gas technicians were able to take over operation of the robot from the developers and work effectively after training. Several exposed leaking joints were visually confirmed as being effectively sealed. Near the conclusion of the trial the two man crew was able to effectively drill and inject sealant into nine joints in a single day.

1996 Problems

Several problems were identified during 1996. Approximately 50% of the mains included in our trials contained significant quantities of dirt. This dirt accumulated in front of the video camera and lighting system and reduced the operators ability to correctly position the robot prior to drilling. The robot frequently was limited to a maximum insertion distance of approximately 130 feet. The current design has no provision to determine when the drilling and chamfering operation is complete. The camera angle needs to be repositioned to allow for more precise positioning prior to drilling.

1996 Observations

Anaerobic sealant suppliers recommend that for external repairs we inject 500 cc of sealant per joint. Internally we are only capable of injecting approximately 180 cc. before significant leakage around the joint occurs. Joints previously sealed by the external injection procedure exhibit the same leakage around the joint. Should Utilities reduce the recommended injection quantities?

Most of Consumers Gas mains are located in city owned property near customers front lawn. Over time, homeowners (or the city) have planted trees, installed driveways, etc. Repair costs in these situations are significantly higher than average. Homeowners are increasingly reluctant to be inconvenienced by our leakage problems. The robot will be a valuable tool in these areas.

PROGRAM STATUS

The internally sealing robot will be redesigned based on the observations of the 1996 field trials. The problems that we have encountered have centred around mains with significant quantities of dirt and rust in the bottom of the main. To date approximately 50% of our trials have had problems with debris. Improving the effectiveness of the robot in dirty mains is one of our prime objectives. A second concern is correct robot positioning. We are able to stop joint leakage when the robot successfully drills into the jute material. If we drill into lead, we obviously cannot inject any sealant and the robot is repositioned to another point on the same joint and the joint re-drilled. Work on finding the best location to drill into a joint at the jute is continuing.

ACKNOWLEDGEMENTS

Consumers Gas wishes to acknowledge the creative design, innovation and adaptability of Dr. Andrew Goldenberg, Pawel Kuzan, and Jacek Wiercienski of Engineering Services Inc.

TRENCHLESS TECHNOLOGY FOR CULTURAL RESOURCE PROTECTION CITY OF SAN DIEGO NORTH MISSION VALLEY INTERCEPTOR SEWER TUNNEL INSTALLATION

Jamal Batta, M.S., P.E.[1], and Mark Giandoni, P.E., M.ASCE[2]

The City of San Diego's North Mission Valley Interceptor Sewer (NMVIS) is the main collector sewer between the San Diego Metropolitan Wastewater System and six contributing wastewater agencies to the east of San Diego. Within the City of San Diego there are eight communities that contribute sewage flows to the NMVIS. The original NMVIS pipeline was constructed in 1960 and was reaching it's maximum capacity limit in the mid-1980s.

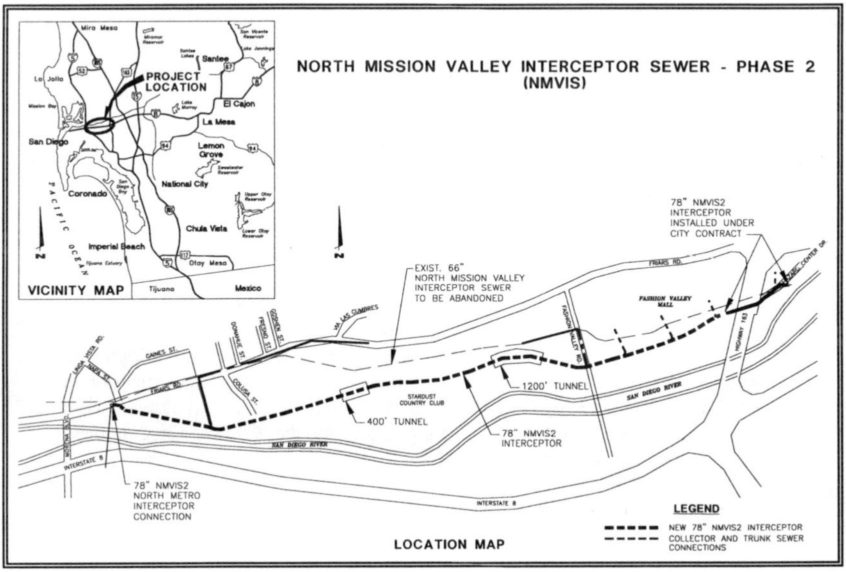

---

[1]Associate Engineer/Project Manager, City of San Diego, Public Works, Water and Wastewater Facilities Division

[2]Senior Project Engineer, Hirsch & Company Consulting Engineers, 3550 Camino Del Rio North, Suite 305, San Diego, CA, (619) 563-4545.

Trenchless Technology for Cultural Resource Protection

In a 1986 study it was determined that the NMVIS needed to be replaced due to lack of capacity and deteriorating conditions on several reaches.

A new NMVIS would be designed and constructed to convey flows for the ultimate buildout of the sewage basin with peak flows of 135 MGD.

The NMVIS pipeline is located in the Mission Valley area of San Diego. The alignment passes through heavily urbanized commercial developments, riparian habitats of the San Diego River and through relatively undisturbed areas in two golf courses. It was in these golf courses where cultural resources, determined to be significant by California standards, were found.

In the preliminary cultural resource investigations completed for the project it was determined that there was a potential for disturbing Native American artifacts during the excavation process of construction. In an attempt to quantify the potential for disruption, additional studies and field work were completed during the final design of the project. With the new information obtained in the final design studies and additional information shared from studies completed by an adjacent transit district light rail train project the resources were deemed significant and any disruptions were to be avoided or mitigated.

With this new information regarding Native American cultural resources in the project alignment the City of San Diego completed an alternative alignment study. It was during this alternative alignment analysis that the potential use of trenchless technology was found to be feasible and economical.

The North Mission Valley Interceptor Sewers Relationship with Other Projects:

The Mission Valley corridor in San Diego is one of the major east/west routes for transportation in southern California. Interstate 8 begins at the Pacific Ocean to the west and continues past the California border in the East. With the major transportation corridor comes dense urban development. While the Mission Valley area was farmland thirty years ago it has seen major developments constructed in the past years and is currently in a stage of rapid development. There are currently two major shopping centers, a convention center, over twenty major hotels, major office buildings and San Diego's Jack Murphy Stadium in the area. All of this development is in a relatively narrow four mile long valley with the San Diego river in the middle.

The NMVIS pipeline was being proposed to be constructed adjacent to the new commuter light rail train (LRT) system. The pipeline was to be constructed by the light rail train prime contractor under an agreement between the City of San Diego and the local transit board, the Metropolitan Transit Development Board (MTDB). To do this the NMVIS pipeline schedule was now tied to the light rail train construction schedule. No major changes to the pipeline schedule would be allowed in the pipeline construction since this would then impact the larger LRT construction contract schedule.

The pipeline passed through one of the major shopping centers in the area that was expanding to nearly double in size. One of the golf courses that the pipeline passed through was planned for development and the pipeline designers were required to address these future development issues of the site. All of these developments needed to be addressed in any pipeline redesigns required to mitigate cultural resource impacts.

In addition to the above work in the area and adjacent to the NMVIS pipeline the NMVIS connected to the 108" North Metro Interceptor Sewer (NMI). Under an EPA Consent Decree, which the City of San Diego had previously negotiated, the NMI was required to be in operation before January, 1997. The NMVIS pipeline was to provide the flows for the NMI to meet this deadline.

With all of the scheduling parameters and physical restriction on the NMVIS construction project, any delays to the schedule would have serious negative impacts to the City of San Diego. It was imperative that any pipeline redesign necessary to mitigate impacts to the cultural resources not delay the construction of the NMVIS pipeline.

Trenchless Technology -

It was during the NMVIS realignment alternative analysis that it became apparent that some method of trenchless construction would be necessary to complete the project on time. The alternative alignments considered this fact and included trenchless technology in the potential realignment alternatives. The trenchless technologies recommended for use were two pass tunnel construction and pipe jacking. These specific construction techniques were recommended due to the size of the pipe (78-inches) and soil conditions to be encountered. Though these are not new methods of trenchless construction it was believed that the rational for their use, cultural resource protection, was unique.

Once the alternative pipeline alignments were evaluated, the feasibility of the tunneling alternatives was investigated. The alignments through and soil condition below the cultural resource rich areas were studied and it was determined that no one method of construction would suit both areas to be protected. The soil condition below the largest resource area was a clay/cobble conglomerate. Simple direct jacking would not be feasible in this area. The second and smaller site was underlain by a saturated silty/sand. This proposed tunnel was shorter and direct jacking would be feasible.

Trenchless Technology for Cultural Resource Protection

With the two general trenchless construction methods determined, pipe jacking and two pass tunneling, the designer recommended that it may be desirable to help the competitive bidding process and allow different variations of the tunneling methods. This would increase the number of potential tunneling contractors by not limiting the work to one specific tunneling or jacking method.

For the longer two pass tunnel wood rib and lagging and steel segmental liner plate primary liner systems would be accepted. The designer felt that these variant methods of two pass tunneling would both be acceptable in the harder clayey/cobble conglomerate soils. It was obvious that a jacked pipe tunnel would not be appropriate since line and grade would be difficult maintain and the jacking head would not be able to cut into the hard soils at the face of the tunnel. The tunneling contractor chose the use of a wood rib and lagging construction method behind an open face tunneling shield.

For the shorter tunnel in a saturated silty/sand the direct jacked pipe tunneling would be feasible. For the shorter tunnel both the two pass tunnel and the direct jacking were acceptable. Once the project was bid the contractor chose to use the pipe jacking method of construction.

The most important process in the use of trenchless technology is the determination of what methods are appropriate for the specific construction project. In the NMVIS project, two tunnels less that one thousand feet away from each other, required different trenchless construction methods. One in soft soils allowed for the use of direct pipe jacking construction methods. The second in a hard clay/cobble conglomerate required another method and would not have been successful if designed for direct pipe jacking.

City of San Diego Decision Process -

The City evaluated numerous alternatives for the installation of the pipeline to avoid impacts to the identified cultural resource sites. The alternatives evaluated were to complete a major cultural data recovery program or realign the pipeline.

Because of the potential delays to the project schedule and the associated initial and long term costs of the data recovery alternative it was determined that this would not be feasible. The NMVIS construction had numerous constraints relating to other projects in the area and an EPA Consent Decree. Prior to the discovery of the significant cultural resources the schedule had been negotiated with the other projects in the area and it was determined that the installation of the NMVIS 78" pipeline must be complete by December 1996. The implementation of the data recovery alternative would have increased the cost of the project beyond the project budget and would have delayed the completion of the project beyond an acceptable time period.

Trenchless Technology for Cultural Resource Protection

The City directed the design consultant to investigate alternative alignments for the project. Four alignments were investigated and it was determined that the alternative to tunnel under the major cultural resource site would be the most efficient alignment and would not excessively delay the project. The realignment alternatives were as follows:

Alternative 1

Alternative 1 would construct the NMVIS pipeline under the MTDB contract only through the Fashion Valley parking lot. At Fashion Valley Road, the NMVIS would be realigned to proceed north of the original alignment along Fashion Valley Road then west along Friars Road to the North Metro junction box. The new alignment would be constructed under a City of San Diego construction contract.

Alternative 1 Benefits

- Eliminates archaeological impact within the Stardust Golf Course.
- Eliminates right-of-way requirements through Stardust.
- Minimizes cooperative construction effort with MTDB's general contractor.
- The project could proceed without further delays.
- Eliminates the majority of the collector sewer work currently specified under the City contract.
- Eliminates all wetland impacts and the corresponding restoration and 5 year monitoring program from the NMVIS project.

Alternative 1 Disadvantages

- The revised alignment would require 7,800 lineal feet of additional tunnel construction at depths up to 65 feet.
- Tunnel construction at the intersection of Friars Road and Fashion Valley Road will create severe traffic control impacts.
- The additional cost to the NMVIS project under this alternative is $6.7 million.

Alternative 2

Alternative 2 would construct the NMVIS pipeline under the MTDB construction contract from station 45+00 to 105+75, which includes all of the Fashion Valley section and the easterly 4,000 feet of the Stardust Golf Course section. Station 45+00 is the intersection of the NMVIS and D Street as proposed on the Riverwalk tentative Map. At D Street the pipeline would turn to the north and continue up to Friars Road. At Friars Road the pipeline would turn to the west and follow Friars Road to the North Metro Interceptor junction structure.

Alternative 2 Benefits

- Minimizes impacts to cultural resources and eliminates the impacts to any new cultural resources located in the areas to be studied.
- Reduces right of way requirements through Stardust Golf Course.
- Reduces cooperative construction effort with MTDB's general contractor.
- Eliminates the most of the collector sewer work specified under the City contract.
- Eliminates all wetland impacts and the corresponding restoration and 5 year monitoring program from the NMVIS project.

Alternative 2 Disadvantages

- The revised alignment would require an additional 4,200 feet of tunnel construction.
- The additional major construction in Friars Road will have additional impacts to traffic circulation.
- The additional cost to the NMVIS project under this alternative have been estimated to be $4.8 million.

Alternative 3

Alternative 3 would construct the NMVIS pipeline under the MTDB contract from station 40+00 to station 105+75, which includes all of the Fashion Valley section and the easterly 4,500 feet of the Stardust Golf Course section.

Alternative 3 Benefits

- Minimizes the archaeological impacts within the Stardust Golf Course.
- Reduces right-of-way requirements through Stardust.
- Reduces cooperative construction effort with MTDB's general contractor.
- Eliminates the majority of the collector sewer work currently specified under the City contract.
- Eliminates all wetland impacts and the corresponding restoration and 5 year monitoring program from the NMVIS project.

Alternative 3 Disadvantages

- The revised alignment would require 4,200 lineal feet of additional tunnel construction.
- Additional construction in Friars Road would have an impact on traffic circulation.
- Additional new right-of-ways would be required.

- The additional cost to the NMVIS project under this alternative is $4.6 million.

TABLE 1 - MAJOR DECISION MAKING PARAMETERS

ALTERNATIVE	COST IMPACTS	SCHEDULE IMPACTS	PROS	CONS
1. Realign entire pipeline to the north	+$2,400,000	Minimal	No cultural impacts, no R.O.W. acquisitions required	7,000 lf of pipeline in busy streets
2. Realign 4,500 lf of eastern section of pipeline	+$4,800,000	Minimal	Reduced R.O.W. acquisitions, eliminates wetland impacts	Additional 4,200 lf of tunnel construction
3. Realign 4,000 lf of eastern section of pipeline	+$4,600,000	Minimal	Reduced R.O.W. acquisitions, eliminates wetland impacts	Additional 4,200 lf of tunnel construction
4. Realignment and separate project into 2 phases	+$1,400,000	Adds 1 year to project completion	Minimizes impacts cultural resource impacts	Additional studies needed

Realignment Alternative 4

Alternative 4 would split the NMVIS project into 2 phases; east and west. The east section would include all the current NMVIS work from the eastern limits of the archaeological site through Fashion Valley. The west section would include the remainder of the NMVIS project to the west. The east section would be constructed under the MTDB contract. The west section would be constructed under the City Contract.

The reason for the split is to accommodate the anticipated field work, data recovery and mitigation and permit review by the Army Corps of Engineers. This work was expected to take up to 12 months. The LRT project will move forward without the NMVIS project during this time.

Alternative 4 Benefits

- Minimizes archaeological impact within the Stardust Golf Course.
- Allows for the majority of the project to be constructed as planned.

Alternative 4 Disadvantages

- The alternative would require approximately 1,400 feet of additional tunnel construction.
- Work in the tunnel section could not begin without additional archaeological investigations.
- The additional cost to the NMVIS project under this alternative is $2.4 million.
- Increased project cost due to additional mobilization.
- Delays to City Contract of up to one year.

Recommended Alternative

The recommended alternative was a modified realignment alternative 4. This alternative required several minor realignments of the pipeline and two tunnels for a total of 1,600 feet would be required. The final alternative included a small data recovery program completed during construction. The modification to the design did not require additional reviews by the Army Corps of Engineers or other resource agencies. The additional cost were limited to approximately $1.2 million and there were no delays to the NMVIS project or any of the adjacent projects.

Conclusion -

By following a rational engineering process the City of San Diego and their design consultant were able to use trenchless technology construction techniques to cost effectively satisfy cultural resource protection requirements for a major pipeline project. By completing a detailed cost benefit analysis for the NMVIS pipeline realignment alternatives the City was able to determine the most efficient alignment alternative. The selected alternative using trenchless construction was able to eliminate impacts to the cultural resources, and minimize scheduling impacts to adjacent and connecting projects.

Trenchless technology provided alternative construction methods for the City of San Diego's North Mission Valley Interceptor Sewer that helped in the successful completion of this major project on schedule and within budget.

CIPP Selected for Cast Iron Force-Main Rehabilitation

M. Najafi,[1] Member, V.K. Varma,[2] Fellow, and K. Stutterheim[3], Member, ASCE

Abstract

This report provides a feasibility study of various trenchless rehabilitation methods for an existing 508-mm (20-in.) diameter cast iron sewer force-main, approximately 2,500 meters (8,000 lineal feet) in length and located along a railroad right-of-way. The failing cast iron pipe is approximately 35 years old and has a history of ruptures with longitudinal and circumferential cracks. Several trenchless pipeline rehabilitation methods can restore the structural integrity of the force-main, however, the specific conditions of the project, congestions of the surface area, and the necessity to bypass the flow, have narrowed down the applicable technique to cured-in-place pipe (CIPP) method. For comparison purposes, an approximate estimate of replacing the pipe with conventional trenching is presented in this report. Based on the analyses of this investigation, the authors conclude that the CIPP method would be the most appropriate method to restore the structural integrity of the cast iron pipe. The CIPP would be designed as a "stand-alone" pipe to meet the internal and external load parameters without any support from the existing cast iron pipe.

Introduction

The history of the City of Atchison, Kansas, situated on the high bluffs overlooking the Missouri River, goes back to 1804. According to 1990 census, the population of the city is approximately 10,600. In recent years, the city has experienced

[1]Assistant Professor, Department of Engineering Technology, Missouri Western State College, St. Joseph, Missouri 64507

[2]Professor, Department of Engineering Technology, Missouri Western State College, St. Joseph, Missouri 64507

[3]Assistant Professor, Department of Engineering Technology, Missouri Western State College, St. Joseph, Missouri 64507

several pipe breaks in the 508-mm (20-inch) cast iron pipe that connects the pump station to the waste water treatment plant some 2,530 m (8,300 ft) away at the south side of the town. Several classes of cast iron pipe have been used for the construction of the pipeline, ranging from class 22 to class 25. The pipeline is more than 35 years old.

In June 1995, the authors were asked to prepare a feasibility study to rehabilitate the pipe using trenchless technology methods. The objective of the study was to develop the most cost-effective and practical strategy for the long-term repair of the pipe. The selected method must restore the structural integrity of the pipe.

Background

Cast iron pipe is considered a rigid pipe. Three main types of cracks may form the preliminary stages of pipe fracture, namely, longitudinal cracks, circumferential cracks and multiple cracks (starting from a point with formation of fragments). The type of crack and the cause of cracking are closely connected. The shape of a crack, its dimensions and its course usually allows conclusions to be made regarding the causes. Longitudinal cracks represent an indication of instability of the pipe. Circumferential cracks are created where the allowable longitudinal bending tensile strength of the pipe, or the pipe's shear strength is exceeded. These cracks run mainly over the whole circumference of the pipe, but they do not represent an immediate danger to stability, as long as there are no changes to the bedding. Multiple cracks start from a single point and are star-shaped, or which have an irregular course.

Of the main types of cracks referred to above, longitudinal cracks occur most frequently in flexurally rigid pipes and are generated when the ring bending tensile strength is exceeded, such as, a line displacement because of construction defects. They can also be created by positional displacement caused by changes to the bedding as a result of leaks or ground movements, as well as incorrect pipe connections. They usually occur in the top, bottom and sides and stretch along the overall length of the pipe.

Table 1 presents a chronological list of break locations reported by the Department of Public Works. The repairs before 1990 were conducted by the Street Department and no records were available. The pipe sections of the break dated July 24, 1995, were visually examined. The cracks were longitudinal with a 45-degree angle at the end of the pipe section. The longitudinal cracks, as said earlier, suggest that the ring bending tensile strength of the pipe has been exceeded. They can be caused by changes in the bedding aggravated by heavily fluctuating operating conditions (frequent overloads) or high pressure.

Table 1
Chronological List of Pipe Breaks

Break Number	Station	Break Date
1	30+00	November 9, 1990
2	31+60	August 29, 1991
3	47+80	August (?), 1993
4	38+90	May 10, 1994
5	39+19	August (?), 1994
6	11+62	December 6, 1994
7	10+50	July 24, 1995
8	30+25	December 16, 1995

Trenchless Pipeline Rehabilitation Methods

The basic trenchless pipeline rehabilitation methods can be categorized into the following types: (1) Cured-In-Place Pipe (CIPP), (2) Sliplining, (3) In-Line Replacement, (4) Close-Fit Pipe, and (5) Point Source Repair. Table 2 provides a summary of these basic methods. For more information on capabilities and limitations of different trenchless pipeline rehabilitation methods, the reader is referred to Iseley and Najafi (1995).

Table 2
Comparison of Different Trenchless
Pipeline Rehabilitation Methods (Najafi, 1994)

Method	Diameter Range (mm)	Maximum Installation (m)	Liner Material	Applications
CIPP:				
Inverted in Place	100-2700 (4-108 in.)	900 (3,000 ft)	Thermoset Resin/Fabric Composite	Gravity and Pressure Pipelines
Winched in Place	100-1400 (4-54 in.)	150 (500 ft)	Thermoset Resin/Fabric Composite	Gravity and Pressure Pipelines
Sliplining:				
Segmental	100-4000 (4-158 in.)	300 (1,000 ft)	PE, PP, PVC, GRP (-EP & -UP)	Gravity and Pressure Pipelines
Continuous	100-1600 (4-63 in.)	300 (1,000 ft)	PE, PP, PE/EPDM, PVC	Gravity and Pressure Pipelines
Spiral Wound	100-2500 (4-100 in.)	300 (1,000 ft)	PE, PVC, PP, PVDF	Gravity Pipelines Only

Table 2 (Continued)

Method	Diameter Range (mm)	Maximum Installation (m)	Liner Material	Applications
In-Line Replacement:				
Pipe Displacement	100-600 (4-24 in.)	230 (750 ft)	PE, PP, PVC, GRP	Gravity and Pressure Pipelines
Pipe Removal	up to 900 (36 in.)	100 (300 ft)	PE, PVC, PP, GRP	Gravity and Pressure Pipelines
Close-Fit Pipe:				
Modified Cross Section	100-400 (4-15 in.)	210 (700 ft)	HDPE, PVC	Gravity and Pressure Pipelines
Drawdown	62-600 (3-24 in.)	320 (1,000 ft)	HDPE, MDPE	Gravity and Pressure Pipelines
Rolldown	62-600 (3-24 in.)	320 (1,000 ft)	HDPE, MDPE	Gravity and Pressure Pipelines
Point Source Repair:				
Robotic Repairs	200-760 (8-30 in.)	N/A	Epoxy Resins/Cement Mortar	Gravity
Grouting	N/A	N/A	Chemical Grouting	Any
Link-Seal	100-600 (4-24 in.)	N/A	Special Sleeves	Any
Point CIPP	100-600 (4-24 in.)	15 (50 ft)	Fiberglass/Polyester, etc.	Gravity
Spray-on-Lining	76-4500 (3-180 in.)	150 (500 ft)	Epoxy Resins/Cement Mortar	Gravity and Pressure Pipelines

Cured-In-Place Pipe (CIPP). CIPP is a liquid thermoset resin-saturated material inserted into the existing pipeline by hydrostatic or air inversion or by mechanically pulling with a winch and cable. The material is heat-cured in place. The primary components of the CIPP are a flexible fabric tube and a thermosetting resin system. For typical CIPP applications, the resin is the primary structural component of the system. These resins generally fall into one of the following generic groups: (1) Unsaturated Polyester, (2) Vinyl Ester and (3) Epoxy. Each resin has distinct chemical resistance and structural properties. All have distinct chemical resistance to domestic sewage.

Sliplining. In this method, a new pipe of smaller diameter is inserted by pulling or pushing into a deteriorated host pipe and the annulus space between the old pipe and new pipe is grouted.

In-Line Replacement. In this method, the existing pipe is removed and a completely new pipe is installed. In-line replacement can be categorized as pipe displacement and pipe removal. Pipe displacement, also known as pipe bursting, is a technique for breaking out the old pipe by use of radial forces from inside the old pipe. The fragments are forced outward into the soil and a new pipe is pulled into the bore formed by the bursting device. Pipe removal uses microtunneling machines with the capability of crushing rocks and stones to excavate existing pipelines for replacement.

Close-Fit Pipe. This method uses coiled deformed new pipe before it is installed, then expands it to its original size and shape after placement to give a close-fit the existing pipe. Most close-fit pipes are first deformed in the manufacturing plant, then shipped to the job site, where inserted and finally reformed by heat and pressure or naturally.

Point Source Repair. Point source repair covers a broad range of techniques such as robotics, grouting, CIPP, link-seal, shotcrete, coatings, spray-on linings, etc. These methods are used when local defects, due to cracking or joint failure, are found in a structurally sound pipeline,

Method Selection Process

For evaluation of the appropriate rehabilitation alternatives to renew the 508-mm (20-in.) force-main, the following selection criteria were established.

Capital Cost. Capital cost includes the cost for implementing the rehabilitation project. Life-cycle-cost, considered for the "do-nothing" option, includes capital cost, annual operations and maintenance (O&M) cost, and loss of revenue due to emergency repairs.

Applicability. Applicability represents qualifications of rehabilitation methods in three areas: reliability, application to sewer force-main, and application to the required diameter and old pipe material. The sewer force-main is critical for the City of Atchison. Therefore, long-term reliability of the rehabilitation method is one of the more important evaluation criteria. The ability to prevent formation of new cracks and the eventual failure of the pipe will be examined under this criterion. Only rehabilitation systems can be recommended that accommodate the design pressure of the force-main.

Constructibility. The construction requirements of each rehabilitation alternative are evaluated based on shutdown requirements; pit requirements; easement requirements; production rates; requirements on curing; ability to accommodate bends; risk of damage to adjacent structure (railroad track); availability of local expertise, and quality assurance and quality control and inspection requirements. Due to specific requirements of the project site, the insertion pits must be spaced on 600-m (2,000-ft) increments.

Life Expectancy. Another very important criterion is the life expectancy of the rehabilitation method. The sewer force-main was constructed more than 35 years ago, and may be close to the end of its useful design life. Construction of a new replacement pipeline outside existing sewer force-main right-of-way would not only be cost prohibitive, due to the existence of high bluffs and possibility of ground movement, and existence of the railroad tracks, it also may be impractical. Therefore, the rehabilitation methods that will revitalize and renew the structural conditions of the pipe (and thus will extend the useful life of the force-main) will be more favorable than a new replacement.

Pipeline Hydraulic Capacity. Some rehabilitation alternatives such as sliplining may improve the interior surface frictional characteristics of the force-main but also reduce the cross sectional area. These changes may increase the head loss and decrease the pipeline capacity.

Structural Integrity. These criteria address structural issues of the rehabilitation system including improvement to the structural integrity, ability to accommodate future pipe settlement or ground movement, offset or displacements, seismic resistance, and ease of repair in case of a pipe break.

The "Do-Nothing" Alternative

To verify the validity and cost effectiveness of the recommended rehabilitation system and for comparison purposes, the costs of "do-nothing" are estimated. The costs of "do-nothing" alternative are the costs that will incur if the recommended rehabilitation method is not carried out. They include the costs of emergency repairs, environmental impact of pipe breaks, costs of periodic major repairs, and the replacement cost of the force-main at the end of its useful life. The cost breakdown of "do-nothing" alternative is presented in Table 3. The cost of pipe replacement with trenching method is based on Means Site Work & Landscape Cost Data, 14th Edition, 1995, for a 3.6-m (12-ft) trench and a 508-mm (20-inch) ductile iron pressure pipe. Any increase in depth of excavation will increase the actual construction costs. It should be noted that no monetary value is given to environmental impact of "do-nothing" option, however, this impact should be considered when "do-nothing" option is compared with other alternatives.

In the replacement with trenching method the following cost items should be considered: (1) preparing bid documents that include substantial field survey work and preparation of contract drawings, engineering, legal and administration; (2) trench shoring; (3) contingencies for possible existence of boulders and dewatering requirements; (4) access roads; and (5) construction of manholes.

Recommended Rehabilitation Strategy

Based on the selection criteria mentioned above, the rehabilitation method requirements listed in Table 2, and information provided in previous section, the following rehabilitation methods are rated as following:

- The CIPP inverted-in-place meets all the project requirements based on the criteria selected.

Table 3
Estimated Cost of "Do-Nothing" Alternative

Item	Type of Work	Cost ($)	Present Worth ($) with a Discount Rate of 3.5%
1	Emergency repair costs (2 locations per year at $15,000 per year for 20 years)	424,200	213,203
2	Replacement by remove and replace after 20 years (year 2015), after the cast iron force-main has been in service for a total of 55 years for 2,530 m at $1,394/m (8,300 ft at $425/ft)	3,527,500	1,772,921
Total		3,951,700	1,986,124
3	Environmental impact	?	?

- The sliplining (continuous) method is eliminated because of limitation on the maximum installation distance that is 300 m (1,000 ft), and because a "stand-alone" sliplining pipe would require a thick-wall pipe that combined with the annular space, would decrease the hydraulic capacity of the pipe significantly.

- The in-line replacement (pipe bursting) method is eliminated because of the maximum installation distance that is 230 m (750 ft), and a very low production rate. The pipe removal is ruled out because of maximum installation distance of 150 m (500 ft) and high unit costs.

- The close-fit methods are ruled out because of the maximum diameter size is 400 mm (15 in.) and maximum installation distance is 213 m (700 ft) (for Nu-Pipe and U-Liner). The maximum installation distance is 300 m (1,000 ft) for drawdown and rolldown.

- The point source repair methods are eliminated because none of these methods would restore the structural integrity of the cast iron force-main pipe.

Based on the foregoing, the authors conclude that the most cost-effective, environmentally responsible long-term repair strategy consists of CIPP (inverted-in-place) installation for the entire pipeline. A summary of the analyses is presented in Table 4.

Table 4
Applicability of Design Considerations for Each Rehabilitation Family

Rehab. Family	Design Considerations						Practical
	Capital Cost ($)	Applicability	Constructibility	Life Expectancy	Pipeline Hydraulic Capacity	Structural Integrity	
CIPP	1,694,000	Yes	Yes	Yes	Yes	Yes	Yes
Sliplining	N/A	No	No	No	No	Yes	No
Close-Fit Pipe	N/A	Yes	No	Yes	Yes	Yes	No
Pipe Replacement (Bursting)	N/A	Yes	No	Yes	Yes	Yes	No
Point Source Repair	N/A	No	No	No	No	No	No
Open-Cut Replacement	1,772,921	Yes	No	Yes	Yes	Yes	No
Do-Nothing	1,986,124	No	No	No	No	No	No

Additional Benefits of the Recommended Rehabilitation Method

Implementing the rehabilitation project will provide the following additional benefits:

- Reduce the risk of the potential catastrophic structural failure of the force-main.
- Facilitate the O&M operations by eliminating the need for unplanned emergency repairs.
- Reduce damage to environment by eliminating leakage and bypassing to the river during the pipe breaks.
- The design life of the rehabilitated pipeline will be 50 years. This prevents replacing the pipe after 10 or 15 years.

Cost Estimate for CIPP Method

The estimated costs of selected rehabilitation (CIPP) include the costs of constructing installation pits, CCTV before and after CIPP installation, furnishing and installing the CIPP pipe, grouting the ends of the CIPP and installing access vaults, bypassing of the existing flow during the construction, and the cleaning of the existing cast iron pipe before CIPP installation. The breakdown of costs is presented in Table 5.

Table 5
Estimated Cost of Recommended Project (CIPP)

Item	Type of Work	Cost ($)
1	Preliminary CCTV Inspection	18,000
2	Pipeline Cleaning	13,000
3	Furnish and Install CIPP	1,560,000
4	Water Pressure Testing (four sections)	14,000
5	Final CCTV Inspection	14,000
6	Install Access Vaults and Pipe Access Work (five each)	75,000
7	Bypass Pumping to Missouri River During Construction (will be provided by the City of Atchison)	-----
Total 2,530 m at $669/m (8,300 ft at $204.00/ft)		1,694,000

Comparison of CIPP with Trenching and The "Do-Nothing" Options

Table 6 provides a comparison of three alternatives, CIPP, replacement with conventional trenching and the "do-nothing."

Table 6
Comparison of Costs for Three Alternatives

CIPP Present Costs to Reconstruct the 8,300 ft of Force-Main	**Conventional Trenching** (Present Worth of Replacing After 20 Years)	**Do-Nothing** (Present Worth of Annual Repair Costs and Eventual Replacing After 20 Years)
$1,694,000	$1,772,921	$3,951,700

Conclusions and Recommendations

Based on the analyses of this report, it was recommended that the CIPP method be used to rehabilitate the entire pipeline. The other rehabilitation methods do not provide the constructibility required by the specific conditions of the pipeline nor do they meet other criteria as explained in the report. Due to railroad right-of-way, and existence of high bluffs near the force-main, the work area for the rehabilitation work is very congested, and only those trenchless methods can be used that provide an approximate installation range of 610 m (2,000 ft). Another main factor considered in this report is the production rate of the applicable method, since the rehabilitation process must be completed within the time limits set by the EPA for bypassing to the Missouri river.

Other options would be the "do-nothing" and replacement by trenching method. The "do-nothing" option will be more expensive as it would require annual emergency repairs and an eventual total replacement of force-main after 15-20 years. In spite of similar capital cost estimates, the trenching method would involve, longer construction duration, problems with congested area of railroad right-of-way, or re-routing of the pipe to top of the bluffs. Existence of ground water, rocks, shoring requirements, terrain and access roads, legal, engineering and administration issues would present some very expensive problems to overcome for trenching method. In addition, preparing the bid documents for an open-cut option would require substantial field survey work and plan preparation. After construction the new pipe may have the risk of ground movements and future pipe failures. The trenchless methods have inherent benefits that provide higher construction productivity and involve fewer risks.

Acknowledgment

The authors would like to thank Mr. John Hixon, P.E., Director of Public Works, City of Atchison, for the opportunity to work on this project.

Definition of Acronyms

PE - Polyethylene
PP - Polypropylene
EPDM - Ethylene Polypelene Diene Monomer
PVC - Poly-Vinyl-Chloride
PVDF - Poly-Vinylidene Chloride
GRP - Glassfiber Reinforced Polyester
HDPE - High Density Polyethylene
MDPE - Medium Density Polyethylene

References

Iseley, D.T. and Najafi, M. (Editors). (1995). "*Trenchless Pipeline Rehabilitation,*" The National Utility Contractors Association (NUCA), Arlington, Virginia.

Najafi, M. (1994). "*Trenchless Pipeline Rehabilitation: State-of-the-Art Review,*" Trenchless Technology Center, Ruston, Louisiana.

U.S. Environmental Protection Agency (EPA). (1991). "*Handbook of Sewer System Infrastructure Analysis and Rehabilitation,*" Office of Research and Development, Cincinnati, Ohio.

Resistance of High Alumina and Portland Cement to H_2SO_4

M. L. Kerr[1] and J.M. Pitt[2]

Abstract

Recently, there has been an insurgence of new technologies used in rehabilitating existing sanitary sewer systems, which have experienced hydrogen sulfide corrosion, and reducing the cost of maintaining these systems. "No dig" technologies are on the rise in the United States and include such innovations as synthetic liners, PVC inserts and grouting systems. The mortars, used in the grouting systems, are the main focus of this paper.

A series of tests designed to assess the resistance of different mortars in dilute sulfuric acid solution, simulating the harsh environmental conditions found in sanitary sewer systems, were performed. The mortars used for this study were High Alumina Cement (HAC), Portland Cement and sand (PC), and Portland Cement, sand and silica fume (MSPC) mortars. The specimens made from these mortar mixes were submerged in water and cured for 28 days. After curing, the specimens were immersed in three different dilute sulfuric acid solutions, 1.0%, 5.0% and 10.0% by volume. Comparisons were made by observing the loss in mass of the specimens. Specimen masses were recorded at 7, 28 and 56 days. At 56 days the tests were discontinued. A 25% decease in mass was considered as an indication of failure for the test specimens. After a review of the data gathered, it was found that the HAC mortar appeared to perform better than the PC or MSPC mortars. The HAC mortar maintained an average of approximately 75% of its mass throughout the test period, in all three of the acid solutions. Alternatively, the PC and MSPC mortars placed in the 5.0% and 10.0% solutions lost more than 25% of their mass, within the first 20 day of the test. The PC and MSPC mortars maintained greater than 75% of their mass, in the 1.0% solution.

[1] Research Assistant, Spangler Geotechnical Laboratory, Iowa State University, Ames, Iowa, Student Member ASCE

[2] Associate Professor of Civil Engineering, Department of Civil and Construction Engineering, Iowa State University, Ames, Iowa, Member ASCE

Introduction

The intent of this paper is to compare the resistance of high alumina cement (HAC), Portland cement with micro-silica (PCMS), and plain Portland cement (PC) mortar to attack by sulfuric acid solutions of different, but constant, concentrations. The mortars studied in this research were designed for rehabilitation of manholes and sewer pipe and the acid concentrations were selected to represent the highly corrosive environments often present in manhole and sewers. Concrete manholes and pipes may be exposed to varying levels of hydrogen sulfide and sulfuric acid. Thistlewayte (1972) reported that, in extreme conditions, pH levels at concrete surfaces can be as low as 1. Corrosion of sewer structures has also been attributed to microbes that produce sulfuric acid and are active as long as hydrogen sulfide is present. Sands et al. suggests that hydrogen sulfide is converted into sulfuric acid by thiobacillus and possibly other sulfide-reducing bacteria. Evaluation of these materials, intended for use in rehabilitation in acid environments, is the focus of this research.

Literature Review

High Alumina Cement

The composition of high alumina cement (HAC), the hydration products, and the fact that formation of certain products depend on the temperature during hydration is represented in Figure 1. The definitions of cement chemistry shorthand notation are in Appendix II. Two hydration products, CAH_{10} and C_2AH_8 are dimensionally unstable when subjected to temperatures greater than 30° C (86° F). This conversion or transformation in crystalline structure is known to result in loss of strength, increase in permeability, and possibly a decrease in acid resistance in that a more porous structure of less stable material is the result (Neville, 1975). However, as hydration and service in manholes and sewers will probably occur over a limited temperature range, this research will concentrate on those products formed and maintained within the intermediate temperature range of 10 to 25° C; this is the temperature range in which C_2AH_8 is formed.

$$CA + H \begin{cases} \xrightarrow{<10^\circ C\ (50^\circ F)} CAH_{10} \\ \xrightarrow{10-25^\circ C\ (50\ -\ 77^\circ F)} C_2AH_8 + AH_3 \\ \xrightarrow{>25^\circ C\ (77^\circ F)} C_3AH_6 + 2AH_3 \end{cases}$$

Figure 1. Hydration of Calcium Aluminate Cement, adapted from Mindess et al.

Portland Cement

The predominant compounds and hydration products of Portland cement are summarized in Figure 2, equations 1 & 2. Calcium hydroxide (CH) is the most vulnerable of the hydration reactants in that hydrogen ions accelerate leaching of calcium which later combines with sulfate radicals to form gypsum. This leaching enhances further acid attack because the cement matrix is made more porous and greater surface contact area with internal sources of CH are available for reaction.

The conventional view is that the calcium silicate hydrate ,$C_3S_2H_3$, is less vulnerable to acid than CH. However, if the acid is highly concentrated, the calcium silicate hydrate may be attacked and transformed into a water soluble silica gel (Mindess and Young).

It has been proposed that an admixture of reactive silica or a pozzolan which combines with the CH of hydrating Portland cement results in a more resistant mortar. This is due to the fact that highly soluble CH components are replaced with a more stable calcium silicate hydrate as represented by equation 4, Figure 2.

Kienow (1991) proposed an alternative hypothesis which presumes that acid attack can be abated by the sacrificial contribution of CH to neutralize the attacking acid. In theory, an unlimited source of acid should consume all of the sacrificial component and with enough acid availability to continue the attack on the calcium silicate hydrates.

$$2\,C_3S + 6\,H \rightarrow C_3S_2H_3 + 3\,CH \qquad \text{Eq. 1}$$
$$2\,C_2S + 4\,H \rightarrow C_3S_2H_3 + CH \qquad \text{Eq. 2}$$
$$CH + SO_4^{-2}\,(aq) \leftrightarrow C\bar{S}H_2 + 2OH^-\,(aq) \qquad \text{Eq. 3}$$
$$CH + S + H \rightarrow C\text{-}S\text{-}H \qquad \text{Eq. 4}$$

Figure 2. Hydration Equations and Gypsum Production

Experiment Design

Three factors evaluated in this study are as follows:

1. Determine the relative resistance of PC, PCMS, and HAC mortars subjected to identical sulfuric acid treatments.
2. Measure the effect of acid concentration on corrosion rate.
3. Evaluate the effect of unlimited availability of sulfuric acid at constant concentration.

The temperature for the experiments was maintained at 20° C. This is based on the presumption that conversion of HAC in the intended application will not happen.

One hundred and sixty 50 mm diameter by 100 mm long cylindrical specimens were cast for each of the three types of mortar included in the study. The specimens were moist cured for 28 days and the initial mass of each cylinder was recorded just prior to immersion into buckets of different acid solutions. The acid concentrations used in this study were 0% (distilled water), 1%, 5%, and 10% concentrated acid by volume measure. Corresponding pH values for the test are in Figure 3. Forty specimens of each type of mortar were placed in each of the four solutions previously described. Each specimen was removed from solution, rinsed, drip dried, and weighed at 7, 28 and 56 days after immersion.

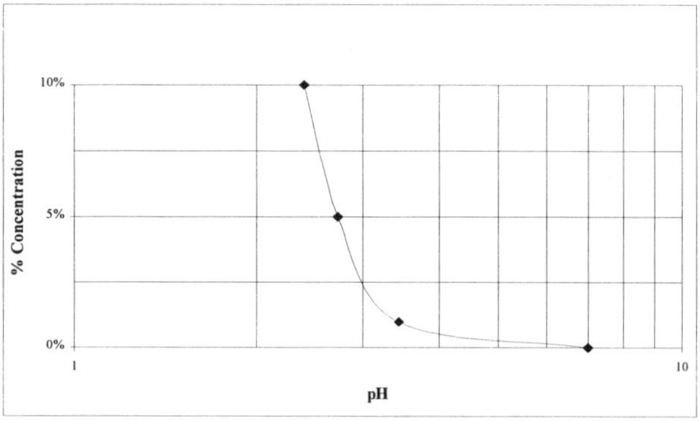

Figure 3. pH Vs Percent Concentration of Sulfuric Acid

The acid concentrations were maintained throughout the experiment by sampling each solution and determining hydrogen ion depletion by volumetric titration with sodium hydroxide on a daily basis. Hydrogen ion concentration was then replenished with addition of concentrated sulfuric acid. Test results reported herein are for 58 days of immersion.

The HAC mortar used in this experiment was a proprietary product in which the sand sized aggregate was ungrounded high alumina clinker. The mortars with Portland cement included silica sand. The PCMS mortars contained silica fume and superplasticizer. The batch quantities for the locally blended mortars are in Table 1. Consistency for all of mortars was similar with the cone penetration value being recorded between 50 to 60 mm per the Cone Penetration Method, ASTM C 780.

	PC mortar	PCMS mortar
Cement	7257.5 grams	7257.5 grams
Sand	9979.0 grams	9979.0 grams
Silica Fume	x	1596.5 grams
Superplasticizer	x	72.5 grams
W/C	0.37	0.45

Table 1. Mix Design for PC and PCMS Mortars

Figure 4 shows the relationship between the HAC, PC and PCMS, mortar subjected to the 1.0% sulfuric acid solution. HAC and PC mortar lost a maximum of 5% and 15% mass, respectively at the conclusion of the 58 day exposure The same exposure for PCMS specimens resulted in an average loss of 32% of mass.

Loss in mass vs. time after exposure to 5% concentration of sulfuric acid follows a similar trend to that of the lower concentration. Corrosion occurs faster and the amount of material lost from the specimens is greater than that of the 1 % case. Still, the HAC offers the most resistance at the conclusion of the test and the PCMS offers the least. This trendcontinues for the highest acid concentration, 10 %, but the rate of loss for the HAC appears to trend asymptotic to the level 25 % loss of mass.

By setting a limit for acceptable loss of mass, the time to failure for each type of material decreases in acid concentration. The longest life can be obtained with HAC and the shortest with PCMS.

Conclusions

The most obvious conclusion that can be reached from these experiments is that the HAC mortar was more resistant to sulfuric acid for the conditions of this study and that this material should be best of the three materials included in this study for sanitary sewer or manhole rehabilitation.

A second observation is that regular Portland cement concrete is less affected by acid than Portland cement with a pozzolan. This is in agreement with Kienow's observation but is in disagreement with the supporting concept. There was not enough CH in the specimens to neutralize the available acid, yet the corrosion process proceeded at a slower rate than did corrosion of the less soluble CSH in the pozzolan mortar. An alternate explanation may be in the fact that reaction products of sulfuric acid and calcium hydroxide or calcium sulfate (equation 3, Figure 2) may precipitate on the surface and serve as a barrier causing further reactions to be controlled by diffusion. Whether such diffusion retardation would be effective in an environment where sulfuric acid is introduced by bacteria depends on whether these

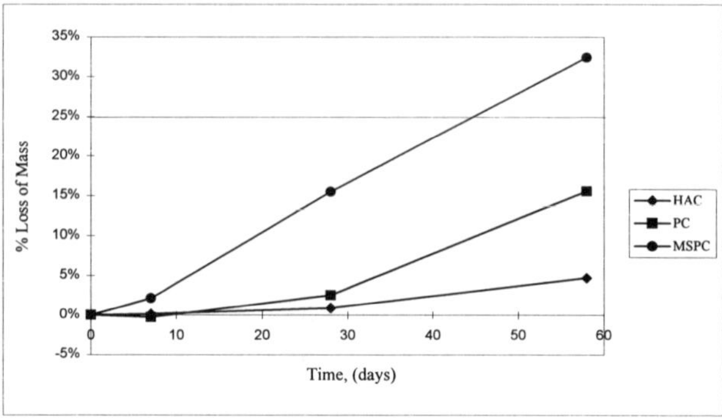

Figure 4. Percent Loss in Mass Versus Time, For Samples Placed in 1.0% Sulfuric Acid Solution

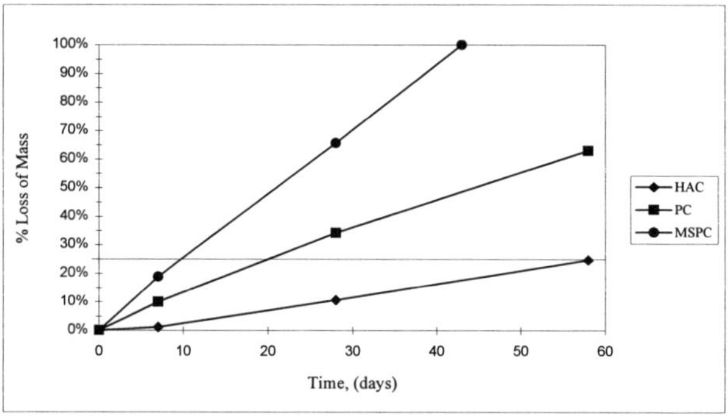

Figure 5. Percent Loss in Mass Versus Time, For Samples Placed in 5.0% Sulfuric Acid Solution

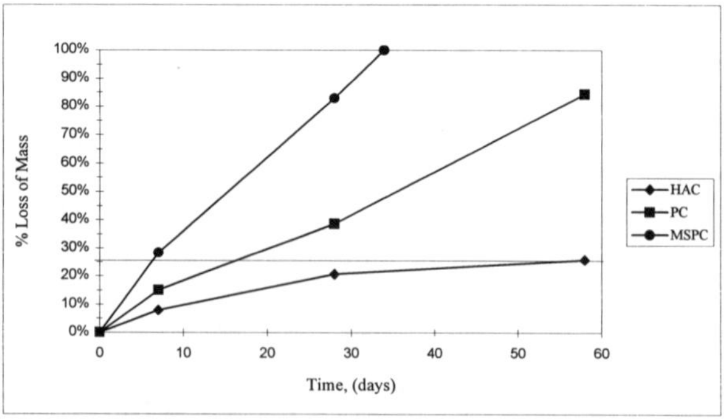

Figure 6. Percent Loss in Mass Versus Time, For Samples Placed in 10.0% Sulfuric Acid Solution

bacteria can live beneath the barrier. This cannot be evaluated by this type of experiment.

Lastly, it appears that calcium-silicate-hydrates are soluble in concentrated sulfuric acid while the hydrates of HAC are not. Thus it appears the best choice of a material where highly concentrated sulfuric acid is present is the HAC. The least suitable material is PCMS and should not be considered as an alternative to regular Portland cement mortar.

Acknowledgment

The authors thank the National Science Foundation and APM/Permaform, Johnston, Iowa, for support of this research through the NSF Small Business Innovative Research program.

References

Durning, T. A. and M. C. Hicks, *Using Microsilica to Increase Concrete's Resistance to Aggressive Chemicals*, 1991

Kienow, Kenneth K. and Karl E. Kienow, *Corrosion Below: Sewer Structures*, 1991

Mindess, Sidney and J. Francis Young, Concrete, 1981

Neville, A. M., Properties of Concrete, 1963

Neville, A. M., High Alumina Cement Concrete, 1975

Sands, Dumas, Marcdargent, Pugliese and Cabiron, *Tests for Biogenic Sulfuric Acid Corrosion in a Simulation Chamber Confirms the On-site Performance of Calcium Aluminate Based Concrete in sewage Application*, ASCE Material Engineer Conference, 1994

Thistletwayte, D. K. B., The Control of Sulphides in Sewerage Systems, 1972

Appendix I

Chemistry Notation

Oxide	Shorthand Notation
CaO	C
SiO_2	S
Al_2O_3	A
Fe_2O_3	F
H_2O	H

Table A1.1. Shorthand Notation for Typical Oxides

The Workshop on Pipeline Research Needs- Where Do We Go From Here?

John G. Bomba, F. ASCE

Abstract

Early in 1996 (March 28-29), a group of about 50 pipeliners got together for two and a half days, at the Landsdowne Resort Hotel just out of Leesburg, Virginia to discuss pipelines -- water, sewer, oil, gas, and freight; what each saw as problems with existing pipelines, installation techniques, materials, design methods, code restrictions/requirements, and potential solutions to these problems. Where research of one kind or another was thought to be necessary, the extent of the required research, potential researchers, and potential funding sources were identified.

The identified research needs were collated and, to the extent required for consistent format, were edited, and printed as an ASCE Publication, "Proceedings of the Workshop on Pipeline Research Needs".

The workshop was hosted by ASCE's Pipeline Division and co-sponsored by the ASCE and seven US and Canadian organizations with an interest in pipelines. The workshop was the brainchild of the ASCE Pipeline Division's (PLD) Pipeline Research Committee (PRC), and was put together principally by two super ASCE staff, Ahmed Habibian and Gretchen Hyde, and Bill Quinn, from PLD's ADCOM, Henry Liu, PRC, and John Bomba, PRC.

Introduction

Buried pipelines are the most environmentally friendly and the safest means of freight transportation. Nations are relying on pipelines to deliver freight- in either liquid, gas, or solid form. The aging of major pipelines which were mostly installed following the economic growth of post World War II period has created a need to develop methods of assessing the remaining useful life, and of means of extending the life of these facilities. New areas into which pipelines are being built are constantly expanding. Several organizations are involved in sponsoring or conducting pipeline-related research activities.The workshop provided a forum to bring research groups together to identify and prioritize major pipeline research needs.

[1] Chief Engineer, Kvaerner RJ Brown, 1253 North Post Oak Road, Houston, Texas 77055

The workshop grew out of a series of discussions at the Pipeline Research Committee's (PRC) 1993, 1994, and 1995 annual meetings. The history of "research" activities in the Pipeline Division was sparse--a 1965 publication "Civil Engineering Research Needs 1966 - 1975" and a conference held at the University of Massachusetts, in 1988, with no significant follow up other than a "proceedings" document. It was apparent that unless a concerted effort was made to follow through with some of the recommendations, another conference or workshop would be a waste of time. It was decided to <u>push</u> to get the research done, <u>and</u> to get the results of the research published in the trade journals and magazines, as well as in ASCE publications, where the people who need to know get their information.

PRC developed the idea of an intra-industry workshop on pipeline research needs, to be attended by 50 invited industry leaders/experts in both onshore and offshore water, sewer, oil, gas, and water pipelines.

The defined purpose of the workshop was: "*TO IDENTIFY AND PRIORITIZE PIPELINE-RELATED RESEARCH NEEDS. THE WORKSHOP WILL COVER ALL PIPELINE APPLICATIONS, INCLUDING CRUDE OIL, NATURAL GAS, WATER AND SEWER, PRODUCTS, PNEUMATIC, AND CAPSULE*".

Seven specific research areas common to most of the industry were identified. Contractors would be represented. Regulatory agencies would be included (Department of Transportation, Minerals Management Service, and the National Energy Board of Canada). Design/Consulting Engineers would participate. Trade agencies would participate. Cross training was encouraged.

PRC invited civil, mechanical, and electrical engineers. It turned out that one of our "civil" engineers actually had a degree in Nuclear Engineering--and he was a pipeline contractor! An Honorarium to each participant was proposed. Sponsors would be solicited. A good site would be selected. A workshop team would be formed. An editor for the Proceedings was chosen.

<u>Why Another Workshop?</u>

An earlier workshop on buried pipeline research needs was held in 1987 at the University of Massachusetts, and was sponsored by National Science Foundation. Since then, significant advances in the field of pipeline materials, engineering, construction techniques, and integrity inspection (including leak detection) methods have occured.

Pipelines are routinely being installed in deeper and deeper water. New problems have arisen onshore. There appeared to be a strong need to reexamine the recommendations of the 1987 workshop, particularly because of the strong onshore emphasis of the 1987 workshop, a review of offshore pipelining aspects was felt to be necessary.

<u>The Workshop</u>

Fifty experts from the industry, government, and academia were invited to participate. An honorarium was offered to each as a way of encouraging participation by knowledgeable pipeliners.

After arriving at Landsdowne and, after attending the mandatory plenary session, the participants spent some time assessing the state-of-the-art in the seven general areas into which the workshop was divided and each group developed a list of needed research topics within its own area. This workshop also served as a forum for the water and sewer industry engineers to communicate with the oil and gas industry engineers.

The pipeline industry has traditionally been segregated into oil and natural gas pipelines and water and sewer pipelines. In order to overcome this "segregation", the workshop was organized according to function - design, construction, safety, operations, and automatic control - rather than by transported media. This integrated approach succeeded in bringing together pipeliners from different industries together to share information and exchange ideas. It also greatly enhanced the dialogue among various pipeline groups that normally do not communicate with each other (even within some of ASCE's own committees).

The program for the workshop was organized under the following seven general areas described below:

(1) **Pipeline Safety and Protection**-- includes pigging for safety and integrity, pipeline leak detection and monitoring, pipeline spills, effects of earthquakes, hurricanes, and floods on pipeline safety, cathodic protection systems, and third-party damage prevention.

(2) **Pipeline Design**-- new design approaches for onshore and offshore pipelines (including earthquakes, flooding, etc.), possible revisions to design codes, and use of expert systems in design.

(3) **Pipeline Operations**-- any non-safety related operational issue such as use of drag-reducing additives to reduce power consumption, handling of emergencies and spills, economics of pipelines, pumping operation procedures, and maintenance of aging pipeline systems. Includes design for and use of pigs for pipeline cleaning, sizing, and entry ports, use of various instruments to detect pigs, leaks and corrosion. How to cope with hydrate formation and paraffin build-up problems, particularly in deepwater oil and gas flowlines.

(4) **Fluid Mechanics/Hydraulics of Pipelines**-- dynamic analysis of pipeline transients, water hammer and column separation, cavitation in pumps and valves, rheology of slurry, hydraulics of capsule flow, and multi-phase flow problems.

(5) **Construction and Rehabilitation of Pipelines**-- new construction techniques for pipelines, and construction under extreme conditions - cold regions, mountainous terrains, swamps and wetlands, and offshore; in-situ lining, replacement of corroded pipe segments, retrofitting existing pipelines to comply with earthquake design, renovating decommissioned oil pipelines and natural gas pipelines for other purposes.

(6) **Automatic Control and Instrumentation** -- PLC's, computers and other new

technologies used for automatic control of pipelines, control strategies, and communication systems; includes measurement of flow, pressure, and temperature of the transported fluids.

(7) **Freight Pipelines**-- slurry pipelines, pneumatic pipelines, capsule pipelines, and tube transportation systems.

Identified Research Needs

The seven groups met, reviewed the list of submitted research ideas, discussed the merits of each, and settled on two, three, four, or five research needs. These were presented at a wrap-up session and turned over to the Workshop team. The editor then went to work, reformatting, editing, picking out representative photographs to include, and encouraging people to send in their final versions.

A listing of the research needs which were written up and justified at the workshop, follows below. Selected items are described in more detail after the listing.

Working Group Reports

A. **Pipeline Safety and Protection**
Facilitator: Tom Steinbauer, The Gas Research Institute
Group A reviewed the topics proposed by the workshop participants deemed to be relevant to pipeline safety and concluded that the two most pressing needs were the development of controls for systems to monitor and detect natural gas pipeline breaks and of a real time tool for monitoring the integrity of a pipeline system.
1.0 Controls for Full Scale Natural Gas Line Breaks
2.0 Real-Time Pipe Integrity Monitoring Research Needs

B. **Pipeline Design**
Facilitator: Raymond Sterling, Trenchless Technology Center
The members of Group B evaluated all of the pipeline design related proposals submitted by workshop participants. After considerable discussion, three major pipeline design research needs were identified. These are listed below.
1.0 Design Guidelines for Trenchless Technologies in New Construction
2.0 Limit State/Reliability-Based Pipeline Design
3.0 Rapid Joining:

C. **Pipeline Operations**
Facilitator: Thomas Hoelscher, Technical Manager Field Operations, Division III, Transco
The Pipeline Operations Group members represented the oil, gas, water and sewer sectors of the pipeline industry, and included engineers from construction contractors, design engineering firms and operating companies. The 30 plus pre-submitted ideas, as well as those developed during the workshop, were carefully reviewed. The ideas worthy of further consideration and possible research are given below.
1.0 Assess the Need for Deep Water Pipeline Repairs

2.0 Prevent Pipeline Blockage Due to Hydrocarbon Solidification.
3.0 Develop a Cost Effective and Non-intrusive Pipeline Inspection Method

D. Fluid Mechanics/Hydraulics of Pipelines
Facilitator: E. B. Wylie, Department of Civil and Environmental Engineering, University of Michigan

The Fluid Mechanics/Hydraulics Group, which included representatives of academia, research organizations, consulting engineers, and gas and electric utilities, recommends the following areas for further investigation and or research. These areas encompass water quality, long term degradation of interior pipe wall, and transient phenomena in both the water and the oil and gas industry. Both onshore and offshore pipelines are included.

1.0 Water Quality Issues in Water Distribution Systems
2.0 Unsteady Flow in Pipeline Systems
3.0 Calibration of Prediction Techniques For Slug Flow and Slug Length

E. Construction and Rehabilitation of Pipelines
Facilitator: B J Schrock, President, JSC International Engineering

Group "E", which developed the research needs statements in the area of pipeline repair and rehabilitation, included consulting engineers, user/agency representatives, pipeline contractors, a plastic pipe materials supplier, and the ASCE Pipeline Division's Technical Liaison representative. The Group represented an extensive background in the rapidly growing area of pipeline repair and rehabilitation.

1.0 Degradation Impact on the Structural Adequacy of Pipelines and Degradation Control
2.0 Long Term Performance of Plastic Pipe
3.0 Determining the Integrity of Cast Iron Pipe
4.0 Structural Rehabilitation Techniques for Existing Utility Pipelines

F. Automatic Control and Instrumentation
Facilitator: William A. Hunt, MSE-HKM Engineering

The Automatic Controls and Instrumentation Group was made up of recognized experts in the application of computer and other new technological advances for automatic control of pipelines, control strategies, and communication systems. From the twenty or so research areas submitted, the group selected the following for further study.

1.0 Instrumentation ***(For automatic control of pipeline systems)***
2.0 Training Simulators ***(For automatic control of pipeline systems)***
3.0 Application of Expert Systems ***(For pipeline systems)***

Group G: Freight Pipelines
Facilitator: Henry Liu, Director, Capsule Pipeline Research Center

Research and Development (R&D) necessary to bring about increased use of freight pipelines should be encouraged by government agencies, especially the U.S. Department of Transportation, the Federal Highway Administration, the U.S. Department of Energy, the U.S. Department of Energy, the U.S.

Department of Interior, and the U.S. Environmental Protection Agency. Five types of freight pipelines deserve attention in research:

1.0 Tube Transport
2.0 Coal Log Pipelines
3.0 Hydraulic Capsule Pipelines
4.0 Pneumatic Pipelines
5.0 Slurry Transport.

Additional Detail

Group D, Fluid Mechanics/Hydraulics of Pipelines:
Calibration of Prediction Techniques for Slug Flow and Slug Length
Justification
Accuracy of current techniques for predicting hydrodynamic slugs and the length of these slugs is not good--somewhere between plus or minus 100 percent for one commercially available software package and plus or minus 1000 percent for another.
Focus of Research
Companies with multiphase flow pipelines in the Gulf of Mexico have developed large operating data bases for their systems over a number of years. What is needed is collection, sorting, and collation of this data to make it available to the industry for calibration of existing prediction algorithms.
Funding Sources
AGA, API, Operating Companies.

Group E, Construction and Rehabilitation of Pipelines:
Structural Rehabilitation Techniques for Existing Utility Pipelines
Justification
In the midst of our aging infrastructure, there are about 55,000 miles of gas lines and some 850,000 miles of water mains in various stages of deterioration. The replacement cost of these pipelines is astronomical--in the billions of dollars. And some of these are in congested areas which are now virtually inaccessible. Over $250 billion is estimated to be needed over the next two decades just to rehabilitate the aging water and sewer lines in the United States alone! Structural rehabilitation, using trenchless technology, can extend the useful life of buried pipelines and minimize failures. Cost effective, safe rehabilitation techniques are urgently needed.
Focus of Research
1) An evaluation of existing rehabilitation methods in the oil, gas, and water industry;
2) Adaptation of some currently in-use rehabilitation methods from the oil and gas industry to water lines;
3) The development of new rehabilitation methods for these pipelines,
Sources of Funding
There are several governmental agencies and industry groups such as AWWA and API/GRI which are interested in developing these techniques.

Finding Research Money

Several Agencies have already expressed interest in putting up some money to get the critical needs started. The PRC has made an "official" contact with one agency and several unofficial contacts. Some of the unofficial contacts have indicated that they were ready to provide matching funds, similar to a joint industry study, such as the current DeepStar study sponsored by Texaco, and in which, most of the offshore producers take part.

Publishing the Results of the Research

As was mentioned earlier, one of the key elements evident during the planning of the workshop, was the need for getting the results of the research out to the working engineer. Articles are being planned for Civil Engineering, the Oil and Gas Journal, AWWA Journal, Water and Waste Digest, and similar publications. An update on the "follow-through" activities will be prepared for the next Pipeline Division Specialty Conference scheduled for Austin, Texas, in 1998.

Where Do We Go From Here?

The Proceedings have been sent to ASCE for printing. "Off the record" contacts have been made with several agencies who have expressed interest in co-sponsoring some of the research relevant to their particular sphere of influence.

The Pipeline Research Committee plans to contact the Agencies/Organizations identified in the discussions of the research needs in a more formal manner. The "Off the Record" contacts will be made official, hopefully getting together enough sponsorship to get the research started.

Need more Details?

Further details on the workshop, plus the findings of the workshop are documented in the "Proceedings of the Workshop on Pipeline Research Needs" published as an ASCE publication.

Appendix A

PARTICIPATING SPONSORS

The following organizations provided funding for the workshop:

American Society of Civil Engineers (ASCE)
Office of Pipeline Safety, U.S. Department of Transportation (OPS)
Minerals Management Service/ Department of the Interior (MMS)
National Energy Board of Canada (NEB)
National Science Foundation (NSF)
Gas Research Institute (GRI)
American Water Works Association Research Foundation (AWWARF)
North American Society for Trenchless Technology (NASTT)

DERIVATION OF LONG TERM DESIGN PARAMETERS FOR PVC PIPE SUBJECT TO EXTERNAL LOADS

Ronald R. Bishop*

Abstract

Applying long term stress design properties to the design of buried plastic pipe subject to external loads due to soil or hydrostatic overburden considerations has long been difficult due to the seeming absence of test data which results from the proper stress state. In this paper, data taken from Hydrostatic Design Basis (HDB) protocol, initial flattening tests, environmental strain corrosion tests and results from pipe that has been in sanitary sewer usage for 25 years have been combined to look at minimum long term safety factors. Test data has been plotted on a Log (stress/strain) vs. Log (time) coordinates. Results of measurements taken from pipe that has been in service as buried sanitary sewer lines is imposed on the regression line. Two states of stress (tensile and bending) are combined with two types of loading (creep and stress relaxation) and are discussed relative to the available test data and the applicability of those results to the design of buried pipe used for rehabilitation of pipelines. The specific data presented is taken from tests on various PVC compounds but the overall rationale is applicable to other polymers with visco-elastic properties. Observations are made regarding life expectancy of past and future pipelines installed per recommended construction guidelines.

Misleading Desgn Premise

During the late 1960's and early 1970's some questions were raised concerning long term design considerations for buried pipe which was subject to external earth loads but not subject to internal pressure. Some individuals representing competitive products circulated a design rationale that applied long term creep data from tensile tests to predict life expectancy for pipe subject to bending.

*Vice President, Lamson Vylon Pipe, 25701 Science Park Drive, Beachwood, OH 44122, (216) 766-6644, Fax (216) 766-6577

The design rationale would have predicted failure of pipe that were installed with more than 0.5% initial strain. (Moser, 1981)

Misconceptions about strength and time.

In order to understand why these pipes endure stress and strain beyond what one would expect based on the rationale above we need to review some basic concepts. First, we may look at two different states of stress. Second, we recognize two different types of loading. This creates several possible combinations as shown in Table 1. For pipe in pressure applications, indeed the HDB creation process is based on tests of pipe under a constant tensile stress and creep loading condition which is depicted in Figure 1. For a pipe that is not subject to internal pressure the state of stress is primarily flexure or bending of the pipe's circumferential ring. There is a component of hoop compression or thrust in the pipe wall also. We need to understand the relative magnitude of these two components of stress.

TABLE 1: PRIMARY TYPES OF LOAD AND STATES OF STRESS FOR BURIED PIPE

	Buried Non-pressure	Internal Pressure
Primary Type of loading:	Stress Relaxation	Creep
Primary State of Stress:	Flexure or Bending	Tensile

A pictorial description of the stress and strain for these cases is given in Figure 1. This figure displays a non-linear relationship between stress and strain for bending. However, the simplified equations in use today assume a constant modulus of elasticity. We may calculate values of flexural and hoop compression stress for an assumed height of cover ranging from 5 feet to 80 feet for pipe stiffness ranging from 25 to 115 psi and soil modulus (E', E prime) ranging from 200 psi to 2000 psi. Calculations based on the Spangler or Iowa Deflection Formula as modified by Dr. R. K. Watkins commonly called the "Modified Iowa Formula". The results for such a wide range of burial conditions and pipe stiffness illustrate that the bending stress exceeds hoop stress for all conditions by a factor of about 2:1 to 14:1. Strains were calculated based on OD/t values of 26, 35 and 41 corresponding to stiffness values of 115, 46 and 25 psi. We conclude that flexural or bending stress dominates. In average cases hoop compression can be neglected. The exception to this would be very high E' values combined with deep burial depths.

Buried non-pressure pipes are primarily subject to flexural stress which is in a loading state of "stress relaxation" which causes a reduction in stress

FIGURE 1: STRESS FOR PRESSURE vs. NON-PRESSURE

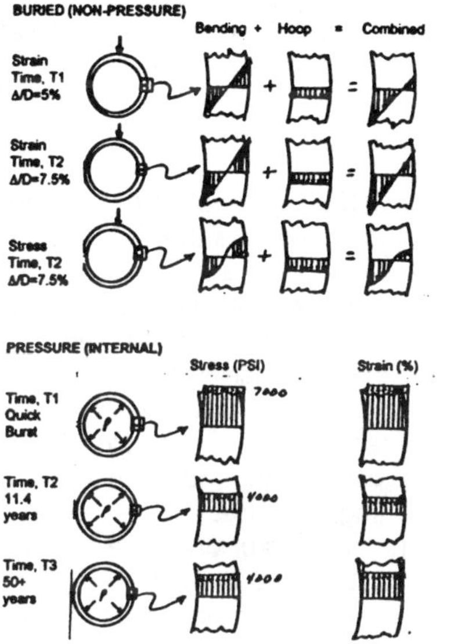

FIGURE 2: LONG TERM DATA (PVC) - (Linear log log, initial stress vs. time)

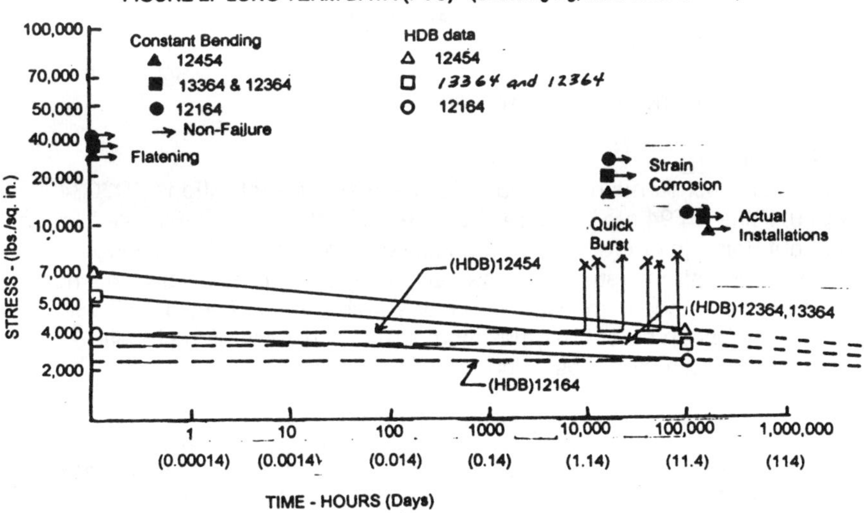

with time. This means that in the absence of some chemically induced reaction the most critical stress (highest stress) occurs at the time the pipe reaches equilibrium with the surrounding soil and deflection stabilizes. This time frame varies depending upon the soil and ground water conditions. (Bishop, 1979; Spangler, 1970 and Howard, 1975).

Available Data

HDB

A wealth of data is available from HDB testing of PVC, PE and fiberglass composite pipe that has been conducted since the 1960's. ASTM D1598 was first published in 1958. Many plastic pipe materials are HDB qualified under ASTM standards as was briefly described earlier. The process of establishing the HDB for any plastic pipe results from pipe subject to internal pressure creating a tensile stress across the pipe wall section as depicted in Figure 1. The internal pressure provides a creep strain (rising strain over time under constant load or stress). The plotted HDB test results of a typical PVC pressure pipe compound are given in Figure 2. This is the result of statistically fitting the (stress - time) failure points of many pipes under test at different stresses or pressure levels. A common misunderstanding is to look at the statistical data fit line (regression analysis) and conclude that the strength is reducing with time. The line does not represent the strength of a single pipe but rather the life of many pipes under known and differing stresses. This has been shown conclusively by R. T. Hucks (Hucks, 1973 and Hucks, 1983) who raised the pressure on pipe to the point of failure just before the predicted life was to expire. He repeated this over a ten year period. The results of these burst tests are plotted on the HDB regression line of Figure 2. The pipe's expected ultimate strength is actually increased after time under stress rather than reduced. This is exactly what is expected by plastic materials which undergo molecular orientation.

Flattening Test

ASTM product standards for buried non-pressure applications (D3034, F679, F789, F794 etc.) require circumferential flattening to 60% deflection without damage or cracking as a quality control test. This test imposes a high flexural strain - stress in pipe samples that is approximately 8 times those experienced in the typical designs which limit deflection to 7.5%. This imposes an outer fiber maximum stress in excess of the tensile strength of the material as calculated below.

Strain (bending) = 4 (1/35)(.6) = 0.0686 or 6.86%
Stress (bending) = 400,000 (0.0686) = 27,428 psi for cell class 12454
Stress (bending) = 500,000 (0.0686) = 34,285 psi for cell class 12364 and 13364
Stress (bending) = 600,000 (0.0686) = 41,428 psi for cell class 12164

Calculations are for a pipe with an OD/t of 35. A summary of these calculations is given in Table 2.

TABLE 2: STRESS/STRAIN IN CIRCUMFERENTIALLY DEFLECTED PIPES.

D1784	SDR	Pipe	7.5% Design Deflection		60% Quality Deflection	
Cell Class	(OD/t)	Stiffness	Strain (%)	Stress (psi)	Strain (%)	Stress psi)
12454	26	115	1.154	4615	9.231	36,923
	35	46	0.857	3429	6.857	27,429
	41	25	0.732	2927	5.854	23,415
13364 or	26	115	1.154	5770	9.321	46,605
12364	35	46	0.857	4285	6.857	34,285
	41	25	0.732	3660	5.854	29,270
12164	26	115	1.154	6924	9.321	55,926
	35	46	0.857	5142	6.857	41,142
	41	25	0.732	4392	5.854	35,124

These maximum stress levels calculated by assuming a linear modulus of elasticity exceed the pipe material tensile strength by significant margins. This is possible because of the high ultimate strain capacity (30% to 100% strain) of these materials. Non-failure points for these quality control levels are added to Figure 2.

Actual Installations

ASTM D3034 was first approved in late 1972 and included cell class 12454. In early 1973 cell class 13364 was added and in 1978 cell class 12364 was added. Cell class 12164 was included in F789 in 1982. Selected properties of these materials are compared in Table 3. In each case the PVC compounds were in use for active sewer pipe applications for one or more years prior to their inclusion in the various ASTM PVC pipe product standards.

Chemical Resistance

ASTM D1784 defines requirements for PVC compound cell class properties for impact strength, tensile strength, modulus of elasticity, heat distortion temperature and chemical resistance. The chemical resistance test exposes unstressed samples to sulfuric acid at 93% and 80% concentration and to ASTM Oil No. 3 for 14 days. Limits to increases and decreases in sample weight are given for each reagent and compound classification. For sulfuric acid exposure, limits on changes to flexural yield strength are also specified. This is a particularly suitable consideration for use in sanitary sewers where the formation of hydrogen sulfide gas gives the opportunity for sulfuric acid generation and is the cause of much pipeline deterioration. The testing results in the classification of three different levels of chemical resistance which is

TABLE 3: COMPARISON OF STANDARD PVC PIPE COMPOUNDS

Product Standard	D1784 Cell Class	First Included	Tensile (psi)	Modulus of Elasticity	Years In use
D3034	12454	1972	7000	400,000	24
	13364	1973	6000	500,000	23
	12364	1978	6000	500,000	18
D679	12454	1980	7000	400,000	16
	12364	1980	6000	500,000	16
F789	12454	1982	7000	400,000	14
	12364	1982	6000	500,000	14
	12164	1982	4000	600,000	14

designated by a letter suffix (A, B or C) added to the cell class number sequence. All three designations have been suitable for the low acid concentrations typically found in sanitary sewer lines. Some significant contributions have been made (Hawkins, 1992) through scanning electron microscopy of PVC pipe exposed to sulfuric acid. Limited surface chemical attack of compound ingredients has been seen but no chemical attack is observed at other than the exposed surface and even then not to the PVC resin. These observations verify the usefulness of the chemical exposure weight change requirements in ASTM D1784.

Strain Corrosion

The combination of stress or strain and chemical environment is another consideration for long term exposure of pipe. ASTM D3681 was created to test fiberglass pipe under the fixed deflection or strain conditions that buried non-pressure pipe are subject to while exposed simultaneously to the chemical environment in question. As with HDB testing the resulting failure points are analyzed statistically. By fitting the data with a log log linear regression of initial strain level versus time the results are extrapolated to the desired design life. The author has applied this test to a number of plastic pipe materials and PVC pipe compounds over the course of the past 24 years. The major PVC compound cell class materials 12454A and B, 13364A and B, 12364A and B and 12164A have each been tested with reagents of sulfuric acid (10% and 5%), sodium hydroxide (10%), ASTM Oil No. 3, tap water, air, Igepal and vegetable oil. No failure points have been recorded in these reagents for strains up to 4.57% corresponding to 40% deflection in SDR 35 pipe. Attempts to induce failures at higher strains in specially made thicker wall (SDR 12) pipe have also led to non-failure points for test times of two years. Even higher strains have been induced in tensile bar samples and then exposed to environments of sodium hydroxide (10%), ASTM Oil No. 3, sulfuric acid (10%), tap water and air. Again, no failures have occurred in these tests now extending to 18 years in some cases. (Moser,1979 and Moser, 1992)

Predicting Life

Non-failure points for data obtained from constant ring bending strain applications have been added to Figure 2 to compare with HDB data. The strain level has been converted to stress based on the minimum modulus of elasticity of PVC compounds in typical sanitary sewer use. Each point falls higher than the log log regression of stress versus time. This is not surprising because one expects flexural stress and the accompanying stress relaxation state to be much less severe than the constant stress and creep conditions that go with HDB determination. What perhaps is surprising is the degree of severity difference. Strains or stresses many times those that would cause short term failure even without chemical exposure in HDB tests, do not cause failures at all for the same PVC material compounds subject to constant flexural strain even when exposed to chemical environments. If there is no driving force (constant stress with unrestrained strain) even mechanically initiated cracks do not propagate. (Moser, 1981 and Moser, 1990) Compared to the strain and stress levels allowed under design conditions generally used in the United States (5% initial deflection and 7.5% long term deflection) one can expect very long useful life expectancy.

Applications to Buried Pipeline Design

The three major design considerations for flexible plastic pipe are 1) ring deflection control, 2) ring compression strength and 3) buckling resistance. For PVC pipe with pipe stiffness values of 46 psi or greater; ring deflection is usually the controlling parameter. For other plastic materials, any of the three may control the design. The much higher specific strength and modulus of fiberglass composite materials allow a lower wall thickness. Ring compression strength is a direct function of both the wall cross-section area and material strength. PE on the other hand has a lower tensile strength and modulus of elasticity than PVC. This requires a higher wall moment of inertia to get the same pipe stiffness. This can be accomplished for all materials by creating profiles that increase the geometric moment of inertia. Maximum strains are a function of the total wall thickness and deeper sections impose proportionally higher strains. Pipe with the same pipe stiffness deflects the same in an identical installation. When constant strain data is available it can be imposed over HDB data to determine how critical those strain levels are. When the constant strain data falls above the HDB regression line one can have confidence in using life predictions based on the combination of data available. Looking at Figure 2 we can pencil in a line below non-failure points for flatening, strain corrosion and actual installations. This is a kind of upper safety limit for pipe subject primarily to circumferential bending in the presence of sanitary sewer flow.

For some rehabilitation and other applications where buckling due to external hydrostatic loads is the controlling design parameter, it is the hoop compression stress that governs the long term pipe material performance against cracking. But the real concern is buckling stability. The several equations for buckling in use today imply the need to use a long term modulus of elasticity for buckling calculations. This is only true for cases where there is a continuing driving force or load over the life time. The pipe's ability to respond to momentary external loads is based on it's stress - strain response to any new load application (Bishop, 1990 and Janson, 1990). For these situations the published typical values of modulus of elasticity may be used in design.

References

Bishop, R. R., "The Structural Performance of PVC Pipe Subjected to External Soil Pressure", Unpublished Masters Thesis, Utah State University, 1973.

Bishop, R. R.,"Time Dependent Performance of PVC Pipe", Proceeding of the ASCE International Conference on Underground Plastic Pipe, 1981.

Bishop, R. R., "Design and Performance of Buried Fiberglass Pipes - A New Perspective", ASCE - NACE Conference on Pipeline Materials and Design, 1984.

Bishop, R. R. ,"Retention of Pipe Stiffness for PVC Pipe samples exposed to Various Environments and Constant Strain", ASTM STP 1093, Buried Plastic Pipe Technology, 1990.

Hawkins, T. W . and Mass, T. R.,"The Effects of Sulfuric Acid on Calcium Carbonate Filled PVC Sewer Pipe", ASTM STP 1222, Buried Plastic Pipe Technology: 2nd Volume, 1994.

Janson, L.-E., "The Relative Strain as a Design Criterion for Buried uPVCGravity Sewer Pipes", Proceedings of the ASCE International Conference on Advances in Underground Pipeline Engineering, 1985.

Janson, L.-E.,"Short Term vs Long Term Pipe Ring Stiffness in the Design of Buried Plastic Sewer Pipes", Proceedings of the ASCE International Conference on Pipeline Design and Installation. 1990.

Moser, A. P., "Strain as a Design Basis for PVC Pipes ?", Proceeding of the International Conference on Underground Plastic Pipe, 1981.

Moser, A. P, Shupe O.K. and Bishop R. R."Is PVC Strain Limited After All These Years?", ASTM 1093, Buried Plastic Pipe Technology, 1990.

Increasing the Reliability of Concrete Pressure Pipe

David H. Marshall, P.E. [1], Member ASCE
Will Worthington, P.E. [2], Member ASCE

Abstract

The Tarrant Regional Water District conveys water to the Ft Worth area through two pipelines constructed of prestressed concrete cylinder pipe (PCCP) installed in 1972 and 1989. Delivery through these lines has been interrupted on eleven occasions by corrosion-induced ruptures. The District has assessed the pipeline condition using both cell-to-cell, and internal visual and sounding methods. More than thirty miles of pipeline has been inspected internally, with some success.

In this past year the district also explored new technologies to assess the condition of these vital pipelines. Acoustic emission detection equipment was installed in the pipeline on an experimental basis. This equipment consists of underwater microphones, or hydrophones, installed in the pipeline, connected to data recorders and signal processing equipment. Data from the acoustic emission testing allows the District to excavate areas of distress to confirm the physical condition of the pipe, and to take appropriate measures to preclude failure. The result is an increase in reliability and useful life of the pipeline.

This paper discusses the current inspection methods, pipe repair and rehabilitation techniques, and acoustic emission inspection.

[1] David H. Marshall, Engineering Services Manager, Tarrant Regional Water District, 1022 North Calhoun, Fort Worth, TX 76106; phone 817 626 2888; fax 817 625 9112

[2] Will Worthington, President, Pipeline Technologies, Inc. 1435 North Hayden Road, Scottsdale, AZ 85257-3773; phone 602 451 3500; fax 602 451 3165

Introduction

The Tarrant Regional Water District is a raw water supplier to eleven counties in North Central Texas. The District serves 29 municipal customers that supply a population of just over 1.2 million people in addition to industrial and irrigation customers. The District's largest customer base is Tarrant County, which includes Fort Worth and the surrounding metropolitan areas.

The District has two sources of water, the West Fork System and the East Texas System. Dependable yields for these two systems are 79,000 and 385,000 acre feet per year, respectively. Flow is delivered directly to three water plants and to Lake Arlington.

In water year 1996, the District delivered over 200 million gallons per day (225,000 acre feet per year) to the municipal customers. About two thirds of the water was delivered via the pipeline system. The water demand increases on average about three percent per year.

Pipeline Description

The pipeline system began with construction of the Cedar Creek Pipeline, completed in 1972. The line is a 72 inch, embedded cylinder prestressed concrete pipe. The lake pump station draws from the reservoir at an elevation of 322 feet. The flow is boosted downstream by the Waxahachie pump station, and if demands warrant it, by the Ennis pump station. Once the flow leaves Waxahachie, it is delivered to two above ground reservoirs (balancing reservoirs) 24.8 miles downstream. From there, flow is gravity to Tarrant County via a 5.9 mile, 84" diameter embedded PCCP. A maximum of about 146 million gallons per day (mgd) may be delivered. Water is pumped a total of 73.8 miles through this line.

The Richland Chambers Pipeline was completed in 1989 and consists of a 90 inch diameter embedded cylinder PCCP. The lake pump station draws from the reservoir at an elevation of 315 feet. Flow can again be boosted by the downstream Waxahachie pump station. Water is eventually delivered to Tarrant County via a 5.9 mile 102 inch diameter embedded cylinder PCCP from the balancing reservoirs. Maximum flow through the line is about 148 mgd. Water is pumped a total of 78.1 miles.

The two balancing reservoirs serve to supply peak demand requirements. They also serve as emergency supplies. Winter demands average only about 100 mgd, and summertime demands average over 135 mgd. Due to the demand requirements, the pipeline system has a limited window of opportunity for maintenance. Close coordination with the customers directly on the pipeline must occur for either pipeline to go down for maintenance. Maintenance activities can occur only during the winter.

Both pipelines are embedded cylinder prestressed concrete pipe. The pipe is a composite pipe: the innermost part of the pipe wall is a 1-1/2" thick concrete core that is cast inside a steel cylinder (typically 0.06" thick). The steel cylinder acts as a waterproof barrier and is also used to restrain forces directed along the length of the pipeline. A 2" to 6" thick outer concrete core is cast outside the steel cylinder.

High strength wire is then spirally wound on the concrete core at a tension equal to 75 percent of the yield strength of the wire, with a wrap spacing as close as ½" on higher pressure class pipe. The wire wrap is then covered with a 3/4" mortar coating, which serves as the outside of the pipe.

Pipeline Failures

The 72" Cedar Creek pipeline has had eight failures to date, with all but one having been corrosion related. Corrosion occurred primarily on the prestressed wire wrap, which is the primary strength member in the design of PCCP. Once the wire wrap (1/4" in diameter) corrodes significantly in a two-foot square (approximate) area, an abrupt rupture of the pipe occurs, tearing a hole in the pipe at least four feet by two feet. The force associated with the abrupt failure creates a large crater (6-8 feet wide, 4-6 feet deep) and all water in the pipeline exits through the hole.

Initially corrosion failures of any pipeline are infrequent, but over time, the frequency increases until the pipeline is unusable. The cumulative number of breaks increases exponentially. If the pipeline is cathodically protected, additional failures are generally stopped or retarded by decades. Failures on the Cedar Creek line follow this behavior. The graph below shows the timing of the failures.

TIMING OF CORROSION FAILURES ON THE CEDAR CREEK PIPELINE

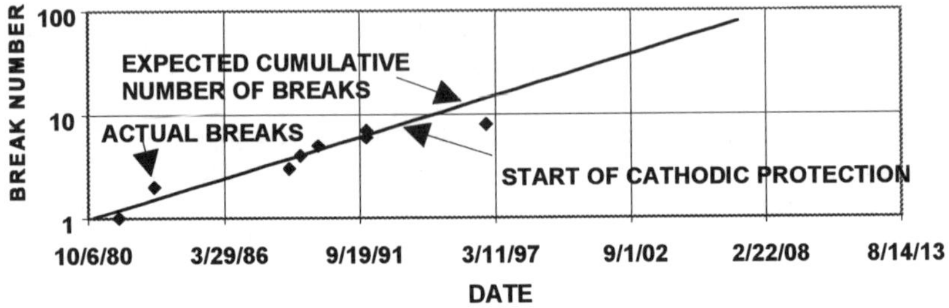

A study of the failures on the Cedar Creek Line was completed in early 1993 by Corrosion Consulting Service Corporation. They concluded that the failures were due to mechanical damage of the exterior mortar coating, permitting chlorides found naturally in the soil to migrate to the wire wrap and corrode the wire. The mechanical damage was attributed to either mishandling during installation or from cracking due to water hammer pressures. Water hammer is mitigated by backflow through the pumps with a delayed, slow-closing control valve.

The one break on the Cedar Creek Line that was not a corrosion failure was likely a hydrogen embrittlement failure. One section of the Cedar Creek Line was put under an impressed current system until 1992. At some point, voltage exceeded 1.1 volts, which tended to create hydrogen embrittlement. Metallurgy tests are now being conducted on samples from the break to test this theory.

The Richland Chambers Line has suffered eight failures since its completion, only two of which were corrosion failures. Four of the failures were due to a failing thrust restraint. One was due to operational problems, and the cause of the most recent failure is yet to be determined. Metallurgy tests on the wire of the latest failure are being conducted to determine if the wire suffered strain aging.

Repairs

Defective segments that are found either by visual inspections or by hammer testing are repaired when the line is taken out of service during periods of low demands. Once the segment is excavated, the exterior coating is examined to determine the extent of the damage. On segments that have limited corrosion damage, a repair clamp is installed over the exterior of the pipe and grouted. On segments that have severe corrosion damage, the entire segment is replaced as is done during an emergency repair.

Pipe segments that fail catastrophically usually have to be replaced in an emergency mode due to a lack of terminal storage in Tarrant County. With part of the pipeline system down, water shortages could occur in just a few days if demands are high. When a break occurs, the system is shut down until the break is located. The other line may be restarted when the break area is isolated and dewatered.

Traditional Inspection Procedures

The District has been monitoring for corrosion since 1989. Interpretation of the data was difficult since the Cedar Creek Line was not electrically continuous. During the winters from 1992 to 1994, the Cedar Creek Line was taken out of service to bond the joints, so the pipeline could be tested properly and cathodically protected.

To stop corrosion damage, the District has begun installing a sacrificial anode system, beginning at the lakes and working toward Tarrant County. Design is based on an average 100 microamps per square foot and uses zinc anodes to eliminate any potential problem with hydrogen embrittlement.

The bonded portions of the Cedar Creek Line have been monitored using a technique that measures the electrical potential difference between the pipe and soil. The graph below highlights readings taken from a short section that was

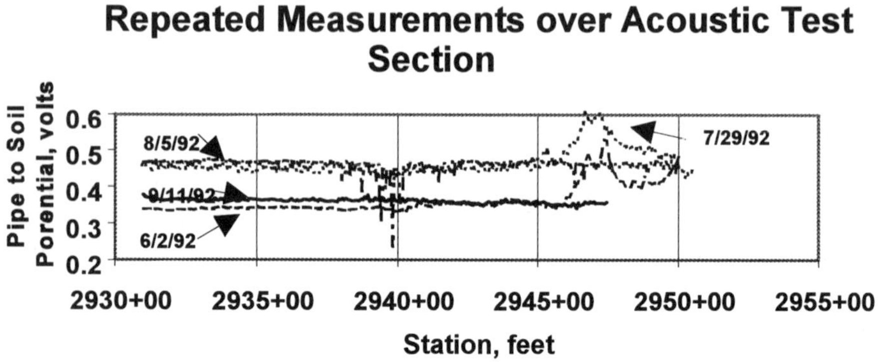

monitored on four different dates.
The differences in the readings may be due to soil moisture, which acts as an electrolyte, permitting the flow of electrons and corrosion to occur. The moisture also blocks oxygen from reaching the pipe. As soil moisture decreases, the corrosion activity increases.

Soils saturated with moisture have only about 10 parts per million (ppm) of oxygen, while soil that is not saturated may allow the atmosphere to migrate into it, and increase levels to 20,000 ppm. Measurements taken when soil moisture is not saturated tend to show higher corrosion rates. Therefore, the best time for a survey to be conducted is right after sufficient rain has fallen to provide moist, but not saturated soil. Unfortunately, this severely limits the time opportunity when surveys may be taken.

During the bonding program of the Cedar Creek line, and during all dewatered repairs, an internal inspection was performed. A team of two would enter to visually inspect the inside of the pipeline. During this phase each pipe segment was carefully examined, numbered and hammer tested.

During the hammer testing the pipe was struck on top, or one side approximately every foot. The team prefers a claw hammer, using the claws as a tuning fork. If the pipe wall is solid, the forks ring. But if the prestressed wire has failed in that area, the sound is hollow.

Acoustic Inspection Procedures

Pipe inspections by voltage potential measurement and interior visual methods have been valuable. However, they have not prevented catastrophic failures of the line with their associated costs and periodic suspension of service to customers. A more efficient and reliable method of pipeline condition assessment is clearly needed.

Ideally a pipeline inspection method can be performed without taking the pipeline out of service and would detect areas of distress early enough to permit refurbishment. One of the methods under consideration by the District is the passive acoustic emission detection technology recently adapted to concrete pressure pipelines.

This method of inspection is based on the acoustic emissions made by the prestressing wire as it releases its energy. As noted previously, prestressed concrete pipe is reinforced by spirally wrapping high strength wire around a concrete cylinder. If the pipe is in a state of distress, the prestressing wire will frequently be involved. When this occurs, the wire will break in a relatively brittle fashion, with an instant release of the tensile force of up to 8000 pounds in that strand of wire.

Much of this energy is in the form of sound energy which propagates through the pipe core and into the column of water within the pipe. The broken ends of the wire are immediately re-anchored in the protective mortar due to the Poisson effect.

If the deterioration continues, the protective mortar is further compromised and the stored energy within the prestressing wire is released in a series of discrete events. As more wires in the area of distress are involved, they too will break. The process of deterioration leading up to a corrosion-related failure takes years to run its course, and unknown to most people, it is a very noisy process.

The pipeline acoustic emission technology draws heavily from the field of anti-submarine warfare. It contains three primary components:

- Two or more hydrophones
- A signal processor
- A telemetry system permitting communication between the hydrophones and the signal processor

The hydrophones are placed in a series along the pipeline, ideally no more than 2000' apart. They may be inserted into the pipeline through existing valves or through additional wet-tapped valves installed for this purpose. Both piezoelectric and all fiber optic hydrophones have been successfully used. As few as two, and as many as twelve have been simultaneously installed on a pipeline.

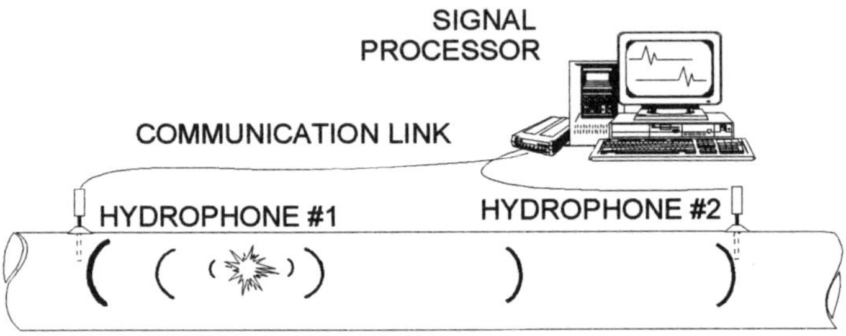

FIGURE 3 - ACOUSTIC SYSTEM SCHEMATIC

The signal processor is a Pentium™ based computer. It has the capability to continuously monitor all noises picked up by the hydrophones and to segregate the wire-related noises from the background noises and from the many other noises within the pipe. It also has the ability to correlate the noises as received on the array of hydrophones and determine its point of origin.

The telemetry system can be fiber optic or shielded metallic cables, or radio frequency transceiver systems.

The District contracted with Pipeline Technologies, Inc. of Scottsdale, Arizona, for a demonstration of acoustic inspection of a test section of the Cedar Creek Pipeline. For purposes of comparison, a test section was selected which had been monitored by potential methods previously. In November 1996 a pair of piezo-ceramic hydrophones was installed, one through a 4-inch air relief valve, and the second through a 12-inch blow off valve approximately two thousand feet apart. The District staff assisted in the installation of the hydrophones, the shielded copper telemetry system, and secured the housing for the signal processor.

The testing was performed during the period November 21 through December 9, 1996. A total of 157 hours of pipe monitoring was accumulated. During that period, a total of 63 distress-related events was detected in the test section and shown in figure 4.

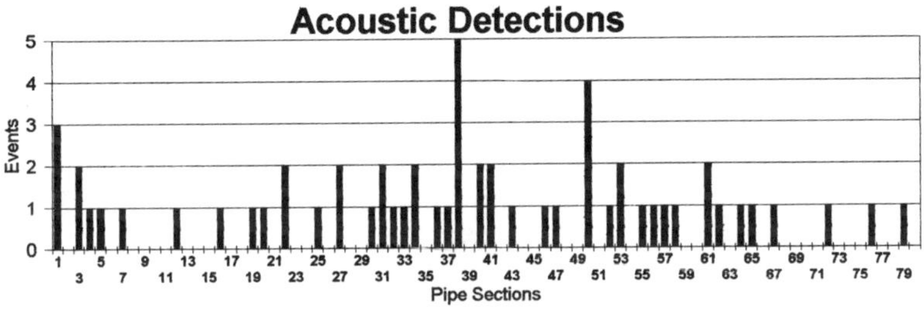

FIGURE 4 - HISTOGRAM OF RESULTS

The result of the acoustic emission testing indicates that pipe sections 31 through 41
and section 50 are in a state of active deterioration and should be excavated. Numerous acoustic events were also detected outside of the original test site. As such, Pipeline Technology recommends a through acoustical inspection of a ten mile section of the Cedar Creek pipeline centered on the original test site, followed by appropriate visual inspection and repairs. As of press date, the above sections of pipe have not been excavated to verify the results of the study.

Conclusions

To improve reliability, the District has undertaken an aggressive maintenance program that relies on continued inspection of each pipe segment. Visual inspections spot segments that are in imminent danger of failing and are repaired as soon as practical. Hammer testing isolates segments that may not exhibit any outward signs of failing, but the corrosion is severe enough to disbond the mortar coating. Both tests are less than ideal due to the requirement of dewatering the line.

Pipe-to-soil potential measurements have not proven to be a reliable method of determining corroded segments. Segments that have failed due to corrosion have shown sharp changes in the pipe-to-soil potential, but any buried metal tends to create spikes in the readings. Many times the readings will be false.

A new technology, acoustic emission technology was recently employed and appears to hold promise. It allows the pipe to be inspected without the requirement of dewatering the pipe. Also, the test can identify the specific pipe segments that require repair and maintenance.

EFFLUENT OUTFALL REPAIR OR AVOIDING A TRENCHLESS PROJECT

F. Stuart Seymour, P.E. MSCE[1]

Abstract

The City of San Diego wished to postpone a major Outfall Tunnel Project due to its high cost and the recent improvements that had taken place on the existing outfall. New treatment plants under construction were also straining budgets.

A recent construction project had shown numerous corroded sections on the steel portions of the onshore outfall pipe. The projected life without repair was estimated at best to be 3 to 5 years.

Alternatives being examined involve completely encasing portions of the steel pipe in concrete. These alternatives, construction methods and the investigation will be presented. Maintaining the structural integrity of the corroded steel pipe which must be kept in service during construction is essential to the project and the wastewater treatment plant operations.

Object

In 1994 a major tunnel outfall disposal system was designed for the City of San Diego to discharge the effluent wastewater from the new North City Water Reclamation Plant (NCWRP) and to serve as a new discharge system for the City's existing Pt. Loma Wastewater Treatment Plant (PLWWTP). This tunnel outfall would replace the first 2 miles(3.2km) of existing outfall that was repaired after a major failure in 1992. This project was designed in four packages to allow for different contracts and for portions of the project to stand alone if needed. The total cost of this system was expected to exceed 230 million dollars and was not politically popular.

Other methods of disposal were being researched and were proving feasible for the reclaimed water being generated. The offshore concrete pipe had recently been inspected and repaired and was deemed to be in good condition. The first fifty feet (Fig. 2) of this

[1]Associate Civil Engineer, Project Management Metropolitan Wastewater Department City of San Diego, San Diego, Ca

outfall located just inshore of the surf line was steel and was the only portion that had not been thoroughly inspected. This pipe consisted of a 108-inch (274.3cm) steel pipe with an 84-inch (213.4cm) wye structure. If this portion of pipe was in dependable shape there would be no reason to construct any portion of the planned tunnel outfall. To construct just the offshore portion of the tunnel outfall was estimated to cost $97,000,000. There had been past reports of pitting type corrosion in this steel portion, but its extent was not known.

The problem

A project was under way to construct the South Effluent Outfall Channel (SEOC)[1] to allow treated effluent to the bypass the vortex structure, which had been entraining air in the outfall system. This was thought to be the cause of the 1992 outfall failure. This new design would reduce the amount of air that was getting into the outfall, allow for redundancy, and inspection and repair in other parts of the disposal system. The SEOC project required a connection to the exist 108-inch steel outfall pipe. This was the first chance the City had to visibly inspect in dry conditions this portion of the outfall. A 96-inch steel coupon was removed from the steel outfall (Fig. 2) in the spring of 1994 to allow for the SEOC connection. When removed, this coupon showed considerable corrosion on the interior of the pipeline. The depth of the corrosion in the steel coupon showed approximately half of the 5/8-inch(1.59cm) plate thickness had corroded. Corrosion of this severity created a need for further investigation and a search for methods to stop further corrosion.

Investigation

This was the metropolitan areas only wastewater effluent disposal system and had to be kept in service at all times. Its average flow was 180 mgd(818mld). Flows could be reduced to one pump operations or approximately 50 mgd(227mld) for several hours in the early morning. The SEOC project would provide opportunities for further inspection and data gathering of the steel outfall system. The construction methods that would allow for data gathering during the SEOC project consisted of the following procedures:

1) Installing a 84-inch bulkhead (Fig. 1& 2) in the wye located downstream of the vortex. This would allow a dry inspection of the portion of 84-inch pipe that was to be removed and replaced in the SEOC project.

2) Removing a 4-inch coupon in the crown of the 108-inch steel pipe to allow for installation of a sampling manhole (Fig.3). This would allow a visual inspection of the 4-inch (10.2cm) coupon.

3) Installation of a 108-inch bulkhead (Fig. 3) upstream of the 84-inch wye allowing for the planned improvements in the vortex and further inspection of the steel pipe.

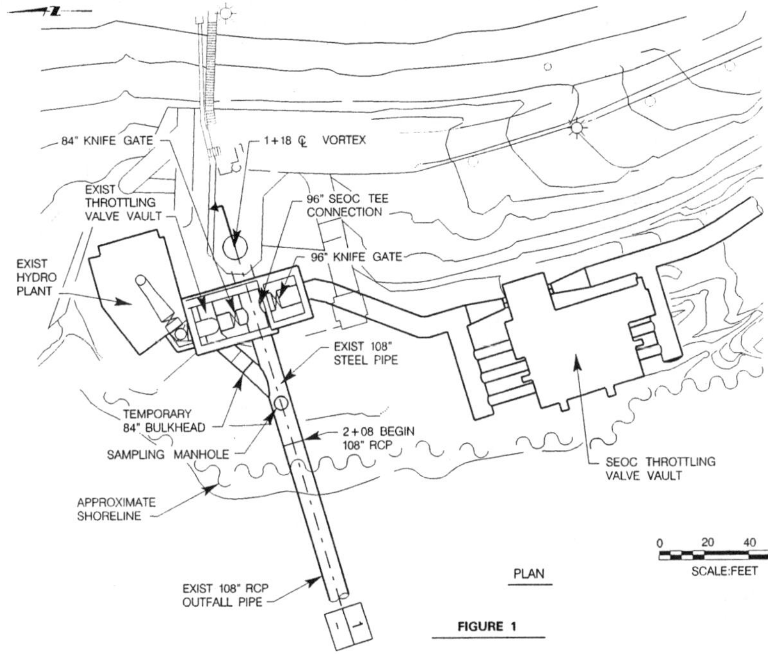

FIGURE 1

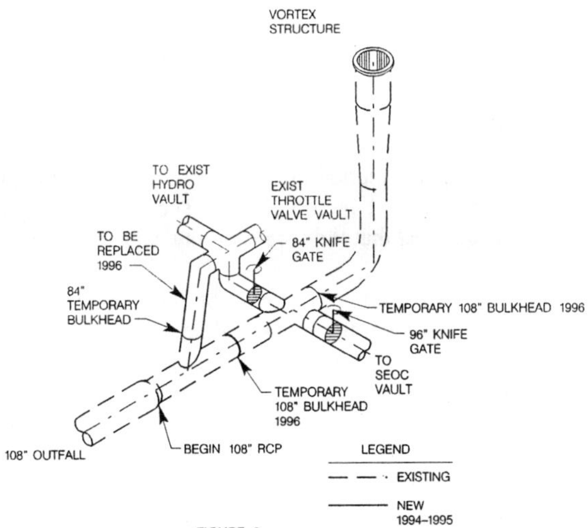

FIGURE 2

The bulkheads would be installed by divers under low flow conditions. The divers would also look for corrosion during the bulkhead installations. The dives would be done in blind conditions with the divers feeling for corroded areas and through radio communication recording the location, position and extent of the pitting. Figure 3 indicates the results of the diver identified pitting.

Shop drawings and as built drawings were also reviewed looking for possible locations of dissimilar steel connections. The divers would feel around the connection of the reinforced concrete pipe (RCP) at sta.2+08 for metal loss in the steel pipe. This was thought to be a possible location for possible dissimilar steel connections between the steel in the RCP and the steel pipe. The divers found no galvanic corrosion or pitting in this area.

Much of the above work was completed but due to construction delays the installation of the 108-inch bulkhead was delayed, however enough work was completed to allow a comprehensive report[2] to be completed. The investigation produced results that estimated the remaining service and structural life of the pipe. From this, improvements and solutions could be planned.

The preliminary investigations indicated a pitting type corrosion. The original design wall thickness of the 108-inch pipe was 0.625 inches(1.59cm). The deepest pit was 0.350 inch(0.89cm), with average pitting of 0.169 inches(0.43cm). This left an average wall thickness of 0.465 inches(1.18cm). The good news was general corrosion was not found. Structural life, meaning catastrophic failure did not appear to be a problem. This was estimated at over 30 years.

The leakage service life, however was another story. The engineers estimated life for the worst case pit in the 84-inch portion to be 1.5 to 3 yrs. This is under a working pressure of 50 psi(2.89kg/sqcm). The 108-inch pipe faired a little better with an estimated service life of 5.5 to 11.5 years.

The further inspections of the interior and exterior of the 84-inch indicated the remaining thickness in some places was 0.1-inch(0.254cm). The original design thickness was 0.5-inch(1.27cm). This indicated that the corrosion rate was faster than anticipated.

As built drawing had shown the interior lining of the thirty-year old steel pipe was coal tar. The engineers assumed that the lining had holidays since the beginning of service and that the corrosion activity had occurred and grown worse over the 30-year life of the pipe. They also assumed after discussions with vendors that a 15-year service life for a lining of this type in pipelines of this size is reasonable.

During removal of the exterior concrete coating, leaks began to appear. These leaks were patched with wooden plugs. Steel plates were then welded over them. This created

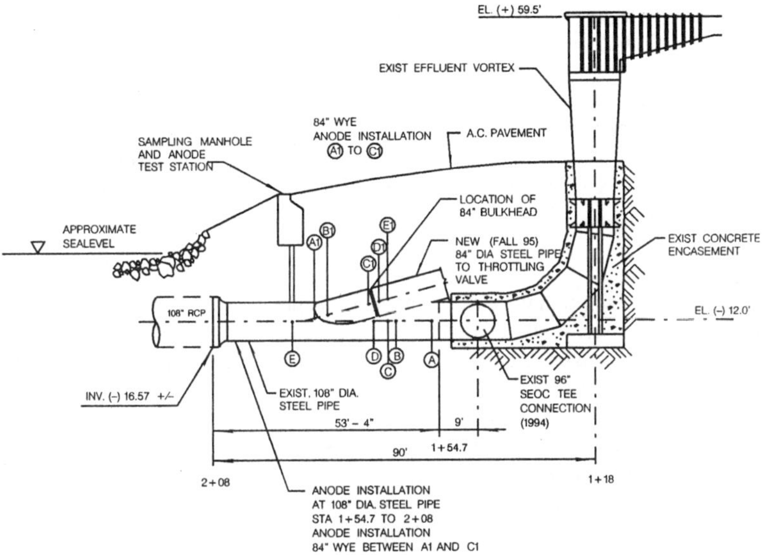

PRELIMINARY CORROSION INSPECTION LOCATION 1

NO SCALE

84" Steel Pipe Wye					
Steel Pipe Wall Thickness *					
Location	A1	B1	C1	D1	E1
Orientation					
12:00 Top	0.500	–	0.225	0.365	0.335
10:00	0.500	–	0.345	0.515	0.345
8:00	0.495	–	0.515	0.520	0.515
6:00 Bottom	0.140	–	0.500	0.525	0.515
4:00	0.490	–	0.520	0.520	0.365
2:00	0.510	–	0.519	0.515	0.520

* ORIGINAL RECOMMENDED DESIGN WALL THICKNESS OF 0.50"

108" Steel Pipe Wye					
Steel Pipe Wall Thickness *					
Location	A	B	C	D	E
Orientation					
12:00 Top	0.640	–	–	0.505	–
10:00	0.640	0.631	0.615	0.630	0.630
8:00	0.630	0.625	0.615	0.625	0.625
6:00 Bottom	0.615	0.645	0.320	0.345	0.345
4:00	0.645	0.625	0.570	0.625	0.625
2:00	0.635	0.620	0.625	0.625	0.623

* ORIGINAL RECOMMENDED DESIGN WALL THICKNESS OF 0.625"

CATHODIC PROTECTION SYSTEM
INSTALLATION DETAIL
FIGURE 3

a need for rapid intermediate repair.

This repair called for a steel plate to be welded from spring line to spring line around the inside of the bottom of the 84-inch pipe. This would be done by divers under low flow conditions and without visibility. The plates slowed the leaks but didn't stop them.

To stop the leaks completely required more innovation. Eight 1-inch(2.54cm) taps were made through the old steel pipe then couplings and 1-inch shut off valves were installed. Through these valves, a polyurethane liquid chemical grout, "SIKAHFIX-HH" was pumped. The valves were then shut and the grout allowed to harden. This chemical grout fix proved successful.

Recommendations

The several alternative solution were recommend as the project progressed. The first two solutions included cathodic protection. The first method was to use an impressed current system. The second was to use a galvanic, zinc sacrificial anode system which could be quickly installed.

The impressed current system allowed the current to be supplied and regulated by a small rectifier. The rectifier could be placed outside of the pipe and the current supplied as required. Anodes would be placed inside the pipe in a PVC tube and connected with an anode wire. They would run the full length of the steel pipe and be suspended in the flow.

The second method required placing galvanic anodes inside the pipe. This method was preferred by the City because no wires would be in the flow. It was felt that the supports in the impressed current system may collect solids on them. The galvanic anode installation went forward. Cost, approximately $40,000, were similar for each alternative.

The galvanic anode design called for 48 pound(21.8kg) zinc sacrificial anodes to be installed in the 84 and 108-inch pipe. There would be 24 anodes to protect the 108-inch and another four for the 84-inch. These would be installed during low flow conditions and the estimated life the anode was expected to be five years. If it worked, the anodes would have to be replaced every five years. They would also have to be inspected by divers twice a year.

Unfortunately this sacrificial anode installation did not prove successful. The electrical potentials never rose to levels that were anticipated. The interior coating condition was worse than anticipated. A greater current output would be required. This could be obtained through the impressed current system.

Four alternatives were examined, three closely. Two demolished the 84-inch discharge wye option. All alternatives would be designed for 50 psi including surge.

Alternative 1 - Concrete Encasement of all Steel Piping
Reinforced concrete would run from the west wall of the turbine box replacement structure to the concrete pipe transition at sta.2+08. This alternative would allow all discharge options to remain available. Internal longitudinal forces in the 84-inch pipe wall would be transferred to the encasement via tension rods that would run from a thrust ring on the new pipe that replaced a portion of the 84-inch pipe. Estimated cost $900,000.

Alternative 2 - Decommissioning of the 84-inch pipe upstream of the Wye
The 84-inch pipe would be removed upstream of the wye structure eliminating a discharge option. This would require reinstallation of the 84-inch bulkhead. The remaining piping would be encased with reinforced concrete and tensioning rods. An air vent would have to be placed at top of the remaining 84-inch stub. Additional head loss would be encountered with the alternative. Estimated cost $962,000.

Alternative 3 - Encasing the 108-inch Steel Pipe Only
This is similar to alternative 2. The 84-inch discharge option would be eliminated. A bulkhead would be placed at the wye structure where the 84-inch and 108-inch intersect. The 84-inch would then be totally removed. This bulkhead would cover the 84-inch opening into the 108-inch pipe. It would be a difficult installation for the divers but it would eliminate the requirement for a thrust restraint. The 108-inch pipe would then be encased in concrete. Estimated cost $1,119,000.

Alternative 4 - Pipe within a Pipe
This alternative called for a pipe within a pipe. A larger diameter 88-inch(223.5cm) pipe would be installed around the 84-inch pipe. The annular space would then be filled with grout. This larger pipe could then be wrapped and coated for protection. This alternative did not address the 108-inch pipe therefore it was not acceptable and no cost estimate was done.

All of the alternatives required a PVC tee lock liner with a mastic adhesive around the outside of the old pipe.

The preferred alterative was No.1 and the City has carried this though to final design. This alternative provides the most flexibility for effluent discharge and eliminates the need for bulkhead installation. Alternatives 2 and 3 required flow from the north to pass through additional bends creating more head loss. The thrust ring is recommended for transferring forces from the 84-inch steel pipe though the encasement. The reinforced concrete encasement would end at Sta.2+08, the concrete outfall pipe, this would maintain joint flexibility. This alternative would also allow for the future outfall tunnel construction, when needed, with a minimum of demolition.

An impressed current system has been designed for installation in the 108-inch steel pipe.

This will be installed prior to the rehabilitation construction. This installation calls for 12 anodes, 3-inches(7.62cm) by 84-inch(213.4cm), to be installed at the 4 o'clock and 8 o'clock position from the vortex elbow to the concrete pipe approximately 90 feet(27.4m). The anode wire will be installed in a 2-inch(5.08cm) schedule 80 PVC pipe. The anodes will have a fiberglass housing which will be bolted to a steel channel and welded to the inside of the pipe by divers. There will also be 3 reference cells located at the 12 o'clock position with leads running to the rectifier on top of the Vortex structure.

Construction Sequencing

The construction techniques and sequencing are extremely important in this project. The outfall must be kept full and in service at all time. The pipe is extremely delicate condition and ocean spills will violate permit conditions.

In order to maintain the structural integrity the contractor will be limited by specification to exposing only 8-foot sections of the pipe below 30-inches (76.2cm) above the pipe invert. The contractor can expose multiple alternating 8-foot (2.438m) sections of pipe. It has been recommended that hand excavation take place as the pipe is exposed below the invert.

Dewatering will be important as the excavation will take place to 22 feet(6.7m) below sea level. It is expected that coffer dams and pressure grouting will be used.

Presently this project is scheduled to go to construction in the spring of 1997. With successful completion of this project the planned tunnel outfall may be able to be postponed until well into the 21st century with a potential savings of greater than $90,000,000.

1.Designed By Black & Veatch Consulting Engineers San Diego, Ca.

2.Corrosion Investigation 108-inch Outfall Pipeline Point Loma Wastewater Treatment Plant San Diego, Ca. by V & A Consulting Engineers, Oakland Ca. July, 1994

NEW TECHNOLOGY FOR ENHANCED PIPELINE EVALUATION

Allen Thomas
Executive Director, National Association of Sewer Service Companies

Greg Laszczynski
Vice President of Operations, Insituform East

Abstract
A review of several pipeline evaluation methods, currently under development, which are designed to increase the amount of information available regarding the condition of vitrified clay and concrete pipe and the soil surrounding them. The usefulness of the information in designing rehabilitation methods is also discussed.

Introduction

A variety of technology is in development in Europe, Japan and the United States that is designed to increase the amount of information available for pipeline inspection and evaluation. This new technology falls into the following general categories:

- Laser scanning and profiling
- Acoustic and ultra sonic testing
- Seismic testing
- Side scan sonar.
- Ground penetrating radar

These technologies are designed to provide additional information (beyond what is provided by conventional CCTV inspection) regarding the condition of clay and concrete pipe which was used extensively for wastewater collection system construction prior to the advent of plastic pipe. In addition to providing an evaluation of the condition of the pipe wall some of these emerging technologies are designed to also provide information regarding the soil surrounding the pipe and to some extent the location of other utilities.

The goal of these new technologies is to provide additional information so it will be possible to more accurately predict the life expectancy of various pipelines and assist in the selection of rehabilitation methods and products and establish rehabilitation priorities. Quality control of the rehabilitation process may also be enhanced. Some of the evaluation methods may provide a way of measuring the thickness of liners and determine if voids or annular spaces exist between the liner and the inside wall of the host pipe.

Laser Scanning and Profiling

This equipment has been on the market for about five years and has been presented at a few of the national trade shows such as NO DIG (North American Society for Trenchless Technology) and WEF (Water Environment Federation). The system consists of a software based measurement system which allows for the measurement of distances and areas on the video image and a battery powered laser to optically project a bright red vertical stripe into the video image. The measurement technique is done by positioning a line or a circle on the object to be measured and the area or length is calibrated by measuring the distance between pixels.

Used together the video measurement software and the laser line allow for the operator of a conventional TV inspection system to locate a defect such as a crack or a hole in the pipe and measure the depth of the hole, the length of the crack or any other object in his field of view.

The precise orientation of the laser projection optics results in an apparent motion of the laser stripe with respect to camera range. When a measurement is to be taken the video is stopped on a single frame and the laser line is positioned on the object to be measured. When the laser line is in position the software based micrometer tools are used to measure areas, distances of flat surfaces and concave or convex surfaces anywhere in the video image.

Ultra Sonic and Acoustic Testing

These two testing methods are based on well known sound testing of ceramic or clay pipes and other non conductive materials and ultra sonic testing which is a standard method of non destructive testing commonly performed on pipes and vessels made of conductive materials. (See figures 1,2,3, on page 7.)

Two devices (acoustic and ultra sonic) were exhibited at the IFAT conference (International Environmental Exposition) in Munich Germany, May 7-11, 1996. The development of these devices was funded by the German Ministry of Research and Technology as part of a program designed to develop new technologies which could be used as a supplement or an alternative to standard video inspection.

The acoustic and ultrasonic testing devices are designed to evaluate the material integrity of the pipe, the permeability of cracks and the integrity of the bell and other connectors. If there is good acoustic contact between the backfill and the pipe, the ultrasonic testing can detect some voids in the backfill.

In general the quality and integrity of ceramic products is assessed by sound testing. Quality control is determined by sound reflection during production and in addition it is a usual procedure to repeat the acoustic test prior to burial.

Sound vibrations are different for individual pipes but when an average is taken for a number of pipes it shows that the frequencies fall into a fairly narrow range which makes it easy to detect pipes that do not meet the standards. Cracks in pipe prevent the propagation of induced sound waves and unsymetric signals are detected, indicating a crack or break in the pipe wall.

In laboratory testing intact and cracked pipes were buried in a sand filled container. The testing determined that the embedment acted to suppress the vibrations of the pipe and produced a higher frequency response than free standing pipes. This situation prevented the application of conventional signal analysis because of the shortened vibration cycles.

Ultra sonic testing is widely used to test steel and other conductive pipes. It has not been used in concrete and clay pipes because the complex composition of poured concrete and to a lesser extent clay pipe. This condition complicates the ultra sonic wave propagation within the pipe wall and makes the interpretation of the signal more difficult. Advanced methods of signal analysis have been developed so that surface waves and other disturbances are identified and eliminated. This allows for a more accurate assessment of the condition of the pipe wall.

<u>Acoustic Prototype</u>
The acoustic and ultra sonic testing devices shown at the IFAT conference were mounted on separate pipe crawlers. The acoustic device consisted of round headed hammers which were activated by solenoids to strike the interior surface of the pipe. The resulting sound waves were picked up by transducers mounted on the crawler and the signals recorded by a data acquisition unit. The signal displays indicated cracks or breaks by an interruption of the wave propagation.

<u>Ultra Sonic Prototype</u>
The ultra sonic device utilized transducers which were positioned against the inside wall of the pipe. One transducer produces the signal and the other transducer receives it. In order for the ultra sonic system to work effectively there must be a direct contact between the transducer and the material being tested. In the case of the prototype unit this contact was made by a film of water introduced between the transducer and the pipe wall. This is the standard medium of contact in ultra sonic testing. Ultra sonic analysis has been tested in Houston in an effort to evaluate the installation of CIPP liners.

Seismic Testing
Seismic methods have been used for several years as a means of determining physical properties of materials beneath the earth's surface. Petroleum exploration is just one area where seismic methods are regularly employed to investigate the subsurface. The velocity of a seismic wave through material depends on the density and structure of that material. Dense materials such as

rock transmit the seismic wave at a high velocity while lower density materials such as loose soils and water transmit the waves at a lower velocity.

The new seismic methods involve a specially generated seismic wave and special detection device, called geophones or hydrophones which receive the wave after it has passed through an area of soil. The technologies work by determining the distribution of seismic velocities through the soil surrounding a sewer. This provides valuable information about soil densities which in turn provides information about passive earth pressures and the ability of the sewer to resist external and internal pressures.

Three different methods of seismic testing have recently been demonstrated in the USA. They consist of the use of a hammer device to strike the invert of a buried pipe and detect the seismic waves on the surface. The analysis of these waves detect gross anomalies in the soil above the pipe. More detailed seismic tests can then be done by placing hydrophones alone the invert of the pipe and introducing seismic waves from the surface down to the pipe. In this instance the seismic testing was developed to provide detailed information about the soils between the sewer and the ground surface where access to the length of the sewer pipe is available. The most detailed seismic evaluations are generated by installing hydrophones in boreholes placed at strategic locations along he length of the pipe. This process allows for the analysis of soil surrounding the pipe at very specific locations and at levels above and below the level of the pipe.

Sonar

Sonar or ultrasound is used to provide direct measurements of the inside diameter of pipelines. These caliper tools are deployed like conventional TV cameras and provide measurements to define the displacement, corrosion, level of debris or other parameters inside the pipe.

Sonar or ultrasound energy is directionally transmitted toward the target. Some of the sound is reflected from the target and detected at the source and the total travel time is measured. For the purposes of the sonic measurements in wastewater collection systems, the only concern is the velocity of the sound in air or water since the surface of the pipe (concrete, clay, plastic, steel) is what reflects the sound. This means that the velocity through the pipe is not being measured but only the interior of the pipe.

Piezo-electric transducers generate and receive the sonar energy which is converted into electrical energy for measurement. Since the sonic velocity can vary with temperature and humidity, the transducers are calibrated for the velocity at the beginning of each inspection. This is done by recording the travel time for several known distances in the pipe radius to be inspected. This time is converted into distance (inches) and becomes the data from which the inspection results are produced. More recently the sonar inspection devices have added a rotating transducer which provides a more detailed interior scan at any point within the pipeline.

Radar

Ground penetrating radar has been used for the detection of underground utilities and other sub surface evaluations for a few years. In the past few months some evaluations of large diameter concrete pipe have been completed by moving the radar unit from the surface into the pipe. This has allowed the radar waves to be sent from within the pipe which provides an analysis of the pipe wall conditions as well as the soil surrounding the pipe.

The emerging in-pipe radar systems utilize a single or multiple radar arrays that transmit the electro magnetic signal which is received by antennae and measured. In this sense the radar evaluation is similar to ultra sound in that they both generate some type of energy which is received and the time or distance of the signal is measured. The key difference with radar is that the electro magnetic signal penetrates non-conductive materials such as concrete and clay pipe easier than any of the other forms of energy.

In general, dry, granular soil has low conductivity and moist, dense soil has high conductivity which means that the electromagnetic signal can penetrate further into the dry soil than the dense moist soil. The other important variable that must be dealt with is the dielectric properties of the objects or materials that are being evaluated and the soil surrounding them. The greater the difference between the dielectric properties the more detail is provided by the radar analysis. For example, metal is not a dielectric material and reflects the electromagnetic impulses better than any other material. This means that metal such as rebar in concrete is very easily seen by the radar system.

Data evaluation and Conclusions

The desired result of having additional evaluation data is to improve and assist in the selection and design of the rehabilitation method used on the collection system pipes. Engineers are often faced with several unknown factors when deciding what method of rehabilitation should be selected and what parameters should be used in specifying and designing the rehabilitation product. Some trenchless rehabilitation products can be designed using two different design conditions, "fully deteriorated" and "partially deteriorated". In a fully deteriorated situation, the rehabilitation product is designed to handle all internal and external loads and no consideration is given for the structural capacity of the host pipe. In a partially deteriorated pipe the host pipe is considered to be structurally sound but deteriorated due to thinning of the pipe wall, cracks, leaking joints, pin hole leaks, etc. and therefore, the rehabilitation product is only designed to resist the external hydrostatic loads, such as ground water, and the internal pressure conditions should the line section experience something other than open channel, gravity flow conditions.

Designing rehabilitation products for the proper host pipe conditions will result in the most cost effective product for the expected life of the pipeline. Specifying a fully deteriorated design condition when not required can result in products being thicker or stronger than necessary which can not only increase the price of rehabilitation but can also diminish the flow carrying capacity of the reconstructed pipeline.

If more accurate data can be obtained on the existing conditions in and around the pipeline such as accurate inside dimensions, amount of ovality, maximum size of offset joints, remaining thickness of the host pipe size and depth of cracks/holes, maximum depth of the pipeline, and surrounding soil conditions, it will assist in selecting the applicable types of rehabilitation to consider and help to design the most cost effective solution.

Some rehabilitation technologies/processes are limited in the range of sizes, and/or condition of the host pipe. For instance, some lining processes are better suited for rehabilitating certain diameter pipelines and changes in diameter or the degree of ovality existing in a given line section will dictate certain types of trenchless technologies over others.

Often times the thickness of the product used to rehabilitate a given pipeline will be based on the pipe's diameter, depth, ground water loading, and the amount of ovality. Accurate design data will help ensure the required product thickness is chosen. Additionally, if these new evaluation technologies can determine the condition and the presence of voids in the soil surrounding the pipe, it would help determine if the pipe needs structural or non-structural repair such as lining or grouting or both.

Improved information and accurate data on the condition in and around an underground pipeline will allow managers and engineers of a collection system to improve their ability to estimate the remaining life expectancy of a line section and to set priorities for the pipes that need repair or replacement. Furthermore, this information and data will ultimately help to select and design the most cost effective rehabilitation solution.

ACOUSTIC AND ULTRASONIC MEASUREMENT DEVICES

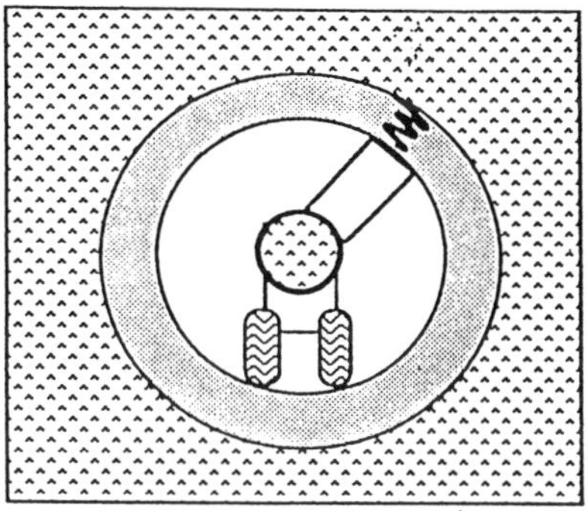

Figure 1. Measurement of wall thickness by ultrasonic impulse.

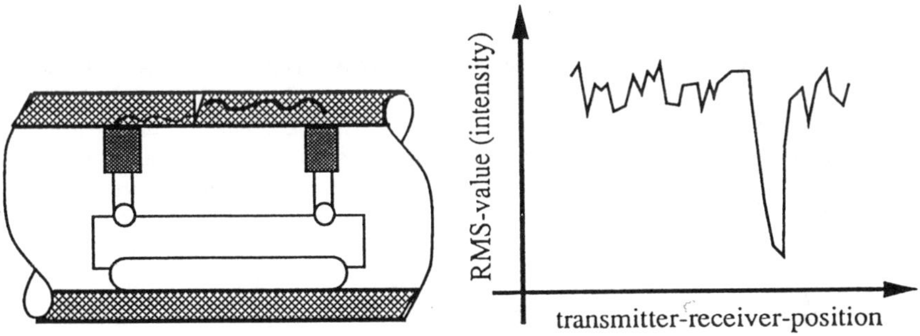

Figure 2. Crack detection with ultrasonic testing.

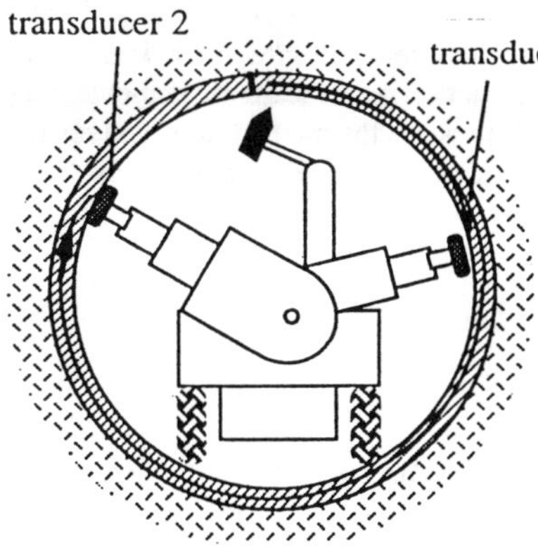

Figure 3. Crack detection with acoustic reflection analysis.

Using Trenchless Technologies and Interceptor Relief for CSO Control in East Boston, Massachusetts

Dennis Doherty[1], Leonard Gillan[2], and Donald Walker[3]

Abstract

The Massachusetts Water Resources Authority (MWRA) is conducting facilities planning for combined sewer overflow (CSO) control in Greater Boston. In East Boston, a total of nine CSO outfalls discharge into Chelsea Creek and the Boston Inner Harbor with frequencies as high as 35 times per year. To allow attainment of water quality standards, a CSO control goal of four or less overflows per year was identified for the nine East Boston CSO outfalls. All nine outfalls are hydraulically associated with a single interceptor system, the East Boston Branch Sewer (EBBS), which conveys flows to the Caruso Pump Station. The overflows occur due to surcharging in the 6,190 m (20,300 lf) EBBS while the pump station operates at less than design capacity. Therefore, it was determined that the surcharging and CSO overflows are caused by inadequate capacity in the EBBS. The EBBS is approximately 100 years old, with its upstream branches consisting of small diameter clay pipe. Given these conditions and the highly urbanized character of East Boston, trenchless technologies were identified as an alternative to traditional open cut construction as a means of implementing interceptor relief for CSO control. This paper presents the methodology and results of facilities planning for interceptor relief alternatives for CSO control in East Boston using trenchless technologies.

---

[1]Metcalf & Eddy, Inc. PO Box 4071 Wakefield, MA 01880-5371

[2]Massachusetts Water Resources Authority, Charlestown Navy Yard 100 First Avenue Boston, MA 02129

[3]Metcalf & Eddy, Inc. PO Box 4071 Wakefield, MA 01880-5371

Introduction

The Massachusetts Water Resources Authority (MWRA) is conducting facilities planning for combined sewer overflow (CSO) control in the four communities within the MWRA service area that have CSOs. Within the communities of Boston, Cambridge, Somerville and Chelsea, a total of 23 projects are proposed, ranging from CSO elimination through sewer separation, to providing floatables control at relatively inactive outfalls. In East Boston, nine CSO outfalls discharge into the Chelsea Creek and the Boston Inner Harbor with frequencies as high as 35 times per year. All nine outfalls are hydraulically associated with a single interceptor system, the East Boston Branch Sewer (EBBS).

The existing EBBS includes approximately 6,190 m (20,300 lf) of interceptor pipe ranging in size from 300-mm (12-in) diameter clay pipe to 1,000 by 1,150-mm (40 by 45-in) brick oval. The interceptor system consists of an approximately 2,040-m (6,700-lf) main trunk, with four tributary branches, and conveys flow from the westerly portions of East Boston to the Caruso Pumping Station, where flow is lifted into the North Metropolitan Sewer for conveyance to the MWRA's Deer Island Wastewater Treatment Facility (Figure 1). The EBBS was originally constructed in the 1890's. The main trunk line is brick, while the branches are predominantly clay pipe. The main trunk of the EBBS is relatively deep (8 to 11 m [25 to 35 ft]) compared to the branch lines, which are at a depth of 3 to 6 m (10 to 20 ft).

During wet weather events, surcharging occurs in the EBBS. For storms up to approximately the 3-month, 24-hour storm, the surcharging is due to insufficient conveyance capacity in the EBBS. For storms larger than the 3-month, 24-hour storm, the total flow potentially tributary to the Caruso Pump Station exceeds the pumping capacity, and backwater from the pumping station contributes to surcharging. Surcharging in the EBBS causes activation of the nine CSO outfalls associated with the EBBS.

Based on evaluations conducted during MWRA system-wide master planning, reducing the activation frequency of CSO outfalls to approximately four times per year through interceptor relief was determined to be the most cost-effective alternative for CSO control in East Boston. During facilities planning, alternatives for implementing the proposed interceptor relief project were developed and evaluated. This process included consideration of alternative routes as well as alternative sewer construction and rehabilitation technologies. Since East Boston is a highly urbanized area with the potential for contaminated soils, the analysis indicated that trenchless construction technologies would be advantageous in terms of reduction in construction costs and construction-related impacts. The alternatives development and evaluation process is described below, followed by a summary of the recommended plan.

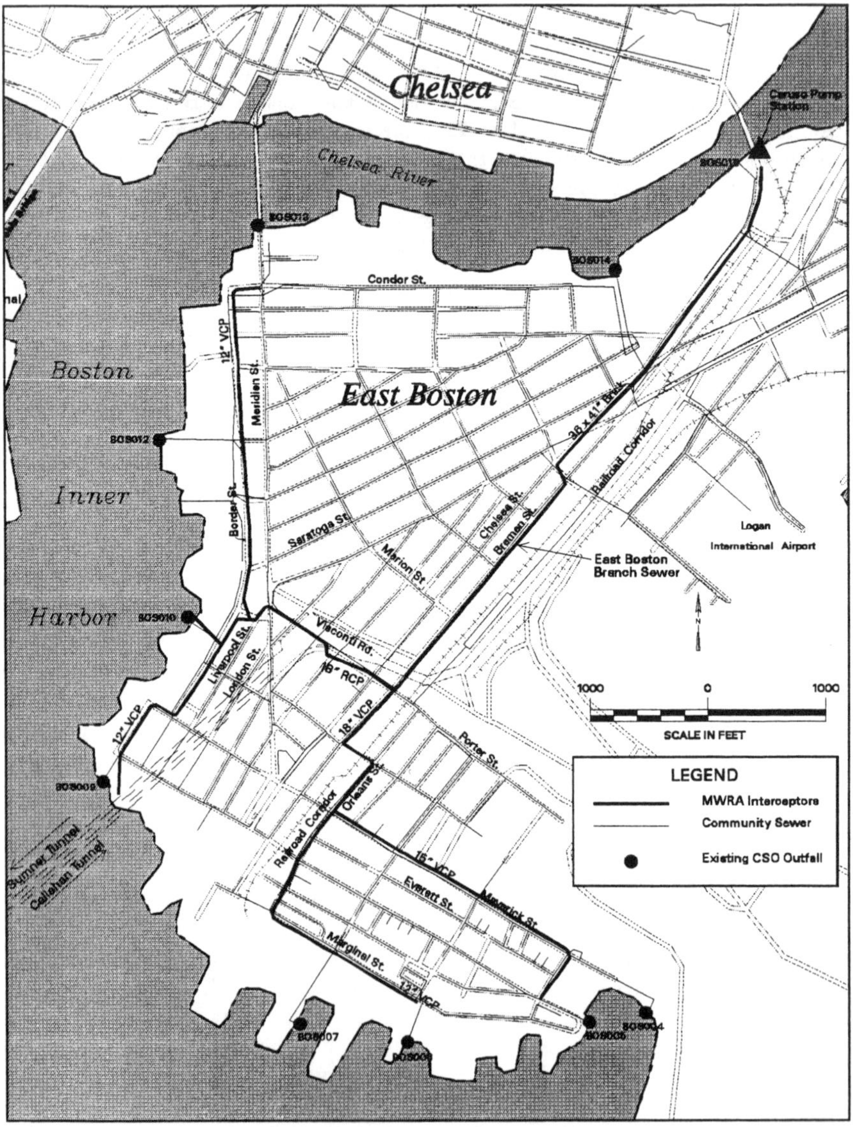

Figure 1. Location Plan for East Boston Branch Sewer

Alternatives Development

The alternatives development process involved the integrated analysis of three general topics: hydraulics, route alternatives, and applicable pipeline construction technologies. East Boston is a highly congested urban neighborhood of Boston, with a large residential population and maritime-related industries along its waterfront. Current and past industrial activities in East Boston have resulted in varying levels of soil contamination throughout the area. In addition, Logan International Airport, which is located in East Boston, contributes significantly to car and truck traffic through the local streets. With these conditions, trenchless pipeline construction technologies present significant advantages over traditional open cut excavation pipeline construction methods, both in terms of construction costs and impacts on the community.

The potential for using trenchless technology construction methods was considered on a site-specific basis, both in terms of hydraulic performance and constructibility. Iterative hydraulic evaluations were conducted using the Stormwater Management Model (SWMM) EXTRAN block. The master planning concept of interceptor relief was developed with the assumption that the existing EBBS would be replaced with larger pipe along the entire route of the interceptor. While this alternative was sufficient to demonstrate the feasibility of interceptor relief for CSO control at the master planning level, it was expected that during facilities planning, more detailed analyses could potentially identify a more cost-effective means for providing the required relief.

The EBBS was evaluated under various hydraulic conditions to assess the pipe sizes necessary for achieving the project objectives. The pipe size requirements helped to determine which construction technologies would be appropriate to meet the project objectives. The evaluations conducted can be categorized as rehabilitation, and a combination of rehabilitation and relief.

Rehabilitation. The first SWMM simulations evaluated the effect of rehabilitating the entire EBBS, including the main trunk and upstream branches. Rehabilitation was modeled either as a liner technology, such as cured-in-place liners, fold-and-form liners, or pipe bursting. The liner technologies would essentially keep the same pipe diameter and reduce the roughness coefficient, while pipe bursting would allow smaller-diameter clay pipe to be replaced by HDPE pipe with an approximately 33-percent increase in pipe diameter (and a lower roughness coefficient). Based on review of existing information, including record drawings, manhole inspection reports and sewer inspection video tapes, the feasibility of using either liners or pipe bursting was established for each section. For example, most of the upstream branches were constructed of relatively small-diameter vitrified clay pipe, which is ideal for pipe bursting. In one of the upstream branches, however, a portion of the existing VC pipe was concrete-encased, which would not be feasible for pipe bursting. This relatively short reach was identified as being appropriate for replacement by open cut excavation. The EBBS main trunk is a larger-diameter sewer of brick construction, which would be more appropriate for rehabilitation using a liner technology.

Using rehabilitation alone, a number of outfalls were predicted to continue to activate more than four times per year. Based on these results, rehabilitation of the EBBS alone was determined not to be sufficient to meet the CSO control objective.

Upstream Rehabilitation and Downstream Relief. The next evaluations involved the hydraulic performance of the EBBS assuming the upper reaches (branches) of the EBBS were rehabilitated by pipe bursting, and the downstream main trunk was relieved by a combination of rehabilitation and installation of new pipe.

This evaluation provided two useful results. First, rehabilitation of the upper branches of the EBBS by pipe bursting appeared to be sufficient to meet CSO goals for outfalls BOS004 to BOS007 and BOS009, if a sufficient level of downstream relief was provided. Since pipe bursting would be less costly than installation of a new pipe by either microtunneling or open-cut excavation, evaluation of alternative routes for relief pipes was not required for these upstream sections. However, intermediate sections between the most upstream reaches and the main trunk required a greater degree of relief than can be provided by pipe bursting. Alternative routes and pipe construction technologies were therefore considered for these sections.

The second useful result was that two alternatives were identified for relief of the main trunk:

- Provide relief and rehabilitation of the branches necessary to convey flow to the upstream end of the main trunk, then provide a new relief conduit parallel to the existing main trunk to convey flow to Caruso Pumping Station. Rehabilitation of the existing main trunk would reduce the size of the new conduit required.

- Provide relief or rehabilitation of the two easterly branches necessary to convey flow to the main trunk, then rehabilitate the main trunk; in addition, provide rehabilitation and relief on the southwest branch, then install a new pipe running north parallel to the northwest branch, then east to a new junction structure opposite Caruso Pumping Station. The new pipe would pick up flows from regulators associated with outfalls BOS009 and BOS012 to BOS014, as well as public connections to the existing EBBS along the northwest branch. This alternative would relieve the EBBS main trunk by diverting flow from the west side of East Boston away from the main trunk.

These evaluations provided the basis for evaluating route alternatives for the EBBS Relief project.

Alternatives Evaluation

Evaluation of interceptor relief alternatives was complicated by the fact that the EBBS consists of a main trunk and four upstream branches. For each branch of the EBBS, alternative routes and construction techniques had to be identified independently of the routes and construction techniques in the other branches. For example, the route and construction technique selected for relief of the southeast branch had no direct impact on the potential routes and construction techniques for relief of the southwest branch. To facilitate the analysis, pipe route alternatives were identified and evaluated by dividing the overall project route into a number of route sections and evaluating route options within each of those sections. Route options included different construction techniques for rehabilitating existing pipe or installing new pipe along the same stretch of road or right-of-way. The applicability of each construction technique was assessed on a site-specific basis. In this manner, preferred route options for each section could be combined to form an overall route alternative. The EBBS relief project was divided into nine route sections, with a total of 36 route options identified for the nine sections.

Evaluation of route options involved comparison of cost, construction risk factors, and environmental impacts for each route option within a section. Costs included allowances for handling contaminated materials based on a planning-level geotechnical and hazardous materials investigation program. Construction risk factors included former land use, traffic, density of utilities, contamination, and geotechnical considerations. These factors were intended to account for certain aspects of construction that could not be fully quantified in terms of cost. The evaluation of environmental impacts included the assessment of impact on a range of specific criteria under the general headings of community and institutional concerns, and natural resources. The costs, construction risk factors, and environmental impacts of the route options within a given section were assembled into a matrix to facilitate comparison. Generally, the low-cost route option was selected, unless the associated construction risk factors or environmental impacts were determined to justify selection of a higher cost option.

As noted above, pipe bursting was determined to provide sufficient relief in the upstream branches. Since the cost and impacts of pipe bursting were less than those for other techniques for installing new pipe, pipe bursting was selected as the recommended alternative in these areas. Where pipe bursting could not provide sufficient capacity, installation of new pipe by open cut excavation was compared against installation by microtunneling. Where microtunneling was determined to be feasible, it generally had a lower cost than open cut excavation. The major cost advantage to microtunneling was in the cost of excavate disposal. The soil in East Boston contains a range of suspected contamination, and a city ordinance requires all soil excavated from beneath city streets for utility construction be disposed off site; i.e. excavated soil cannot be used as backfill. Given this requirement, the significantly lower excavation quantities associated with microtunneling more than compensated for the higher base unit cost of microtunneling as compared with open cut excavation.

In terms of construction risk factors, former land use generally either favored open cut excavation, or was not a differentiator. Where existing information suggested the presence of old seawalls or building foundations along the proposed route, open cut excavation was considered to carry less risk than microtunneling. Where a series of abandoned foundations were known to exist along one particular route, microtunneling was not considered as an option. Where routes were along existing streets, consideration of the impact of traffic on the contractor's operations generally favored microtunneling. Since the construction duration along a given route was significantly less for microtunneling than for open cut excavation, the impact of traffic-related delays on the overall project schedule would be less critical. Density of utilities was generally not a differentiator, although where local pockets of dense utilities were located, microtunneling was favored based on an assumption that the shafts could be constructed with less utility relocation than an open cut pipe trench. Contamination-related risks favored microtunneling, as the excavation quantities along some routes were an order of magnitude less for microtunneling than for open cut excavation. Microtunneling would also minimize the exposure of workers to the contaminated material. Geotechnical considerations were generally not a differentiator. Review of available geotechnical data, including a planning level geotechnical exploration program, indicated that bedrock was well below the inverts of the proposed pipelines. The stratigraphy in East Boston generally consists of a variable layer of fill, underlain by interbedded layers of sand, silt, and clay.

Environmental impacts were generally more of a differentiator among routes, as opposed to differentiating among pipe installation technologies. Where new pipes were to be installed in existing streets, the most common impact was on traffic. If a microtunneling jacking shaft had to be located in a street, the impact of the shaft on lane reduction would be approximately the same as a stretch of open trench excavation. The exception would be where a shaft could be located in a traffic island or some other off street location. Detailed shaft locations were not, however, identified in the facilities planning process. Given general knowledge of the proposed routes, it was assumed for the sake of identifying potential impacts that most jacking and receiving shafts would be located in public rights of way. However, at specific high-traffic intersections, it appeared reasonable to assume that the shafts could be located to minimize disruption to the intersection.

Summary of Recommended Plan

The EBBS Relief project will reduce the activation frequency and volume of the nine CSO outfalls in East Boston, tributary to the Mystic/Chelsea Confluence and Upper and Lower Inner Harbor, to a level whereby water quality standards could be met in those receiving waters. The relief project will increase the capacity of the interceptor system in East Boston to convey wet weather flows to the Caruso pumping station. The increase in interceptor capacity will reduce CSO activations by lowering hydraulic gradients at CSO regulators during wet weather. Trenchless technologies are proposed for the installation of new pipe and rehabilitation of existing pipe. The technologies proposed, pipe bursting, microtunneling, and cured-in-place lining, were found to reduce the overall construction costs and have more favorable non-monetary impacts than open cut excavation. The construction project will involve approximately 3,625 m (11,900 lf) of new pipe installed primarily by microtunneling; replacement of 1,875 m (6,150 lf) of the existing EBBS by pipe bursting; and rehabilitation of 2,040 m (5,700 lf) of the existing EBBS main trunk by cured-in-place lining.

The project will reduce the annual CSO discharge volume from nine CSO outfalls in East Boston from approximately 110,000 m^3 (29 million gallons) to 22,700 m^3 (6 million gallons). The total estimated capital cost of the project is $29.8 million. Design is scheduled to begin in March, 2000, with construction starting in March, 2003. Some of the advantages of trenchless technologies for this project are summarized in Table 1.

Parameter	Benefit of Trenchless Technology
Project cost	Trenchless technology alternatives had lower cost than open cut excavation, due to reduced excavate handling cost
Excavate handling	Reduced volume of excavate for disposal translates to reduced cost and reduced risks associated with contamination
Duration of construction	Higher production rates than open cut excavation
Utilities support/relocation	Less utility relocation generally required, provided shafts can be strategically located
Traffic impacts	Shafts located to minimize disruption to major intersections

Table 1. Summary of Benefits of Trenchless Technologies

Houston's Solutions for Successful Rehabilitation

A.D. "Art" Kidder, Jr.[1]
William H. Sukenik, P.E.[2]

Abstract

The city of Houston is in its final year of a $1.2 billion wet weather overflow control program including sanitary sewer hydraulic relief and rehabilitation. This is in response to administrative orders issued by the US EPA and the Texas Natural Resource Conservation Commission. Full compliance is required by December 31, 1997. Included in this effort is a commitment to structurally rehabilitate the most deteriorated portions of the sewer system.

Houston's investment in rehabilitation will sum to over $300 million in a five-year period ending mid-1998. To date, over $240 million have been appropriated effecting the rehabilitation of over 3.6 million feet of sewer. The success of the rehabilitation effort is attributed to the effectiveness of the partnership of all the stakeholders involved including the city, sewer service customers, contractors, design engineers, products suppliers and construction management personnel. However, that partnership has not been without challenges requiring the formulation of creative solutions in order to assure the success of the program.

Introduction

In the 1980s, it was evident that Houston was experiencing a substantial amount of sewer breaks primarily attributable to the corrosion of unreinforced, unlined concrete pipe. Deteriorating sewers contributed to eroding sewer service quality and

[1]City of Houston, 1100 Louisiana, Houston, Texas 77002
[2]Greater Houston Wastewater Program, 1100 Louisiana, Houston, Texas 77002

customer confidence, inconvenienced citizens and visitors, threatened water quality, and damaged public and private property. Based on late 1970s survey data associated with 62 cities, Houston was experiencing the second highest rate of reported sewer breaks per year per 1,000 miles of sanitary sewer, i.e., 571 (four-year average). Furthermore, Houston was reported as having the sixth highest capital requirements, i.e., $64 million, for the correction of infiltration/inflow and sewer rehabilitation in the nation. (Peterson et al, 1984)

In 1987, Houston was issued an order by the Texas Natural Resources Conservation Commission requiring reduction of wet weather overflows in its wastewater system. Similarly, in 1989, a series of EPA orders were issued dictating the correction of some 200 wet weather bypasses and overflow points by December 31, 1997. Historically, 20 to 30 bypasses were experienced per year. This prompted the City to initiate an ambitious $1.2 billion program to evaluate and rehabilitate its sanitary sewer system and construct wet weather overflow control and relief facilities.

Only an intensive structural rehabilitation program of proportions currently under pursuit by the city could be capable of turning the situation around. Dividends are now being realized as corrective maintenance work orders for stoppages, overflows and collapses have shown significant decline, i.e., a 16 percent decline from the late 1980s to 1994, while customer satisfaction improves. Houston has committed to the U.S. EPA and the Texas Natural Resources Conservation Commission to spend $300 million on its structural rehabilitation program over a five year period through June 1998. It is anticipated that rehabilitation will continue well beyond this period with an average annual expenditure of $50 million or more.

Background

Houston's wastewater collection system includes 6,000 miles of pipeline serving 348,000 customers over a 592 square mile area. The city operates 51 wastewater treatment plants and 337 pump/lift stations. Seventy percent of the system's sewer size is 8 inches or less in diameter, while there are 135 miles of sewers 48 inches or greater in diameter. Approximately thirty percent of the sewer system is constructed of unreinforced, unlined concrete pipe.

To date, over $240 million has been appropriated for sewer rehabilitation. The methods of rehabilitation that have been used in the city's projects with associated quantities are as follows:

Fold & Form	30,000 lf
Cured in place	1,130,000 lf
Remove & Replace	440,000 lf
Pipebursting	400,000 lf
Sliplining	1,600,000 lf
Total	3,600,000 lf

Eighty eight percent of this rehabilitation has been accomplished by trenchless methods.

Annual services agreements contribute $10 to 15 million per year to the overall rehabilitation effort with sewer cleaning, internal television inspection, main line and lateral point repairs, and other related services on a work authorization basis.

Encountered Problems and Identified Causes

Common problems encountered during the rehabilitation effort have been:

1. Failed coatings
2. Liner ovality
3. Pipe joint voids on tunnel sections
4. Wrinkled cured in place liners
5. Poor ground restoration
6. Offset manhole frames
7. Protruding gaskets

The causes of the above mentioned defects are known, and each is considered as attributable to one or more of the following:

1. Inadequate inspection
2. Design deficiencies
3. Substandard workmanship
4. Inadequate product testing
5. Misapplication of products under the ambient conditions
6. Ambiguities and misinterpretation of specifications

Pursuing the contractors to remedy these defects results in the expenditure of already critically limited resources. Furthermore, if the defects go undetected, they could contribute to some future sewer failure, or minimally, deprive the city from realizing the maximum extension of service life that the rehabilitation should have provided.

Solutions

Houston has targeted solutions to these common problems through emphasis in three main areas:

- Enhanced Communications
- Improved Training
- Product Evaluation and Approval

Enhanced Communications. The city has proactively involved the construction contracting and engineering communities in the development of its standard specifications for capital sewer projects and rehabilitation effort. Through regular meetings with the Houston Contractors Association and the Association of Consulting Municipal Engineers, invaluable input has been received and ultimately reflected in Houston's processes and procedures for delivering its construction projects. This spirit of cooperation has undoubtedly contributed to the enhanced bidding of city jobs, resulting in better pricing and minimizing the number of claims and change orders.

Improved Training. Under the theme "good inspectors make good inspections", Houston is striving to raise the skill-level and knowledge of its inspectors through classroom and site training, and on-the-job experience. A tremendous opportunity to develop this resource is afforded by the shear magnitude and scope of the city's current wet weather program. Houston has integrated its inspection forces with those of its consultant program manager to allow the efficient transfer of the latest techniques of the private sector in construction inspection and management, to city forces.

Product Evaluation and Approval. Houston relies on its Standard Wastewater Products Committee (SWPC) to maintain a list of approved standard products, materials and construction methods associated with the installation, maintenance and rehabilitation of sanitary sewers. The Committee is composed of nine voting technical City staff members, and one non-voting member each representing the Houston Contractors Association and the Association of Consulting Municipal Engineers. Manufacturers, vendors, contractors and engineers must submit an application to the city for consideration of their products and processes as the initial step in the approval process. Literature reviews, case studies, manufacturers' printed materials, demonstration projects, field investigations and/or university testing are then utilized to further evaluate the products and processes. The list is continuously updated to add and remove items accordingly.

This approach provides the City with many benefits including:

- Use of the most cost-effective, high quality products
- Application of the newest technologies
- Pricing reflective of a healthy competitive environment for contractors, manufacturer and vendors
- Standardized approach to product evaluation and selection

Summary Conclusions

Houston's success is attributed to the involvement of all the stakeholders in its sewer rehabilitation effort, from the customers to the design engineers to the product vendors, through effective communications, exercising good listening skills, and responding accordingly to all input. The continuous evaluation of new products and construction techniques will allow Houston to remain innovative while enjoying the benefits of the latest available, cost-effective quality products in its projects. Ultimately, this means enhanced competitive pricing of its projects, while efficiently meeting the primary objectives of the rehabilitation effort -- satisfied customers and a reliable sewer system meeting service life expectations.

Acknowledgment

This project is funded by the City of Houston, Department of Public Works and Engineering under the auspices of Jimmie Schindewolf, P.E., Director, and Wendell L. Barnes, P.E., Deputy Director and Program Director of the Greater Houston Wastewater Program. Art Kidder is the Division Manager, Technical Support and Operations in the Department of Public Works and Engineering, Utility Maintenance Division. William H. Sukenik is a Task Manager for the Greater Houston Wastewater Program and an employee of Montgomery Watson.

References

Peterson, G.E.; Miller, M. J.; Godwin, S. R.; Shapiro, C. (1984) Guide to Benchmarking of Urban Capital Condition.

FRICTIONAL RELATIONSHIP OF HDPE LINERS AND STEEL HOST PIPES

John E. Taylor[1], Reda M. Bakeer[2], and Michael E. Barber[3]

Abstract:

Friction coefficients are used to quantify the relative resistance to movement between two bodies in contact. In trenchless rehabilitation with deformed/reformed liners, movement of an installed polyethylene liner due to shear stress applied to the liner can be related to the coefficient of static friction (μ_s). If the shear stress is too large, slippage between a liner and a host pipe may cause damage to service connections. This is especially important inside a pipeline during an earthquake. The frictional resistance is also necessary to evaluate the force required to remove the liner from a pipe at the end of its service life or following an improper installation. Interface shear tests were performed between high density polyethylene (HDPE) and steel; the μ_s was determined to be 0.46. Pullout tests were used to quantify the force required to dislodge a HDPE liner from a steel host pipe. Using the data obtained from these two tests, the maximum normal force exerted by a liner on a host pipe was determined to be 6 kN for Standard Diameter Ratio (SDR) 32.5 liners and 14.6 kN for SDR 26 liners.

Introduction:

Many pipeline infrastructures in the US and around the world are in need of rehabilitation. Public works departments are actively seeking more economical methods for improving the quality of older networks (WEF/ASCE,

[1] Former Graduate Student, Dept. of Civ. and Env. Engrg., Tulane Univ., New Orleans, LA 70118.
[2] Assoc. Prof., Dept. of Civ. and Env. Engrg., Tulane Univ., New Orleans, LA 70118.
[3] Asst. Prof., Dept. of Civ. and Env. Engrg., Washington State Univ., Pullman, WA 99164.

1994). This philosophy has prompted the development of several new "trenchless technologies." This growing field is lead by firms that minimize excavation and pavement replacement costs by offering "no-dig" solutions (Kramer, 1992). Like any new industry, trenchless products have to be supported by tests to quantify their engineering characteristics. This paper presents the results of a study of the frictional relationship between deformed/reformed HDPE liners and a steel pipe. Frictional resistance controls the tendency of a liner to shift within the host pipe during an earthquake. Such shifts may cause damage to service connections. Knowledge of the frictional resistance is also necessary to determine the force required to dislodge a liner from a pipeline after installation. This situation may develop when the liner has to be replaced at the end of its service life or due to improper installation.

Friction coefficients quantify the relative resistance to motion between two surfaces in contact. The coefficient of static friction (μ_s) defines the pullout force necessary to cause a shift between the inside surface of a steel pipe and the outside surface of a liner. The μ_s can be determined from the relationship: $F_{PO} = \mu_s \cdot F_N$, where F_{PO} is defined as the measured pullout (shear) force, and F_N is the normal force between the steel pipe and the liner (Giancoli, 1984). To model the friction relationship, it was first necessary to conduct interface shear tests to determine the μ_s. Pullout tests were then performed to evaluate the pullout force required to dislodge a HDPE liner from a steel pipe.

Interface Shear Tests:

Interface shear tests were conducted to determine the coefficient of friction (μ_s) between an HDPE liner and steel pipe. The steel specimen used in the tests was carefully selected to ensure that it contained roughness and surface texture similar to the steel pipes which were to be used later in the pullout tests. The HDPE specimens were cut from a 250 mm (10 in.) outer diameter liner coil with a Standard Diameter Ratio (SDR) of 32.5. SDR is an industry standard which defines the required thickness of a pipe wall as a function of its diameter. Three nearly flat rectangular HDPE samples; 50 mm x 75.3 mm (2.0 in. x 3.0 in.), 62.3 mm x 87.9 mm (2.5 in. x 3.5 in.), 75.3 mm x 100.4 mm(3.0 in. x 4.0 in.); were used in the interface shear tests. The 7.72 mm (0.31 in.) thick HDPE samples were heated to approximately 93 °C (200 °F) then clamped between two pieces of HDPE to cool down. This process removed most of the remaining curvature from the specimens and simulated the effect of reforming during field installation. In order to model in-situ conditions, two more interface test runs were conducted on the largest specimen (number 3) with fine grained soil or fine sand inserted between the HDPE and the steel specimens. Results of this test could be used to determine the effect of the presence of soil particles in the annular space between the liner and the host pipe.

A direct shear machine designed for testing soils was used to perform the interface shear tests. Accordingly, some modifications to the machine were made to accommodate the specimens. The brass box which would contain the soil sample was replaced with a flat steel specimen, fully restrained to the frame, with a width of 126 mm (5-in.) and a length of 201 mm (8-in.) representing the steel host pipe material. One of the HDPE specimens was then placed over the steel plate and loaded with the appropriate normal load. Under a given normal load, the force required to initiate displacement in the HDPE specimen was recorded. The direct shear machine motor was set at a low constant speed of about 5 mm/min (0.2 in/min). The specimens were tested with the outside surface of the deformed liner pipe placed against the steel specimen. The HDPE specimen was oriented in the same manner as it would be tested in the pullout tests. To test the effect of soil intrusion between the liner and the host pipe, approximately 28 gm (0.1 ounce) of dry soil was spread directly on the steel specimen. The fine grained soil consisted of particles in the typical range of clay, silt and fine sand.

Pullout Tests:

Ten tests were conducted to determine the pullout force needed to dislodge a processed HDPE liner from a steel host pipe. The liners were installed in 151 and 201 mm (6 and 8 in.) inside diameter steel pipes according to the standard field installation procedure. All ten pipes were approximately 2440 mm (8 ft) long. Four tests were conducted in 151 mm (6 in.) pipes using SDR 32.5 liners (Group A), and six tests were conducted using 201 mm (8 in.) SDR 26 liners (Group B). The tested samples were liners which had been previously used in buckling tests (Bakeer et al. 1995). The liners were reformed to closely fit the host pipes following a sufficient period of recovery.

A forklift was used to dislodge the liner from the pipe. The pullout force was measured using a tension spring type scale, with accuracy of 4.55 kg (10 lbs), which was attached to the forklift via a steel chain and to the liner with another steel chain and a hook. The far end of the steel pipe was chained to a steel column to hold it in place. Slack was removed from the chains so that the actual pullout forces would be registered on the scale. The liner was pulled out at a slow rate to determine the force at which the liner material would dislodge from the steel pipe. Pullout rate were varied in order to investigate the effect of the rate on the required force. The pullout rate was measured by the forklift speedometer and was kept under 8 km/hr (5 mph).

Results of Interface Shear Tests:

The normal force and the shear force required to displace the HDPE specimen in the interface shear tests were recorded. The average required shear force was calculated for each normal load. A simplifying assumption was made that the weight of each sample was negligible and was therefore ignored. A plot of the results is contained in Figure 1. The relationship should pass through the origin as there is no adhesion between the two materials.

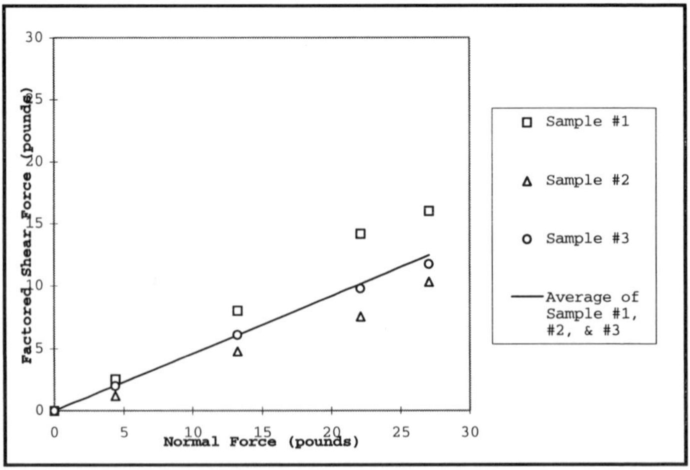

Figure 1. Results of Interface Tests without Soil.

The μ_s between the HDPE liner and a steel host pipe can be approximated by the slope of the shear force versus normal force curve. Because the curve of the sample averages was close to being linear, it was concluded that μ_s between these materials can be estimated by calculating the average slope for the 12 resulting points. The average friction coefficient was calculated to be about 0.46 with a standard deviation of 0.03. The typical friction coefficient (μ_s) between a polyethylene plate and a metal plate is between 0.23 to 0.8, whereas the friction coefficient between two polyethylene plates is typically in the range of 0.1 o 0.6 (Nielsen 1963).

A similar relationship was established for the soil interface tests, as shown in Figure 2. This plot also includes the curve for sample 3 and the average of the three samples.

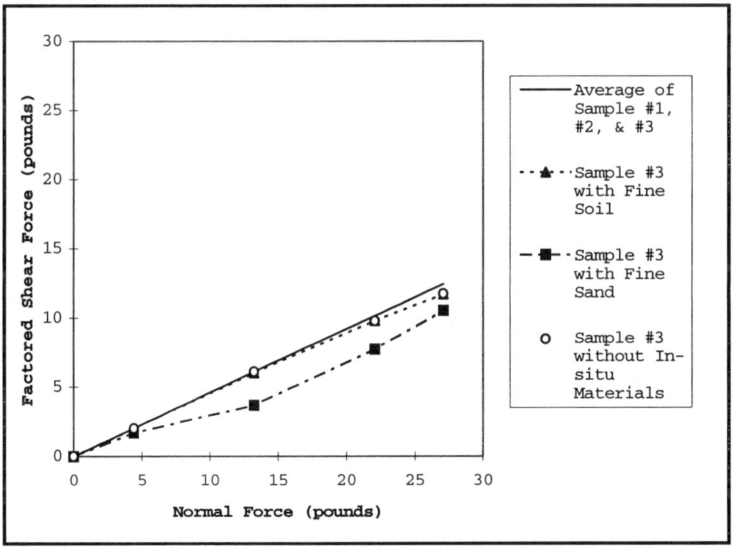

Figure 2. Results of Interface Tests with Soil.

It can be seen that inserting small amounts of fine grained soil between the HDPE liner material and the steel has a negligible effect because the resulting slopes are almost equal. The same average friction coefficient of 0.43 was obtained for specimen 3 for testing with and without the fine grained soil. This behavior can be attributed to the small size of the particles and their relatively low strength. Lower friction, however, can be expected under moist conditions because water will act as a lubricant. Insertion of sand between the HDPE liner and the steel specimens caused the value of the coefficient of friction to decrease slightly by 5% from 0.43 to 0.41. This response is expected due to the relative high strength and size of these particles. The voids between the sand particles would also reduce the contact area between the two specimens.

Results of the Pullout Tests:

Results from the pullout testing of SDR 32.5 liner samples (Group A) and SDR 26 liner samples (Group B) are contained in Table 1. Pullout forces for each group of pipes were found to be relatively consistent. This had not been anticipated since using a forklift and chains to provide the pullout mechanism might have introduced some imprecision into the experiment.

Table 1. Pullout Force Test Results.

Sample	HDPE Liner		Time Sec	Force kN (lb)
	Diameter mm (in.)	SDR		
1-A	150.6 (6)	32.5	30	2.22 (500)
2-A	150.6 (6)	32.5	20	2.67 (600)
3-A	150.6 (6)	32.5	5	3.34 (750)
4-A	150.6 (6)	32.5	50	2.85 (640)
1-B	200.8 (8)	26	10	1.34 (300)
2-B*	200.8 (8)	26	20	4.89 (1,100)
3-B	200.8 (8)	26	25	6.23 (1,400)
4-B	200.8 (8)	26	18	7.79 (1,750)
5-B	200.8 (8)	26	5	8.23 (1,850)
6-B	200.8 (8)	26	5	10.01 (2,250)

* Liner was under-processed.

The average pullout force observed for the four samples tested in Group A was 2.76 kN (620 lb) with a standard deviation of 0.4 kN (90 lb). Sample 2-B was purposely under-processed for a previous experiment and it was therefore excluded from the average of Group B. Under-processing means that the liner sample was not reformed in the steel host pipe according to the standard installation procedure with regard to recommended temperature and/or pressure. The pullout force required to dislodge the under-processed sample was 4.89 kN (1,100 lb). As expected, this value was lower than the other samples as less contact surface exists between the under-processed liner and the host pipe. The reduction in contact surface lowers the frictional resistance and, consequently, lowers the required pullout force. For the five normally processed B samples, the required average pullout force was 6.72 kN (1,510 lb) with a standard deviation of 2.94 kN (660 lb). Sample 1-B dislodged at

1.34 kN (300 lb), a value much lower than the other four samples in Group B. The required average pullout force, excluding sample 1-B, was 8.05 kN (1,810 lb) with a standard deviation of 1.34 kN (300 lb).

NORMAL FORCE IN HDPE LINED CHAMBERS:

Based on a coefficient of friction of 0.46, the average pullout forces was used to calculate the average normal force in each chamber group. The average normal force for the four SDR 32.5 chambers was 6 kN (1,350 lb) based on an average pullout force of 2.76 kN (620 lb). The average normal force for five SDR 26 chambers was 14.6 kN (3,280 lb) based on an average pullout force of 6.72 kN (1,510 lb). If the extremely low value was excluded, the average normal force for the four SDR 26 chambers was 15.08 kN (3,930 lbs) based on an average pullout force of 8.05 kN (1,810 lb). Forces are reported rather than stresses at the interface, which is typically more useful, as stresses can not be calculated with enough confidence. It is practically impossible to determine the exact area of contact between a host pipe and a liner as it will vary from one sample to another. In addition, a liner will not be in full contact with the steel pipe at any given cross section and along the entire length of the steel pipe. Because the annular space between the HDPE liner and the steel host pipe may vary across the liner length, the actual contact area cannot be determined (Pechon, 1996). Chunduru (1996) reported that several factors would impact the size of the annular space including installation conditions and the quality of the host pipe. In addition, tolerances of the inner-diameter of the host pipe and the outer-diameter of the HDPE liner, which are functions of pipe material and diameter, would also impact the size of the resulting annular space.

Summary and Conclusions:

The coefficient of static friction (μ_s) between HDPE deformed/reformed liner material and a steel host pipe was found to be 0.46. Insertion of a dry fine grained soil between the two surfaces had a negligible effect on μ_s. If the fine grained soil had been moist, the coefficient of static friction would decrease due to the lubricant effect of water. Insertion of fine sand particles between the HDPE and the steel decreased the coefficient of static friction by about 5%. This thereby reduces the amount of force required to dislodge the liner once it has been installed.

Pullout tests were conducted to calculate the amount of force required to dislodge a HDPE liner from a steel host pipe. Results of the tests showed that the required pullout force depends on several parameters including pipe diameter and installation conditions. Given this information, it was possible to calculate the normal force exerted by the liner on the inside of the steel host pipe for the chambers tested. Knowledge of the size of the annular space between the host pipe and the liner would yield the amount of pressure exerted on the pipe. Additional interface and pullout tests are needed to determine the effect of moisture, other pipe materials, and temperature on the frictional resistance.

Acknowledgments:

This work was funded by a research grant provided by the Louisiana Education Quality Support Fund (LEQSF) and Pipe Liners, Inc. under contract No. LEQSF(93-96)-RD-3-12. The authors would like to express their gratitude to Ivan Mandich, Keith Herbert, San Hla Aung, Guiqin Yang, Chunduru Suryanarayana, Sean Pechon, Alexander AAlders, and Kimberly Kurtis for their assistance in the laboratory work and analysis of this project.

Appendix, References:

Bakeer and Barber (1995), "Evaluation of U-liner Technology for Trenchless Sewer Rehabilitation, Tech. Report Submitted to LEQSF, Dept. of Civil and Env. Engineering, Tulane University, LA, pp. 1-10.

Chunduru, S. (1996), "Material Performance and Numerical Methods of HDPE Used in Trenchless Technology," Ph.D. Dissertation, Dept. of Civil and Env. Engineering, Tulane University, LA.

Giancoli, D. (1984), General Physics, Prentice-Hall, Inc., Englewood Cliffs, NJ, pp. 68-69.

Kramer, R. S. (1992), An Introduction to Trenchless Technology, Van Nostrand Reinhold, New York, NY.

Nielsen, L.E. (1963), Mechanical Properties of Polymers, 2nd Ed, Reinhold Publ. Corp., Chapman & Hall Ltd, London, UK.

Pechon, S. (1996), "Effects of Installation Procedure on the Behavior of High Density Polyethylene Liner Pipe," Masters Thesis, Dept. of Civil and Env. Engrg., Tulane Univ., LA.

WEF/ASCE (1994), Existing Sewer Evaluation and Rehabilitation, WEF Manual of Practice FD-6, ASCE Manual and Report on Engrg. Practice No. 62, pp. 133-175.

HEAT TRANSFER IN THICK POLYETHYLENE LINERS

Reda M. Bakeer[1], Michael E. Barber[2] and Guiqin Yang[3]

Abstract:

Laboratory tests were conducted to examine the temperature distribution within the walls of thick HDPE and MDPE liners during heating. Hot air was blown inside a pipe section and temperatures were measured at various depths within the pipe wall at different time intervals. After about 3 hrs of heating, the average outside temperature of MDPE SDR 17 liners was 92 percent of the average internal skin temperature and 72 percent of the average internal air temperature. The HDPE liners followed similar trends although heat was transferred at a faster rate. After less than 1 hr of heating, the average outside temperature of HDPE SDR 26 liners was 95 percent of the average internal skin temperature and 81 percent of the average internal air temperature. Test results are currently being used to develop a constitutive model for polyethylene. The model will be used in future finite element analysis.

Introduction:

Leaking pipes may produce catastrophic environmental and economical repercussions including infiltration, inflow, groundwater contamination, ground subsidence and soil erosion. Recently, municipalities have employed new "Trenchless Technologies" for sewer rehabilitation. The term "Trenchless, or No-Dig" refers to a broad class of rehabilitation techniques whose common denominator is

[1] Assoc. Prof., Dept. of Civ. and Env. Engrg., Tulane Univ., New Orleans, LA 70118.
[2] Asst. Prof., Dept. of Civ. and Env. Engrg., Washington State Univ., Pullman, WA 99164.
[3] Former Grad. Stud., Dept. of Civ. and Env. Engrg., Tulane Univ., New Orleans, LA 70118.

that none of them requires excavation or removal of existing pipes. Trenchless techniques include a wide variety of procedures, such as chemical grouting, pipe jacking, plastic lining and sliplining (EPA, 1991). A comprehensive review of trenchless technologies along with their main advantages and disadvantages is provided by Schrock (1992). An updated manual by ASCE/WEF (1994) also helps explain the various trenchless methods.

One such trenchless method utilizes high or medium density polyethylene (HDPE or MDPE) liners. The liner pipe is manufactured round then deformed, or folded, into a U-shape. The installation procedure utilizes a combination of pressure and high temperature steam to reform the liner pipe back to form a close fit with the host pipe. A key element to the performance of the liner is insuring that proper heating occurs during installation. Quality installation is assured by closely following a processing protocol developed specifically for the given in-situ conditions. Installers are required to measure the internal temperature at both ends of a liner as part of their QA/QC procedure. High temperature is needed to soften the material and allow it to conform gradually to the host pipe configuration under low internal pressure. Obviously, the time it takes for the temperature to transfer through the liner (e.g., the time difference), effects the time in which the liner can be processed. Moreover, since the process monitors internal steam temperature in the liner, a theoretical model is needed that correlates measured internal temperatures with outside ground conditions. Moreover, optimizing the time required for heating may reduce the energy cost and time of installation. This information is also essential in developing a material response model for representing long-term performance of thick liners under a variety of possible loading and in-situ conditions.

Testing Procedure:

Laboratory tests were conducted to explore heat transfer properties of thick HDPE and MDPE liners. Two specific types of tests were conducted;(1) pressurized steam tests conducted on confined liners to model a simulated installation; and (2) hot air tests on thick liners to examine temperature gradients within the liner wall. Because of space limitations, only a brief summary of the tests and the results is given. Readers interested in more specific details should see Yang (1995).

(a) Heat Transfer Under Pressure

A 0.92 m long section was cut from a 0.46 m diameter schedule 40 steel pipe. A SDR 26 HDPE liner; 2.44 m long, 0.41 m diameter and 15.6 mm thick; was placed inside the test section leaving approximately 0.76 m of the liner extending outside each end of the setup. Teflon insulted thermocouples were placed inside the liner and on its inside skin to measure the steam and skin temperatures. Each thermocouple was connected to an electronic digital thermometer. The liner was then inflated inside the steel pipe according to the steam injection method used in the field. Pressure inside the liner was increased gradually from 0 to 165.5 kN/m^2 over a 35 minutes period until the liner was completely reformed. Pressure was measured by a mechanical gauge installed on the steam inlet. Temperature measurements were taken every 5 minutes from the two thermocouples. The ambient temperature at the start of the test was 17 °C. Steam temperature was raised gradually to 107 °C in the first 5 minutes of the test and was held between 107 and 124 °C in the next 30 minutes.

(b) Temperature Gradients in Thick Liners

The second test series was conducted to examine temperature gradients within the thick liner wall. Tests were conducted to examine the influence of liner thickness, thermal heating and cooling, type of polyethylene and location within the pipe length. The new liner sections were all less than 2.44 m long so that temperature could be controlled with greater accuracy. One end of the liner was sealed with two plywood circular disks attached with steel angles and screws. A 25.4 mm diameter hole was drilled at the center of the disks.

Six thermocouples connected to electronic digital thermometers were installed in full contact with the liner at different locations. One thermocouple was placed inside the liner to measure the inside air temperature. Temperatures within the liner wall thickness were measured by thermocouples attached on the inside and outside surfaces of the wall and three thermocouples inserted in holes spaced 25.4 mm apart that were drilled from the outside surface of the liner to depths of 0.75t, 0.5t, and 0.25t, where t is the liner thickness. In some tests, this setup was implemented at two sections along the liner length to determine the temperature differences with respect to the cooling of air as it moved along the pipe. Readings were taken from the thermometers at 1 min intervals. An oil-fired turbo heater capable of providing

100,000 BTU was used to blow hot air through the open end of the liner while allowing the air to flow out through the hole at the opposite end. No internal pressure was used in this series of tests.

A total of 11 tests were conducted on the new setups. The setup was allowed to cool down to ambient room temperature of about 27 °C prior to commencing the following test. Air temperature inside the liner was raised or lowered at various rates in order to provide a wide spectrum of measurements. The duration of the tests ranged from 35 to 180 minutes.

Results of the Tests:

Testing conditions were not identical to actual field conditions but were sufficient enough to provide insight on heat transfer characteristics of thick PE liners. Tests were conducted on relatively short sections and the results may differ in a long pipeline. In addition, temperature distribution tests were conducted using hot air rather than pressurized steam which is used in the field. Ambient conditions in the field may also vary significantly from one site to another. In addition, most field installation involve thinner liners.

(a) Heat Transfer Under Pressure

The results shown in Table 1 indicate that temperature of the inside skin of a liner is not the same as the steam temperature. Temperature of the inside skin of the liner continued to increase gradually from 35 °C at the beginning of the test to a maximum of 99 °C at the end. Figure 1 indicates that temperature of the inside skin would have reached a steady-state in approximately 50 to 60 minutes if the steam temperature were held at 121 °C, but the test was halted after 35 minutes when the liner has been completely reformed to closely fit the host pipe. The drop in steam temperature during the test was reflected on the measured liner temperature after about 5 minutes. Accordingly, a time difference can be expected in the response of a HDPE liner to a measurable change in the steam temperature. Time difference is a function of several parameters including material density and liner thickness. Temperature ratio of the inside skin to steam (I/S) increased drastically after 10 minutes to about 0.66. The final I/S ratio reached 0.83 when the liner was completely reformed and the pressurized steam had been released.

Table 1. Results of the Pressure Test

t	P_i	Steam		Skin		I/S
		T	Δ	T	Δ	
1	0	107	162	35	11	0.42
5	69	107	0	38	5	0.44
10	35	102	-4	61	42	0.66
15	103	107	5	69	11	0.70
20	152	121	11	74	6	0.66
25	152	118	-2	86	13	0.76
30	159	121	2	94	7	0.80
35	166	124	2	99	5	0.83

t = Elapsed time in minutes, P_i = Internal pressure in kN/m^2, T = temperature in °C, Δ = Change from previous reading in percent, I/S = Ratio of inside skin temperature to steam temperature.

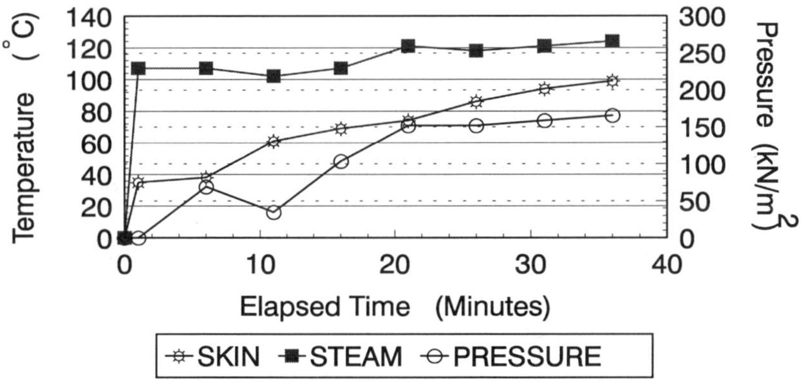

Figure 1. Heat Transfer During a Pressure Test

(b) Temperature Gradients in Thick Liners

Typical curves of internal air temperature and temperatures along the liner thickness are shown in Figure 2. As shown, a steep increase in the internal air temperature after approximately 18 minutes does affect the temperatures along the liner thickness for 10 to 15 min. After heating for 90 to 100 minutes, temperature of the liner walls appear to approach a steady-state at temperatures slightly below the internal air temperature. As expected, temperature was not distributed uniformly across the thickness of the liner in all tests. For the particular test shown, there is a measurable difference between the internal and external skin temperatures of approximately 9 °C. This may not be the case in much thinner liners.

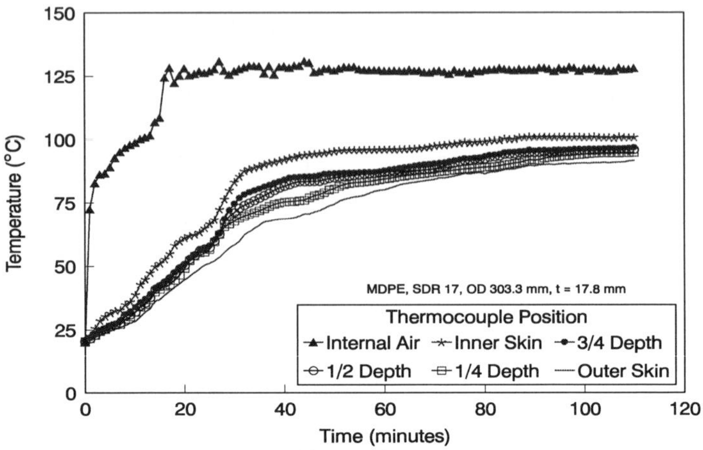

Figure 2. Temperature Gradient in a no Pressure Test

Table 2 summarizes the results of four tests conducted on thick HDPE liners and seven tests on thick MDPE liners. Results of all tests were similar to those presented in Figure 2. Temperatures decreased gradually within the liner wall where the deeper points closer to the inside skin were much warmer that the points near the outside skin. In a short duration test, the temperature difference between the inner skin of the liner and its outer skin was over 38 °C after two minutes, but the difference dropped to about 5 °C at the end of the test. In longer tests, the temperatures differences between the outside and inside surfaces of the liner were much less as a steady-state was reached. The difference in response can be attributed to the slower heating of the liner in the longer tests which allowed the distribution to become more uniform within the material.

An interesting point can also be seen at the internal and external interfaces. While the temperature difference between the 0.75t point and the 0.5t point was less than 1 °C and the temperature difference between the 0.25t and the 0.5t points was 2 °C at the end of the test, the temperature flux across the internal and external surfaces were 6 and 5 °C, respectively. This is a result of film coefficients and is indicative of temperature drops that often occur at phase change locations (Carslaw and Jaeger, 1959). All tests showed the presence of a time difference between the response of the liner to a given change in the internal air temperature. This was more evident with the major drop in temperature experienced during one of the shorter tests. As expected, the inner surface of the liner had a shorter time difference and was more sensitive to the changes in hot air temperature than the other points within and on the outside skin.

Table 2. Thermal Gradients in a no Pressure Test.

	Time	Temperature °C					
	min	Air	I	3/4	1/2	1/4	O
A	50	128	108	105	103	101	99
A	35	127	102	101	97	NA	95
A	51	129	109	108	107	111	106
B	34	NA	122	113	107	102	99
C	180	128	78	76	71	68	66
C	110	128	101	97	97	95	92
C	60	128	91	87	85	84	82
D	75	127	104	102	101	NA	92
D	70	129	103	NA	101	99	96
D	110	132	122	NA	NA	NA	117
C	180	129	112	NA	NA	NA	102

I = Inside skin temperature, O = Outside skin temperature.
A = SDR 24 HDPE liner 0.30 m in diameter and 12.7 mm thick.
B = SDR 26 HDPE liner 0.41 m in diameter and 15.8 mm thick.
C = SDR 17 MDPE liner 0.30 m in diameter and 17.8 mm thick.
D = SDR 17 MDPE liner 0.20 m in diameter and 11.7 mm thick.

Summary and Conclusions:

Tests were conducted to examine heat transfer characteristics of thick HDPE and MDPE liners during reforming process. It was found that the inside skin temperature reaches about 80 percent of the temperature of the steam used in the process upon completion of reforming in a laboratory setup. The tests also indicated that the temperature on the outside skin is lower than that on the inside skin and that the difference becomes smaller with time when the internal temperature is maintained at a constant level. Temperature distribution within the skin of thick liners remains non-uniform throughout the processing period. At near steady-state conditions a smaller temperature variation is present. The time to reach steady-state varies with liner material and its thickness. Disruptions or sudden increases in the internal air temperature will take approximately 15 minutes to effect a thick liner.

Results of the heat transfer tests are currently being used in conjunction with the results of buckling tests on PE liners to develop a constitutive model for the use in finite element analysis of PE liners. A comprehensive paper containing the results will be published upon the completion of the study.

Acknowledgments:

This work was funded by a research grant provided by the Louisiana Education Quality Support Fund (LEQSF) and Pipe Liners, Inc. under contract No. LEQSF(93-96)-RD-3-12. The authors would like to express their gratitude to Ivan Mandich, Keith Hebert, San Hla Aung, Chunduru Suryanarayana, and John Taylor, Sean Pechon for their assistance in the laboratory work and analysis of this project.

Appendix - References:

_____ (1994) Existing Sewer Evaluation & Rehabilitation, Manuals and Reports on Eng. Practice No. 62, 2nd edition, ASCE, New York, NY, Manual of Practice FD-6, WEF, Washington, D.C.

_____ (1991) Handbook - Sewer System Infrastructure Analysis and Rehabilitation, Rep. No. EPA/625/6-91/030, US Environmental Protection Agency, Ctr. for Env. Res. Info., Cincinnati, OH.

Bakeer, R.M., M.E. Barber, R. Subramanian, and I.C. Mandich, (1994). "Effect of Different Fluids on Mechanical Properties of a U-Liner", Proc. Conf. Localized Damage, Udine, Italy, June 21-23.

Bakeer, R.M. and M.E. Barber, (1995). "Evaluation of the U-Liner Technology for Trenchless Sewer Rehabilitation," Interim Report to Louisiana Education Quality Support Fund (LEQSF), Baton Rouge, LA.

Carslaw, H.S. and J.C. Jaeger, (1959). *Conduction of Heat in Solids*, 2nd Ed., Oxford Univ. Press, Oxford, England.

Schrock, B.J. (1992). "Pipeline Rehabilitation: A Guide to a Growing Market," NUCA, August, pp. 14-21.

Yang, G. (1995). "Effect of heat on the Behavior of Polyethylene Pipes," MS Thesis, Tulane Univ., New Orleans, LA.

Deformation Measurement of Liners During Buckling Tests

Alexander C. Aalders[1], Reda M. Bakeer[2], and Michael E. Barber[3]

Abstract:

Several methods were employed to measure the deformation of a polyethylene liner during a buckling test. In the test, a deformed (folded) liner is reformed back to closely fit a steel casing pipe then it is subjected to a uniform radial pressure. The test simulates a field situation where uniform pressure is being exerted on a liner by the groundwater seeping through the defects in a deteriorated host pipe. Simple methods, such as shining a strong light through one end of the pipe or videotaping, were found to be adequate to record failure patterns; but not accurate enough to provide exact measurement of deformation. Volumetric changes were detected by filling the liner with water and measuring the amount of water being squeezed out of the liner under the applied pressure. A mechanical device was specially designed to provide a 3-D scan of the longitudinal profile of the liner during loading. Measurements are taken every 40 degrees along the pipe's axis using an internal electro-mechanical unit connected to a data acquisition system.

Introduction:

The term trenchless technology encompasses a wide variety of methods used for inspection, installation, replacement, or rehabilitation of utility lines; or for leak detection with minimum ground excavation. Rehabilitation of sewer networks using a polyethylene

---

[1] Grad. Stud., Dept. of Civ. and Env. Engrg., Tulane Univ., New Orleans, LA 70118.

[2] Assoc. Prof., Dept. of Civ. and Env. Engrg., Tulane Univ., New Orleans, LA 70118.

[3] Asst. Prof., Dept. of Civ. and Env. Engrg., Washington State Univ., Pullman, WA 99164.

(PE) deformed/reformed liner is one of many methods currently available in the growing trenchless technology market.

This paper gives a summary of an ongoing research on the buckling characteristics of deformed/reformed PE liners. Several buckling tests under uniform radial pressure were conducted on polyethylene liners (Bakeer and Barber 1997). A buckling test simulates the situation of a deteriorated underground gravity sewer line fitted with a flexible liner and subjected to a uniform groundwater pressure. In a buckling test, a liner is reformed inside a steel chamber pipe then subjected to a uniform radial pressure. Three types of buckling tests were performed in this study (Pechon, 1996); namely short-, medium- and long-term tests. These tests may not fully represent the exact field situation due to some factors such as the differences between a typical underground host pipe and the tested specimens with respect to boundary conditions, length and integrity. However, the tests provide useful insight with regard to buckling performance of polyethylene liners. This simulation allows for examining various parameters such as collapse pressure, safe design pressure, volume reduction, cross-sectional changes, and deformation. Different methods were used to measure the deformation of the PE liner during buckling tests. These methods included: light projection, videotaping, laser beams, strain gages, water measurement, and an electro-mechanical measuring device.

Buckling Tests:

A typical buckling test setup consists of a rigid steel chamber pipe fitted with threaded inlet holes. Two circular steel rings are welded to each end of the steel casing pipe, as shown in Figure 1. The testing chamber is then fitted with a PE liner of appropriate length. Liners with Standard Diameter Ratio (SDR) of 32.5 and 26 were mostly used in the tests as these are the most common sizes used in the field. SDR is an industry standard which defines the required minimum thickness of a pipe wall as a function of its diameter. The liner is reformed to closely fit the casing using a process that is as close as possible to the field installation procedure. A second steel ring is then bolted to the welded ring on each end. The two rings hold in-place the flared ends of the PE liner to create a tightly sealed annular space between the outside surface of the liner and the inside surface of the casing. This condition will produce some restraint at the liner's ends which may not be present in the field, but it is necessary to create the tightly sealed space. The threaded holes on the steel casing are used to apply, regulate and measure the air pressure in the annular space. An air compressor is used to supply the required pressure. Mechanical dial gages are used for pressure measurements and are calibrated prior to each test.

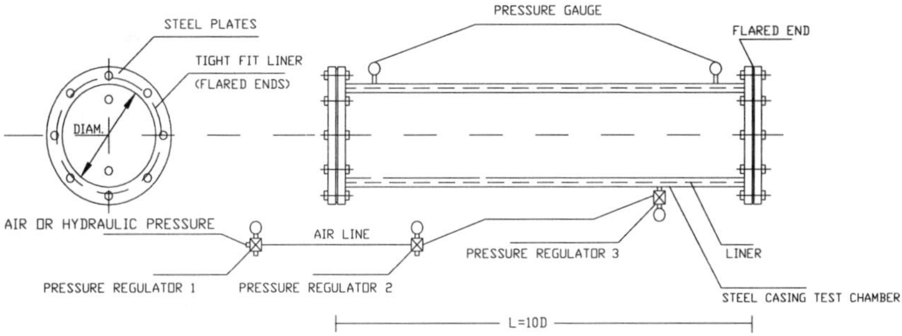

Figure 1. A Typical Testing Setup.

A short-term test is performed by increasing the uniform pressure in equal increments until the liner collapses which occurs typically within an hour or more. In the field, this condition simulates the gradual application of a load; such as a temporary surcharge or a rise of the groundwater table. In a medium-term test, a new pressure increment is applied every 24 hours to allow the liner to adjust to the applied pressure. The medium-term test requires several days to complete and it is terminated when the liner collapses. This test models a slowly applied load such as the construction of an embankment over an underground line. Collapse in both tests is defined by the loss of function of the pipeline due to either excessive deformation or structural failure of the liner. A long-term test is performed over a 10,000 hour period using a low pressure representing a typical service condition. In a long-term test, the liner is not expected to collapse, but it would experience some volumetric deformation and creep under the service pressure.

Several tests were performed on confined liners with different length-to-diameter ratios, deformed cross sections, and SDR's to establish their collapse pressure as well as to design the ultimate testing setup and procedure. The tests revealed that a setup with a length-to-diameter ratio of 10:1 is adequate to eliminate end effects (Bakeer and Barber 1997). The influence of end restraints is severe on the performance of shorter setups, while the tested longer setups yielded similar results as the 10:1 setups. Accordingly, use of a setup with a 10:1 ratio was deemed adequate to represent the much longer pipelines in the field with respect to buckling behavior.

Light Projection:

A strong light source was projected through one end of the pipe on a translucent plastic plate mounted on the opposite end. A sheet of engineering paper was affixed to the plastic plate which allowed for the shadow of the deformed liner to be traced during each pressure increment. The traced shadow was then transformed into a computer image and the deformed area was calculated using an electronic digitizer and a CAD software. This method had its limitations. Failure occurred at a very fast pace in short-term tests which made it extremely difficult to trace the deformed shape. Calculation of the deformed cross-sectional area of the liner using the traces proved to be too simple, but was equally crude. This is due to the non-uniform deformation pattern along the liner span. The projected light always cast the shadow of the most deformed section. Accordingly, when this area was used to calculate the reduction in the volume of the liner pipe it yielded an overestimated value. In addition, the liner's deformation becomes excessive during collapse such that it prevents any light from passing through.

A short-term test was performed on a SDR 26 HDPE liner installed in a steel pipe 1.6 m long with an inside diameter of 0.27 m. The liner was loaded over a 5 minute period with consecutive pressure increments of 36 kN/m^2 until the liner collapsed at 144 kN/m^2 (14 m of water head). Collapse was defined as loss of function due to excessive deformation, but the liner did not rupture. The collapse pattern was non-uniform where the liner bulged inward at the middle of the setup on one side and near both ends on the opposite side. The traced shadows showed that the reduction in the cross sectional area of the liner was less than 2% under a pressure of 72 kN/m^2 (7 m of water). The cross sectional area decreased by about 36% under a pressure of 108 kN/m^2 and 65% at collapse. It should be noted that this test was extremely quick and high pressure increments were used. This procedure was used in the early stages of this research to develop the setups and testing procedures. In later stages, similar liners have shown much higher collapse pressure in slower tests and under lower pressure increments (Bakeer and Barber, 1997).

Video Taping:

A video camera was used to record the liner's deformation during buckling tests. Video taping provided a reliable method by which the rapid change of the liner's cross-section during collapse could be monitored. The process worked satisfactorily with short- and medium-term tests. The camera was also used overnight to record collapse during long-term tests, if it does occur. Some video frames were transformed into computer images and calculation of the area was done with a CAD software. Like the light projection method, video images only show maximum profile of deformation and their interpretation depends on personal judgment.

Laser Measurements:

A setup consisting of several laser beam sources mounted on a guiding frame was used to detect deformations. The setup was found to be impractical. It was difficult to manually adjust the laser beams to follow the progress of deformation especially when using multiple beams to construct a 3-D profile. In addition, the laser beam would only monitor the point with maximum deformation which would mask all other locations along the laser path. It was found that a laser setup can only be used to monitor a long-term test under low pressure where collapse is not expected. However this technology was not utilized in long-term tests since 20 different setups were tested simultaneously and the cost of the required laser equipment became prohibitive.

Strain Gages:

Use of strain gages mounted on the inside wall of the liner at different locations was found to be not feasible. PE does not allow for adhesives or resins to bond with adequate strength to attach a strain gage. Only butt fusion was found to be possible, but was ruled out as it would change the material properties in the fusion areas and accordingly would affect the measured buckling resistance (Bakeer and Barber 1997). It was also feared that drilling or shaving the liner wall to attach a strain gage would also impact the strength of the liner.

Water Measurement:

Change in the volume of the liner pipe was detected by filling it with water then measuring the amount of water being squeezed out during the test. Two lexan or plexiglass plates were affixed tightly to the ends of the steel casing. Six threaded holes were drilled in each plate and pipe sections with valves were affixed in the holes. The pipe sections were used to fill the liner with water, remove entrapped air, and measure volumetric changes in the liner pipe. The liner pipe was then fully filled with water, and air was used to apply uniform external pressure on the liner. Deformation of the liner under pressure caused the water to be squeezed out which was collected and measured.

This measurement system was used successfully in medium- and long-term tests. For example, a medium-term test was performed on a 0.20 m diameter SDR 32.5 HDPE liner installed in a steel casing with 10:1 length to diameter ratio. Change in the volume of the liner pipe was monitored using water. The test was performed over a 26 day period using uniform pressures of 43 kN/m^2 for 6 days, 58 kN/m^2 for 7 days and 72 kN/m^2 for 3 days. Results of the tests showed that volume of the liner pipe was reduced by 1, 1.5 and 2 percent of its original volume at the end of each pressure increment, respectively. In long-term tests, small volumetric changes continued to occur in the liner due to creep. A creep curve was generated by plotting the measured volumetric changes over the 10,000 hr period (Taylor 1996).

LVDT Deformation Measuring Device:

Studies were made at Ohio University (Sargand, 1995) on two large polyvinyl chloride (PVC) buried pipes (457 and 914 mm in diameter). The pipes were fully instrumented and loaded to failure. To measure deflection, a specially designed rotational LVDT measuring system was deployed. This design was found to be impractical for the small diameter pipes used in the buckling tests (152 and 203 mm). An internal measuring device was opted for because it required less space, would remain stationary with respect to the pipe, and was transportable from one setup to another.

The design of the deformation measuring device allows for accurate measurement of the liner's cross-section. The apparatus can be attached directly to the end rings of the same testing setup. Due to the small diameter of the tested liner pipe (188 mm), the device used would only allow for measurements of less than 5 percent of the liner's cross-section (9.4 mm). This value is acceptable based on the tolerance required for typical cleaning and inspection equipment to move freely along a pipeline without jamming. Under actual field conditions, a liner should not experience higher value to ensure proper maintenance. It should be noted that the pressure used in medium- or long-term tests is typically much lower than the collapse pressure of the liner.

The deformation measuring device shown in Figure 2 employs a threaded rod supporting a LVDT (linear voltage displacement transducer) via a specially fitted aluminum carriage. The LVDT is attached to a computerized data acquisition unit for collecting deformation data along the liner pipe. The operator can adjust the sampling rate and duration of the data acquisition via the computer. The threaded rod is driven by an electric motor fitted with a gear reduction box and controlled by an adjustable speed control box. The LVDT carriage is mechanically steered by the rotation of the threaded rod while the carriage is held longitudinally in position at the selected angle. The apparatus is capable of measuring the longitudinal profile at any point along the interior wall of the liner, although it is specifically oriented to rotate at intervals of 40 degrees about the longitudinal axis of the liner pipe. The motor sits atop the L-brackets (parts 23 and 24) and is connected to the threaded rod by means of a coupling (part 37). The testing chamber's other end is identical except for not having parts 1-7, 11, 12, 20-24, 30, 31, and 37. Parts 1-7 are included in this list as there is only one LVDT measuring carriage. This series of tests is currently in progress.

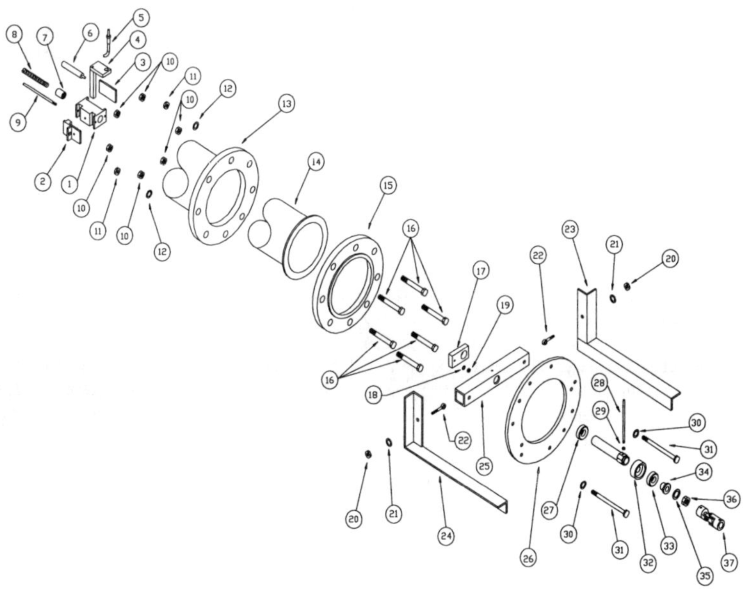

Figure 2. Deflection Measurement Device.

Conclusions:

Several methods for measuring the deformation of a polyethylene liner during a buckling test were utilized in a comprehensive study. Buckling tests were used to model different field situations when the groundwater flows through the cracks in an underground pipeline. Shining a strong light through the liner pipe or videotaping the progress of deformation can be used to record failure pattern, but cannot be used to measure the exact deformation. Water was used to measure the change in the volume of the liner pipe during medium- and long-term buckling tests. This procedure was found to be simple and it yielded valuable data regarding creep in PE liners. A mechanical device was specially designed to measure the longitudinal profile of deformation of the liner during loading. Measurements were taken during medium-term tests along the span of the liner using a comuterized electro-mechanical unit.

Acknowledgments:

This work was funded by a research grant provided by the Louisiana Education Quality Support Fund (LEQSF) and Pipe Liners, Inc. Under contract No. LEQSF(93-96) -RD-3-12. The authors would like to express their gratitude to Ivan Mandich, Keith Herbert, San Hla Aung, Chunduru Suryanarayana, and Sean Pechon for their assistance in the laboratory work and analysis of this project.

Appendix - References:

Bakeer, R.M. and Barber, M.E. (1997), "Evaluation of U-liner Technology for Trenchless Sewer Rehabilitation," Technical Report Submitted to Louisiana Board of Regents, LEQSF, Baton Rouge, LA.

Pechon, S. (1996), "Effects of Installation Procedure on the Behavior of High Density Polyethylene Liner Pipe," Masters Thesis, Dept. Of Civ. and Env. Engrg., Tulane Univ., LA.

Sargand, S.M., Hazen, G.A., Liu X.A. (1995), "Structural Performance of Buried PVC Pipes Under Large Distributed Load," Transportation Research Board, 74th Annual Meeting, Washington, D.C.

Taylor, J. (1996), "Material Performance of HDPE used in Deformed/Reformed Liner System," Masters Thesis, Dept. of Civ. and Env. Engrg., Tulane Univ., New Orleans, LA.

Internal Joint Seals Extend Life of 50-Year Old Pipeline

Michael Cronin[1]

Abstract

There are many trenchless pipeline rehabilitation methods available to the industry. Each one of these methods are suited for their particular application. Relining and the insertion of a small diameter pipe in the existing pipe are the most recognized forms of trenchless technology. There are, however, many other rehabilitation products and procedures that play an equally significant role in the trenchless rehabilitation industry.

The oldest form of pipeline rehabilitation, particularly in the man-entry pipes or structures, is point repairs/spot repairs. In non man-entry, trenchless point repairs are available today in a myriad of methods, materials, and equipment. When man-entry rehabilitation is added to the equation, repair possibilities are endless.

In current economic times, the priority to accomplish more work for the dollars spent must be balanced by the need to satisfy concerns for improved public relations, minimal environmental impact, speed of accomplishment, quality, performance, and permanence. The internal joint sealing process utilizing an elastomeric joint seal with preformed retaining bands meets this criteria. They provide a mechanical seal on the inside of the pipe joint that gives a flexible, noncorrodible, and bottle-tight joint repair. A proven system of pipeline rehabilitation, internal joint seals offer a fast, safe, effective and efficient repair for underground pipelines.

---

[1]District Manager of Miller Pipeline Corporation, 8850 Crawfordsville Road, Indianapolis, IN 46234

Background

Internal joint seals have been utilized as an effective joint repair worldwide in large diameter piping 406 mm (16" and larger) since 1966 with continuous changes and modifications of the seals to improve reliability, durability, and broaden the application ranges. This method of repair requires safe man-entry and, therefore, precludes any pipelines smaller than 406 mm (16").

The internal joint seals have been installed in water, natural gas, wastewater, nuclear, and other industrial piping. The first North American installation in natural gas was in 1974, followed by industrial piping in 1977. The first North American installation in water was in 1982, with wastewater in 1983. Nuclear and fossil fuel power installations occurred first in 1987.

Pipes manufactured from cast iron, ductile iron, reinforced concrete, prestressed concrete cylinder pipe, steel, polyethylene, PVC, vitrified clay, fiberglass, fiberglass composite, and corrugated metal pipe have received extended life spans with the use of internal joint seals. With continuous evolution and product improvements, this allowed installation in not only round pipe but elliptical pipe, square and rectangular conduits, horseshoe shaped pipe, and other nonround structures. The internal seal is a flexible rubber leak clamp that provides a bottle-tight seal around the full inside circumference of the pipe joint area. Its design features a series of lip seals that ensure leakproof sealing on both sides of a joint. The seals for water and wastewater are manufactured from EPDM (Ethylene Propylene Diene Monomer) rubber. This material exhibits superior properties in resistance to ozone, weather, heat, low temperatures, water, chemical, and microbiological bacteria. Other polymers are used for natural gas and other applications.

All metal components for sealing are made from a variety of stainless alloys and are permanent, noncorrodible, and able to handle operating pressures up to 2068 kPa (300 psi). Flexibility of the seal can accommodate normal pipe movement resulting from ground shifting, thermal expansion or contraction, and vibration.

Not all internal joint seals are installed by the manufacturer. However, Miller Pipeline Corporation not only manufactures the

AMEX-10/WEKO-SEAL Internal Joint Sealing System, but also installs the seals with specially trained confined-space technicians. These technicians not only are trained to install the internal joint seals but also undergo extensive training to comply with all the requirements of the United States Federal OSHA's Permit-Required Confined-Space Regulation 29 CFR 1910.146. By design, the internal joint sealing method greatly reduces the number of excavations where manholes cannot be utilized for man-access. Example: 183 meters (600 feet) of 508 mm (20") pipeline can be sealed with only two access openings and 610 meters (2,000 feet) of 914 mm (36") pipeline can be sealed with only two access openings. Predetermined access points can be at excavations with pipe cutouts or through manholes.

Problem

The Massachusetts Water Resources Authority (MWRA), Boston, Massachusetts, which serves 43 cities and towns throughout Eastern Massachusetts recently decided to use this modern trenchless technology to rehabilitate a portion of one of their sewer pipelines that has been in continuous service for 50 years. This particular 914 mm (36") reinforced concrete pipeline had been plagued with leaking joints that allowed considerable infiltration of ground water into the sewer system which then increased the flow into the treatment plant.

A recent closed circuit television inspection revealed that the 914 mm (36") gravity flow pipeline was in excellent condition structurally. The television inspection clearly showed infiltration of varying degrees taking place at the joints which are at 2.5 meters (8-foot) intervals.

Alternative Analysis

MWRA completed an alternative analysis prior to selecting the internal joint seal. This analysis included cured-in-place-pipe (CIPP) repair and chemical grouting. Since CIPP provides a continuous liner at a greater cost than internal joint seals, the MWRA rejected this method of repair because they could not justify lining this entire length of pipe that was in excellent condition with the exception of leaking joints.

An alternative method of joint repair was chemical grouting. This process was not selected because the MWRA was not confident that chemical grouting would withstand the hydrostatic loads in wet weather.

Since joint leakage was the only identified problem with this pipeline, the internal joint seal was selected as the best method of rehabilitation.

Solution

There were a total of 176 joint to be repaired in 427

meters (1,400 lineal feet) of pipe. The contractor/installer indicated that the work could be performed in a short period of time, thereby reducing any inconvenience to the general public and to the utility. In order to safely accomplish the man-entry into the pipeline to install the seals, flow diversion by-pass pumping up to 80,000 m^3 (21 million gallons) per day was required. After the preliminary by-pass pumping was accomplished, the technicians were able to safely enter the pipeline and to begin installation.

The first phase of the internal joint sealing procedure is to inspect, clean, and map the pipeline. Once the project is completed, a map of the pipeline is given to the customer for their records. The pipeline must be as clean as possible to provide a safe working environment for the technicians. During pipe cleaning operations, any gaps between the joints must be cleaned of dust and debris leaving a clean area for joint filling. The joints are filled with mortar and rendered flush with the internal surface of the pipe. The mortar has no joint sealing capabilities and is used only as a backing for the internal seal.

Any imperfections along the pipe wall in the joint area are removed. The equipment that is used varies depending on how much preparation is required (pipe ID grinder, hand scrapers, wire brush, etc.).

Immediately prior to fitting the seal, the joint area must be cleaned with a dry brush and coated with a nontoxic vegetable base lubricate compatible with the composition of the internal joint seal. The lubricant is used as an aid to fitting the internal joint seal tightly up against the pipe wall.

The internal joint seal is placed in position bridging the joint gap. Before the retaining bands are placed in the grooves provided in the seal, radiused shims are placed underneath the wedge area in the grooves to provide a bridge that will transmit the radial load evenly as the retaining bands are expanded. A hydraulic expanding device is used to apply a set pressure to the retaining bands of the internal joint seal. A radiused locking piece called a "wedge" is fitted between the exposed gap of the expanded band ends.

There are two individual pressure tests applied to the internal joint seal. The first regulated pressure test of 69 kPa (10 psig) is applied through the test valve which is located in the body of the seal. The outer edge of the seal is tested with soap and water. In the second test, a regulated 34 kPa (5 psig) test is introduced through the valve and the pressure is sustained while a soap and water test is applied to the entire body of the seal.

After the final test, the test valve is sealed with a counter sunk hex head completion plug using a nontoxic thread sealing compound.

Upon completion, a final visual inspection on the entire length of the pipeline is performed.

The MWRA internal seal project was safely completed in only four (4) working days. Once the project was completed, a post-video was provided to the MWRA for their permanent records.

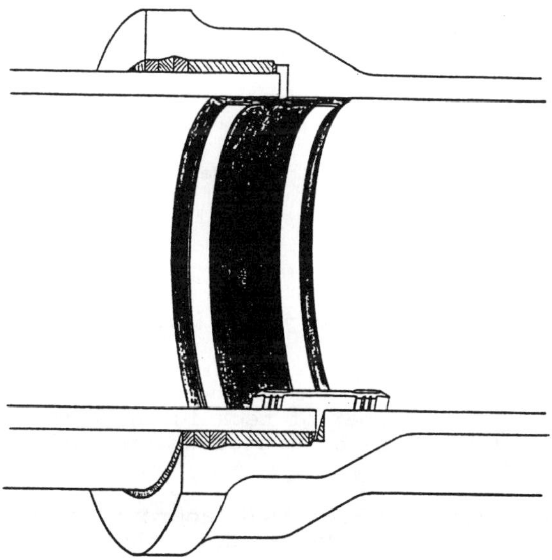

Figure 1. Round Seal Drawing.

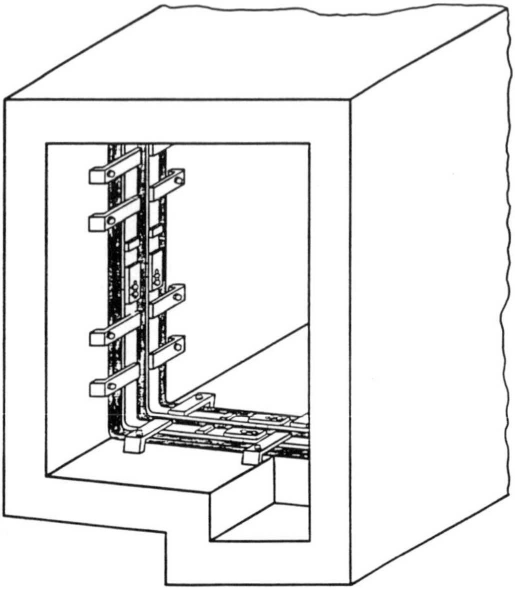

Figure 2. Square Seal Drawing

Cost Analysis

The 176 - 3454 mm (136") joints were manufactured and installed for $800.00 per joint (2.5 meters (8') joints).

All internal sealing projects are unique, therefore, the price of the project is determined by the size of the pipe, number of joints, number of technicians needed to complete a project, difficulty of installation, and mobilization/ demobilization.

CIPP would have cost the MWRA approximately $180.00 per foot for installation compared to the internal seals at $100.00 per foot. CIPP installation would have also meant a greater cost to the MWRA because the pipeline could not have been rehabilitated in the four (4) working days it took the internal joint sealing project, thereby creating additional costs for by-pass pumping, police detail, MWRA employees, and possible excavations as well as additional public inconvenience.

Open cuts and pipeline replacement were not even considered because the price would have been cost prohibitive.

Conclusion

This was the first project completed directly between the MWRA and the contractor/installer. In the past, successful projects had been completed with other layers of engineers and general contractors between the MWRA and the internal joint seal provider.

The MWRA indicated that this project was a good experience for many reasons: (1) The Sewer Division now has first-hand knowledge of what it takes to safely complete a joint sealing project of this size; (2) They now are aware of the many advantages of this trenchless technology system; (3) They believe that there can be future applications not only for capital programs but for spot repair work realizing significant cost savings over conventional repair methods.

Considering the excellent condition of the rehabilitated pipeline, it is conceivable that this pipeline will provide many more decades of service to the MWRA and their customers.

Many utilities now use the internal joint seal as the primary method of repair for man-accessible pipelines.

Grouting Around Trenchless Repairs
The Forgotten Technology

Richard M. Berry, MASCE[1]
Robert R. Farrar, MASCE[2]

Abstract

The last decade has seen the rapid growth of trenchless and no-dig methods to rehabilitate sewer lines, gas lines, and other underground utilities. Starting with slipline, progressing through numerous other complete and partial relining methods, and continuing today with pipe bursting and small scale directional drilling, the trend continues. However, such methods do not always economically address all the problems in a system including required tie-ins, and continuing leakage into or around the system. Eliminating excavation at industrial sites is particularly important considering pollution possibilities. Infiltration can continue, toxic contaminants can still flow, pipeline subsidence can continue, and adjacent structures can be further damaged.

Grouting offers a solution to some of these problems. When it works, its cost is normally a small fraction of the cost of other alternatives. This paper covers a number of projects where grouting was used to remediate after some new form of relining had been completed. In other cases, grouting was done before no-dig remedies were applied to stop settlement and control groundwater.

Introduction

Use of trenchless and no-dig technologies has grown rapidly over the last decade and the phrase "trenchless" is now used broadly to include relining of pipes, placing of new underground lines without any excavation, and repair of underground facilities adjacent to pipelines. Generally, the many new methods involve miniaturization of techniques from the larger construction world of directional drilling, tunneling, and oil drilling.

Small pipelines and conduit are being placed with precision trenchless machines which accurately direct a small drill stem into place, and then back ream to pull the desired size casing into place. These new lines then must be tied into the existing system and its buried structures. Accomplishing that tie-in without leakage can become difficult especially if the area is contaminated in any way as it can be on industrial sites.

----

[1] Richard M. Berry, member ASCE's national Grouting Committee, President Rembco Engineering Corporation, PO Box 23009, Knoxville, TN 37933-1009, 423/690-6917
[2] Robert R. Farrar is Vice President of TRB Specialty Rehabilitation, 777 Annapolis Rd, Gambrills, MD 21054, 410/923-1182

Sectional relining of sewer pipes by slipline or CIPP (cured in place pipe) or pipe bursting methods can renew whole sections of a system, but water tight taps and connections must be made to entering pipes and to manholes and vaults. Adjacent sections of lateral sewers sometimes infiltrate badly particularly in the first few pipe sections as they are often deeper and under the groundwater table. New methods allow the taps to be cut remotely and without dig up, but this does not necessarily solve the leakage problems if the connection is below the water table.

Pipelines in trouble from imminent collapse are often found to have been leaking into the surrounding soil excessively and for long periods of time which can cause washout of soil fines and then settlement. Even after the pipeline is repaired, the pipe trench may have partly washed away causing voids and settlement of the surrounding area and structures. The washed out trench also can be an excellent French drain sewer, causing further damage at best, and carrying chemicals and other contaminants at worst.

Leaks in underground pressure pipelines can do a great deal of damage in a very short time. The pipeline can sometimes be repaired internally but the damage outside can be severe involving plant downtime, extremely costly dig up, and major by-pass constructions. Grouting has been used to remedy many of these situations.

A Major Settlement Problem

Sewer pipe disintegrates and collapses for many reasons. It can be simply that over the years infiltration has washed away enough of the bedding that settlement has stressed the pipe into collapse. But corrosion can have the same effect. A sinkhole-like depression developed in an area with congested electrical and electronic utilities at DOE's Oak Ridge Nuclear Weapons Facility. The circle of depression was some 20 feet (6 m) in diameter. Within that circle was a cable conduit connector box on a 3 by 5 foot concrete pad which had settled nearly a foot at an irregular angle. It turned out that this depression was above a 36 inch (90 cm) diameter concrete storm sewer buried 25 ft (7.5m) deep. Although this sewer had not carried radioactive wastes, it was in an area where radioactive contaminants exits. It was nearly impossible to shut down the utilities, so in-place repair technologies were mandatory.[3] Television inspection showed the concrete sewer to be badly corroded with the bottom out and bedding gravel visible. The pipe was reinforced with steel hoops, now the only thing holding it up. The pipe section under the depression and several others were relined using the *Insituform* CIP process.

What to do with the stressed and settled utilities at the sinkhole was another matter. Figure 1 shows the situation. The selected approach was to compaction grout the depressed area lifting it back to near normal.

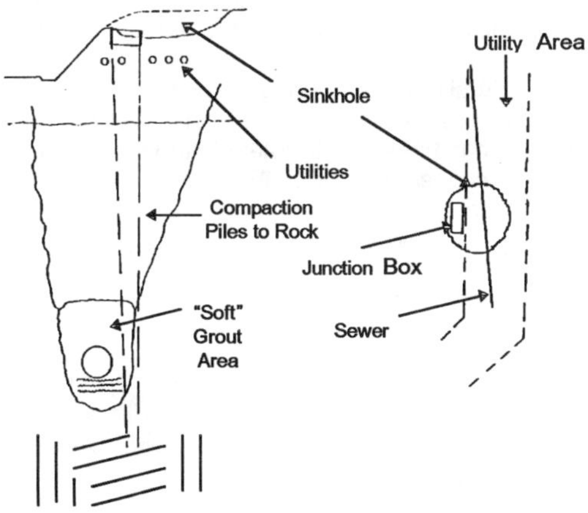

Fig 1. Section & Plan Sinkhole Remediation

Probe drilling of the area showed the clay soil very wet and poor from the pipe up to within 6 to 10 (2 to 3 m) feet of the surface. At the invert, the soil was saturated. Near the surface, the soil was good fill, well compacted with a fine crusher run gravel surface at the top. The junction box foundation pad was to be supported on compaction piles taken down to the rock surface, which was not far below the pipe depth.

Pressure grouting outside a pipe can possibly cause collapse so a combined approach was used. The *Insituform* liner design was increased in thickness to a full depth load bearing liner. Then, the bottom 10 feet (3m) of soil in the trench near the pipe was grouted with a thin non-cohesive grout. Injection pressures were watched closely to see that the line was not damaged and that grout was allowed to set overnight. Then the area above was grouted with a one inch slump but still non-cohesive grout, a mass grouting formulation. During these injections, pressure could be applied without damage below and the whole area including the sagged electrical conduit was lifted. Finally, the pad with the electrical switch box was supported on compaction grout piles seated on rock.

All in all, some forty six cubic yards of grout were injected at 29 points and two compaction piles were placed at the panel box pad. The entire area was brought up to its original elevation without damage to the utilities, and surface drainage was re-established.

Lateral Grouting

Rehabilitation of sewer lines by re-lining has proven to be a popular and effective structural repair method. After lining, laterals are commonly reconnected using internal cutters or grinders monitored by a television camera. If infiltration existed in the sewer pipe prior to lining, it often persists by traveling through the annular space created between the host pipe and the liner and then entering the system either at the manhole or through the newly cut lateral openings. Additionally, infiltration entering the adjacent sections of the laterals will continue unabated.

Over fifteen years ago, methods to access and grout laterals from the sewer were developed, and they have grown in their sophistication. Currently, lateral grouting is used in both lined and unlined sewer pipelines within a distance of about eight feet back from the pipe connection. Since most shallow laterals are above the water table but drop into the sewer main close to it, sealing this section can remove most of the problem.

Lateral grouting is similar to sewer grouting. Using television equipment, the operator maneuvers the lateral packer into position inside the pipe. The middle section of the packer has the ability to rotate in order to center an inversion tube directly over the lateral. Using air pressure, the inversion tube is expanded into the lateral. The end of the inversion tube is flared out to form a seal at its end. The end sections of the packer are then inflated and a three point isolated area is established which encompasses the lateral connection and several pipe joints up the lateral. This area of the pipe can be air tested and grouted if leaky. See Figure 2 for a diagram of the special packer.

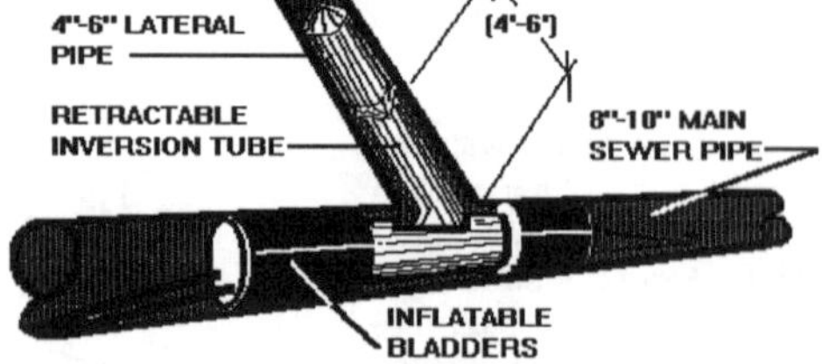

Figure 2

The value of lateral grouting is obvious when two facts are considered: 1) As reported in the ASCE Manual Existing Sewer Evaluation and Rehabilitation, on average, over one-third of system infiltration enters through the laterals, most of which comes in at connections or very near the main line, and 2) The average cost to treat this extra waste water can exceed $3.00 per thousand gallons as reported by WSSC.

If a lateral contributes just one gallon a minute, the yearly cost to treat this extraneous water, using WSSC's cost, will exceed $1,500.00. The cost of lateral grouting typically is in the $400-600 range and the economics are obvious with a less than one year payback. When this excess flow is reduced, both economic and environmental benefits are realized. Treatment plants become more effective when they work on more concentrated feed, there are fewer overflows, system capacity is effectively increased without further capitalization, fewer consent orders are entered, soil stops washing the bedding out, and on and on. Two major system operators have taken advantage of this process early on.

The Washington (DC) Suburban Sewer Commission (WSSC) is one of the largest water and sewer utilities in the USA. With almost 300,000 service laterals, engineers began concerted efforts to address the problems created by water entering the laterals fifteen years ago. Working with an equipment company and TRB, a system to seal laterals was developed which is in regular use today.

Since the program began in the early 1980's the WSSC has spent over three million dollars in lateral grouting, removing more than two million gallons a day of infiltration and inflow. Annual savings far exceed the total program expenditures so there is actually no cost to the program at all. By lateral grouting, the WSSC has removed water from the lateral joints and connections at the main. Extensive follow up tests by WSSC have shown that lateral sealing is long lasting. Of the many laterals tested, many as much as four years after sealing, all have continued to pass the air pressure test requirements set during the original application. Philip M. Hannah, WSSC maintenance reconstruction division head, has indicated that the WSSC intends to continue lateral sealing as an important part of its overall rehabilitation program.

The infiltration reduction quantities were derived from a mathematical estimate. "Given basic information derived from the air test (e.g., initial void pressure, final void pressure, holding time) and assuming appropriate soil property values (e.g., soil porosity and conductivity) in lieu of site specific data, it is possible to estimate the potential leakage of a defective connection. The air test data were collected from reports generated by the contractor during the testing. The soil properties are values taken from literature. This information, taken altogether, is used to build a mathematical estimate of the size of the defect. Although the resultant leakage area is theoretical, it should be stressed that the estimate is conservative and does not, at this point, take into account such amplifying effects as soil washout surrounding the connection which would permit increased infiltration rates. The leakage rates were estimated from an averaging of groundwater and soil conditions in the regions where the testing occurred. These rates could be more severe in other areas."[4]

Miami-Dade Water and Sewer Authority has made lateral grouting an integral part of the sewer line rehabilitation program. Sewer lines in South Florida are generally located below the water table, often in coral rock. The area has very high rainfall amounts, so it is no surprise that the collection system has experienced unusually high infiltration and inflow. Many years ago, Miami-Dade began an aggressive program of sewer rehabilitation directed at infiltration and inflow reduction which made extensive use of grout sealing of sewer joints. Louis Aguiar, assistant director of transmission systems has said, "When you look at what you get for your money, there is no doubt that chemical grouting of the most cost effective repair method available to anyone trying to deal with an infiltration problem."[5] After realizing that a significant part of the infiltration was entering the system through lateral connections, they began including lateral grouting in the program. According to Rod Lovett, chief of the collection division, "There's a lot of stress on lateral connections and that's where you generally get a crack. If a pipe has been lined, you'll still have the leaks at the lateral connections, plus groundwater will migrate through the annular space between two pipes and come in the lateral opening."[6] As a result, Miami-Dade now insists on the sealing of lateral connections when pipes have been re-lined.

Although the Washington Suburban and Miami-Dade systems have received significant accolades for their leading role in introducing lateral sealing, it is now gaining acceptance in smaller systems. Lateral sealing offers a no compromise, negative cost (net gain) solution to the problem of stopping infiltration and inflow from entering through lateral connections, and from the pipe annulus after re-lining.

A Distressed Underwater Effluent Pipeline

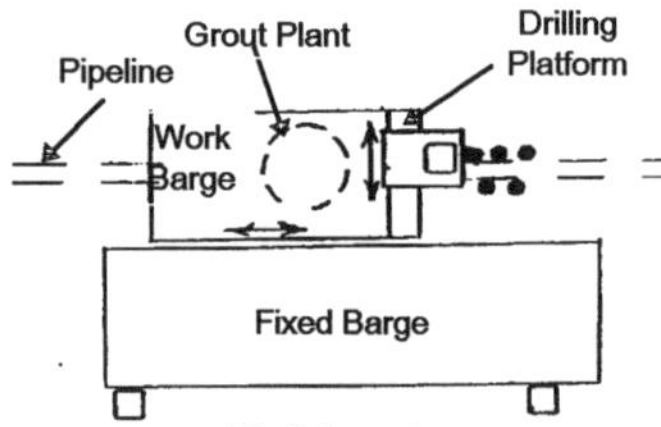

Fig 3. Layout

At the Scott Paper Company's Mobile, AL plant, weakened joints in a six foot diameter polyester fiber glass pipeline were allowing effluent to enter the Chickasaw Creek, a navigable artery of the Mobile River. The single pipeline carried about two million gallons an hour to the company's waste treatment facility across the river. The distressed zone is buried in about twenty five feet of mud and is about fifty feet below the water surface. The line had been repaired during its laying in 1978 but showed leakage from a failed joint in 1991. Several times divers had repaired the leaks from the inside with internal sleeves, but the repairs did not hold. The pipeline had been weighted down to prevent its vertical movement by using saddles over the top, but it still moved horizontally. Core sampling found that some river bottom mud had got into the pipe trench during construction possibly causing and at least adding to the problem.

Scott Paper investigated line replacement whose cost was found to be in excess of six million dollars. Grouting was put up as a potentially lower cost alternative. Soil tests suggested that a water thin grout would be required. Based on its water thin viscosity, suitable curing time, strength of the cured grouted soil, material cost, and environmental concerns, a new acrylic grout, AV-130, was chosen. Hard grouts involving cement are not acceptable for use around plastic fiber glass pipe as any hard spots will eventually cause pipe failure.

The grouting was done from a two barge system where one barge was spudded in parallel to the pipeline and the other winched alongside. Figure 3 shows the layout of the job. As the grouting progressed, the grout barge was moved along the primary barge to allow exact positioning with a minimum of surveying. The drilling and grouting mast is shown in Figure 4.

Injection points were located by probing to find the edges of the pipe. Grout was injected continuously through an open center drill bit in a top-down manner at 72 locations.

A total of 6,500 gallons (24,600 L) of grout (15% polymer content) was pumped with the area having a soil with apparent voids of 45 percent.

Before and after standard penetration tests were performed every ten feet along the pipeline to verify complete grout placement. The project was completed at a cost of $308,000, just slightly more than five percent of the replacement cost.

Figure 4

Storm Sewer Trench Sealing in a Contaminated Zone

A concrete storm sewer at the Oak Ridge National Laboratory was taking on radioactive contaminated seepage from a nearby retention pond. The upper end of the pipe was below the ground water tables and took on water primarily through pipe joints. It was relined with an *Insituform* CIPP liner. This action had the effect of immediately eliminating radioactive materials in the river outfall. However, a slow increase in contaminants began shortly after which gradually increased over the next several months to nearly where it had been before repair. A close look showed the contaminated flow now was coming from around the sewer pipe; the contamination had moved to the sewer trench utilizing its French drain capabilities. The area is generally silty clay with a weathered limestone and shale Karst surface below. The sewer was about 25 feet (7.5 m) deep near the pond and gradually surfaced as it came to the outfall.[6]

Anti-seep collars were considered along the pipeline but rejected because of radioactive dig-up considerations. Instead, grouting of the soil in the trench was the chosen economic solution to cut off the flow routes, seal the leaking the areas of the berm, and force grout back into the leaking cracks and fissures. Acrylamide grout was chosen on the basis of its history of stability in the presence of radioactivity, its water thin consistency for this low permeability soil, its excellent gel time control, and its tested longevity. In-situ flow tests were made to find the maximum rate of grout injection. Grout pipes were 3/4 inch (23 cm) diameter schedule 80 pipe and were wash bored into the pipe trench using a 1 1/4 inch (3 cm) open center bit. The pipes were drilled along each side in an alternating pattern and extended about two feet into the natural soil below the pipe invert. Figure 5 is a cross section of a typical zone of the trench.

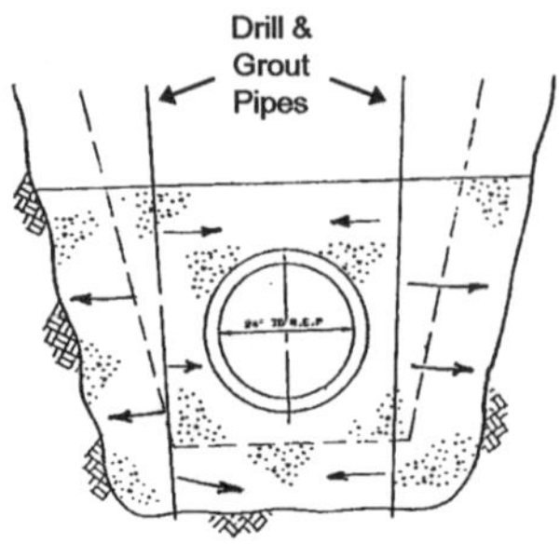

Fig 5. Sewer Trench

Grout consumption was very close to initial calculations based on soil data showing an average soil void of 42 percent. Grout pipes were checked for radioactive materials as they were withdrawn from the soil and about 20 percent were lost to the project. Radiation protective gear was required for all personnel entering a striped off safety zone along the trench. The grout plant was set on a trailer and moved along the pipeline as the work progressed and was generally within about a hundred feet of the injection points. Flows from the outfall have fallen to near zero in dry times; demonstrating the initial effectiveness of the solution.

Grouting to Save a Pipeline

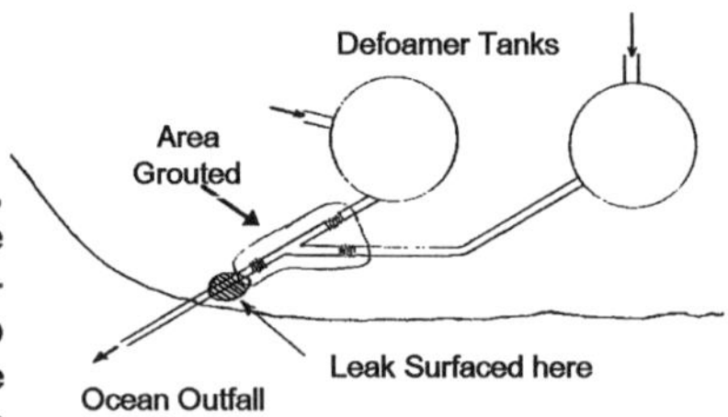

Fig 6.

During a routine facility inspection at a major West Coast paper plant in Canada, a small leak appeared near the shoreline where an underground 42 inch (110 cm) fiberglass reinforced pipe carried effluent to diffusers in the ocean. The leak was above a location where two individual pipes came from defoaming tanks joining in a "Y" to a single line before discharge. Figure 6 shows the facility layout. In a few days, the small leak had increased to a flow of about 500 gallons (1,900 L) a minute at the surface. At the point of leakage, the top of the pipe is some 22 feet (7 m) deep. The pipe carries about two million gallons an hour. The pipeline can be shut down for about two hours a day without slowing plant production. The paper company had the inside looked at with television but could not get a clear picture through the cloudy effluent. However, sonic scanning revealed an offset of nearly five inches at joint #1 in the "Y". This offset precluded conventional grouting with a packer from the inside of the pipe.

With plant shutdown for annual maintenance months away, the fear of catastrophic failure was felt to be right around the corner. The company decided to try grouting on the recommendation of several specialists. The objective was to try to stabilize the area in place so the entire assembly would not wash out causing a long term shutdown with a secondary objective of shutting off as much flow as possible.

The pipe was bedded in 3/4 minus limestone sand. Above the pipe were cobbles, large boulders (some desk sized, a driller's dream), and more sand. The pipe was fiberglass so hard grouts had to be avoided. And drilling is sensitive as one easily can go through a fiberglass pipeline. Since it was probable that much of the fine bedding material had washed away, a two grout program was proposed. A hydrophilic urethane, AV-220 Hydracure, was chosen as the initial grout to fill voids and large areas. Its ability to foam in a closed space without generating high pressure and yet free foam to a closed cell state, was important. The 300 centipoise shooting viscosity (like maple syrup) was thought to help prevent the grout from moving through the breaks or joints in the pipe to any great degree. The area would then be grouted with acrylamide to penetrate low permeability areas. Grouting would be carried out during the two-hour downtimes which could be made available while normal plant production continued.

Materials, equipment, grouting engineer, and technicians were transported by air to the plant. On the job, drilling began through the difficult material over the pipe. With limited access for large equipment, drills were hand held and the operation was one of finesse rather than power. At the same, tests were made on site of what was thought to be material representative of the material in the trench and carried out at in-ground temperature. In-situ groutability tests were performed for both grouts. The layout of grout injections is shown in Figure 7.

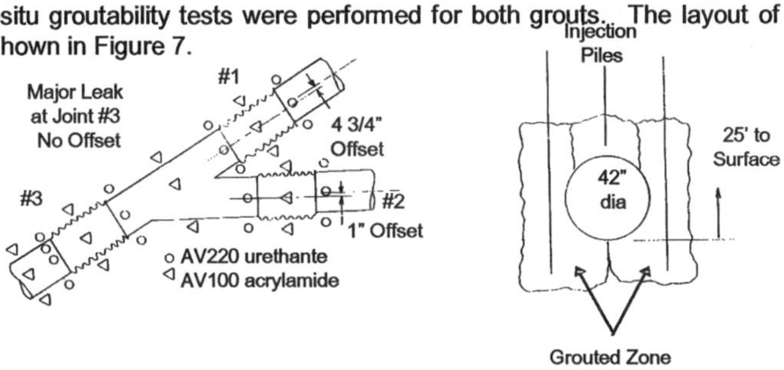

Fig. 7

A total of 350 gallons (1,350 L) of urethane was injected at 18 locations, stabilizing an estimated 670 cubic feet (19 cu. m) of bedding. After the urethane grouting, leakage encouragingly dropped to about twenty gallons a minute at the surface. A total of 2,375 gallons (9,025 L) of 10% acrylamide gel grout was then injected through 15 grout pipes filling another calculated 700 cubic feet (20 cu. m) of bedding. During this secondary grouting, **all** the flow back to the surface stopped. To the surprise and delight of all, a major leak in a large pressure pipeline operating at an internal pressure of 18 psi (125 kPa) had been stopped by grouting from the outside. Grout placement was verified by probe drilling at ten locations. Notified of the project on February 6, 1996, Rembco was on site and at work on February 12, USA and Canadian border officials notwithstanding. Work was completed February 22, ten days later.

Final Thoughts

Grouting in less than conventional situations can be both technically effective and cost effective in controlling leakage and in preventing further damage to pipelines as these examples have shown. Truly a complement to no-dig technology. Governing considerations should include a thorough investigation and thinking through of the problems in the specific situation to make sure they are the *real* problems and the *complete* situation. Then all potential solutions should be considered. It is well to engage both engineer and grouting professional at this point, the more technical and experienced the better.

Method and type of grout go hand in hand in the consideration of alternative solutions. The worst materials will not stand up to the environment. The best materials will not work if applied improperly. Process is important. Testing to obtain maximum flow rates and proper gel and cure times for permeation grouting are essential. Analysis of such special items as the need for soft grouts for plastic pipelines is essential. Verification testing should be made part of the program. However, the fact remains that grouting can often do things other processes cannot. Grouting can be both low cost and long lasting.

[3] John D. Long, Lockheed Martin Energy Systems, private communication, January 15, 1997

[4] WSSC Internal Service Connection Sealing Unit, Pilot Program and Cost Benefit Analysis, March 1983

[5] Trenchless Technology, May-June 1993

[6] Water and Environment and Technology, March 1996

Conquering Glacial Till in Ames, Iowa

Mark Schumacher, P.E.[1]; Mike Ellis[2]

Abstract

This paper will investigate the problems associated with microtunneling in dense glacial tills with hard rocks and cobbles, and the many changes required to successfully complete the project. The paper will discuss the design elements of a new pipe used for the first time on the project to cope with the soil conditions, and its subsequent performance.

Introduction

In 1995, the Public Works Department of the City of Ames, Iowa began planning for the expansion of a trunk sewer line along Lincoln Way. Lincoln Way is the main east-west corridor through the Iowa State University campus extending into the downtown area. The alignment of the proposed new sewer ran along the four lane roadway through the college campus. The road in this area has a carefully landscaped median and is bordered by extensive landscaping on each side of the road. The University wanted the mature landscaping preserved as well as minimizing the traffic disruption. The City then began designing the project around microtunneling to meet the project needs.

The project, consisting of seven drives totaling 929 m (3,048 ft) of 610 mm (24-in) inside diameter pipe, was bid in November of 1995 and awarded to Westcon/Dillingham Microtunneling of Pleasant Grove, Utah. Construction began in the spring of 1996 in what was believed to be dense sandy clays with unconfined compressive strengths of 34 MPa to 62 MPa (5,000 psi to 9,000 psi). A new Herrenknecht AVN 600 microtunneling machine was selected to complete the work.

----

[1]President, PipeForm LLC, 750 South Plaza Drive, Mendota Heights, MN 55120
[2]President, Westcon/Dillingham, 800 South Main, Pleasant Grove, UT 84062

The head was set up with a bore diameter of 787 mm (31 in) was designed for excavating clay material. The equipment injects bentonite into the overcut from the trailing tube behind the machine. No intermediate jacking stations or bentonite ports were used. The jacking frame has a capacity of 1,780 kN (200 tons) of thrust.

Drive Summaries

The following is a summary of the drives and the techniques used to conquer the materials. Figure 1 shows a schematic of the shaft locations, drive lengths, and type of head used.

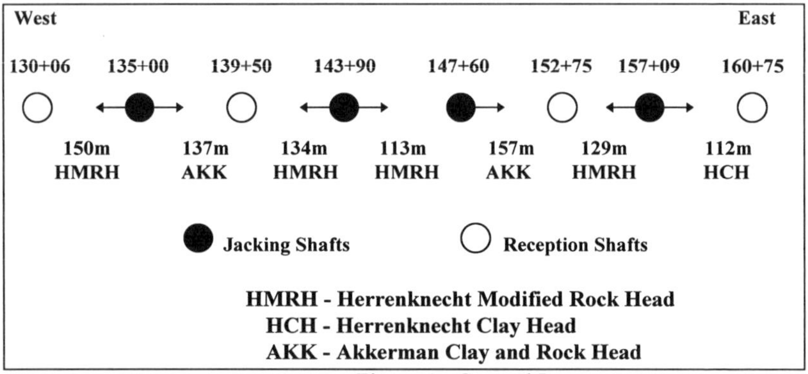

Figure 1. General Layout

The first drive was from station 157+09 heading east toward station 160+75. As the drive progressed, rocks and cobbles were encountered in the clay material as well as a stratum of sand and gravel. The rocks were very hard granite with high compressive strengths and abrasive qualities. Difficulties with the pipe string then caused the contractor to receive permission to open cut the last 7.5 m (25 ft) of the drive. Examination of the cutting head after retrieval showed that all of the cutting teeth were completely worn down or broken off.

The contractor then contacted PipeForm to discuss using their composite concrete and PVC pipe for the remaining drives. The specific project conditions were reviewed and PipeForm began immediately to design and produce a pipe customized for this project. The outside diameter of 749 mm (29.5 in) was selected to match the onsite equipment which was skinned to 762 mm (30 in) and had a cutting diameter of 787 mm (31 in), providing an over cut of 19 mm (0.75 in) on radius.

The joint detail utilizes a flush joint configuration to maximize the surface area for jacking. A step or offset joint was designed to further maximize the end surface area and to still allow for a 2 degree angular movement on each side of the collar.

The collar was constructed of 6.3 mm (0.25 in) zinc metallized steel and was designed to be no more than 1.3 mm (0.050 in) off the outside diameter to help deflect the rocks and cobbles away from the joint area, which was believed to be a problem in the first drive. The pipe was designed with a concrete compressive strength of 48 MPa (7000 psi) to yield a jacking capacity of 1,780 kN (200 tons) minimum to match the capacity of the jacking frame. A drawing of the joint cross section is shown in Figure 2.

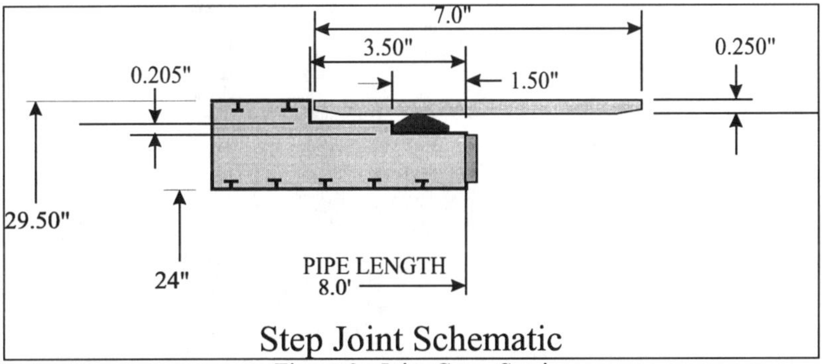

Figure 2. Joint Cross Section

The pipe was delivered to the jobsite four and a half weeks after the decision was made to use the pipe, and excavation on the drive commenced the following day.

In launching the machine, large cobbles were encountered immediately. A decision was made to pull the machine back prior to starting the pipe jacking and remove the cobbles. Upon removal, the head showed that all the cutting teeth had already been broken off. The contractor then decided to bring in a rock cutting head and modify it for the heavy clays that surround the cobbles. The photographs in Figure 3 show the original head and the modified rock head with two rock cutting wheels. The other wheels have been removed to create openings for the clay material and cutting teeth have been welded in by these openings to help bring in the clay material. The tradeoff here is that the total opening in the head has been reduced which will slow down production rates for the clay but will allow the rock to be cut more efficiently.

Representatives of Akkerman Equipment of Brownsdale, Minnesota visited the site and took the original soil head with them to rework into a design they felt would help cut the rock as well as excavate the clay.

Figure 3. Photos of Modified Rock Head and Original Soil Head

Five days after the original attempt the machine was relaunched. Jacking forces at the end of the drive were just under 890 kN (100 tons). Figure 4 shows the plot of jacking forces over the entire drive for four of the drives. When the machine was retrieved at the end of the second drive, a large rock was found wedged into the center opening on the rock head as shown in the photograph in Figure 5.

The contractor then moved to station 147+60 and prepared to launch the machine using the reworked Akkerman head. There was a delay of two days to allow for the rerouting of a 300mm (12 in) sewer line that was found to be in the alignment of the proposed sewer.

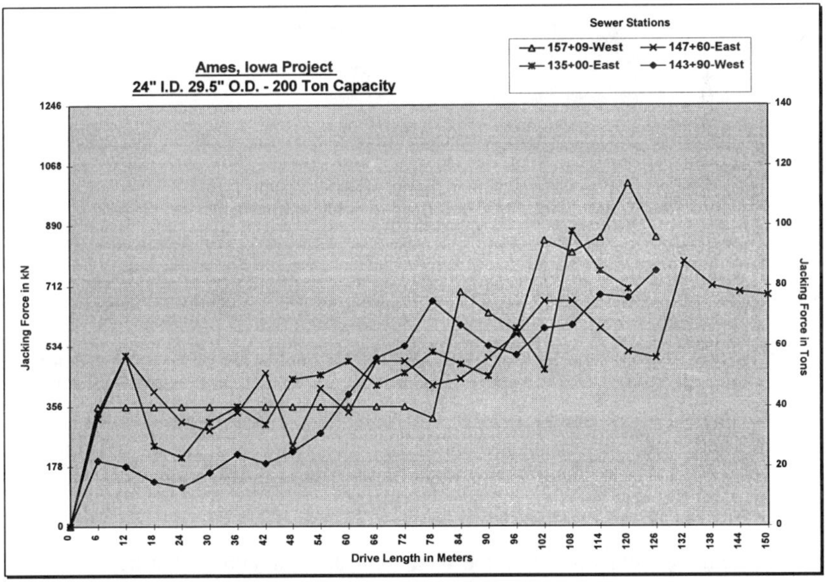

Figure 4. Drive Records

Figure 5. Photo of Rock

When the machine was retrieved at the end of the drive the head was caked solid on the outside with the heavy clay completely covering up the cutting teeth.

At this stage of the operation, a second Herrenknecht AVN 600 machine was brought in to help meet the revised project completion date in November. This machine was set up at station 143+90 and would utilize the modified rock head. The other machine was set up at station 135+00 and would use the Akkerman modified head. Each machine would make a drive in both directions out of these shaft locations.

The drive from 143+90 to the west went through an area that had been an old railroad grade and dump site. Items from tree limbs to bolts were found in the slurry separator.

The drives out of station 135+00 however were again characterized by heavy soils with very hard and abrasive rock. Upon retrieval of the machine from the drive to the east, examination of the head revealed that cobbles had worn away most of the steel cross members of the head including all evidence of the teeth.

In preparing to launch the machine for the final drive to the west from station 135+00, the head was changed back to the modified rock head. In addition, the high pressure water nozzles at the head were changed to a fan type offering lower pressure but greater volume of flow.

Observations/Lessons

The conditions encountered present some of the most challenging conditions for microtunneling due to the mixed face conditions that can include clay, rock, sand, or gravel. As this project shows, microtunneling can be successfully completed in these conditions.

This project also highlights the important role that detailed and comprehensive geotechnical information plays in successful projects. The geotechnical information drives the whole decision making process, from equipment selection to pipe design and selection. Microtunneling and jacking operations are a much more interdependent system as compared to open cut projects. As shown here, once the right combination of excavation head and pipe material were found, the project proceeded to completion.

Several areas for future development to improve the efficiency and success of microtunneling are suggested by the experiences on this project. First, development of a standard geotechnical "package" of tests designed around discovering the information needed by the microtunneling contractor would be extremely useful. A standard package of information that improves the characterization of the actual ground conditions will help in all aspects of the design and selection of equipment and materials. Second, continued development and refinement of heads for mixed conditions will improve the efficiency of microtunneling operations and by extension the economics of the technique. As was the case here, the contractor needs to be flexible and willing to try different equipment setups to try and conquer the glacial tills.

ENGINEERING A CHALLENGING TRENCHLESS UTILITY CROSSING

Brian H. Zelenko, P.E., M. ASCE [1] and James D. Smith, P.E., M. ASCE [2]

Abstract

Not every project warrants the use of the most current technology to be successful. A challenging pipeline installation was designed to allow the use of conventional auger boring and pipe jacking methods in addition to expensive high tech microtunneling equipment to accommodate a limited construction budget. With an appropriate engineering effort, a cost effective and practical tunneling application was developed for this interesting trenchless pipeline project.

Site constraints required that a relatively short tunnel 50.3 m (165 ft.), 1.524 m (60 in.) diameter be mined under a critical 80-year old brick storm sewer and the operating mechanical room of an existing 5-story building in downtown Baltimore, Maryland. The project, designed jointly by Ballinger/RCG Joint Venture and Haley & Aldrich, Inc., was planned to provide the final link for the completion of a new chilled water loop at the University of Maryland at Baltimore (UMAB) Nursing School.

Technical details regarding site limitations, design and construction considerations, and the practical design solutions developed for the project will be discussed. These details include chemical grouting for groundwater control and ground stabilization.

Introduction

Upon initial inspection, this project appeared destined to be designed for construction by microtunneling. Microtunneling has become a popular choice for tunneling methods in recent years for tunneling in difficult (soft, saturated, hazardous) ground conditions. However, there was more than one way to construct this project.

---

[1] Senior Engineer, Haley & Aldrich, Inc., 12200 Tech Road, Silver Spring, MD 20904
[2] Vice President, H&A of New York, 189 N. Water Street, Rochester, NY 14604

The project team evaluated the design and construction considerations and developed a design which allowed a more conventional tunneling approach in addition to microtunneling. By allowing a range of tunneling methodologies, the owner was more likely to receive competitive bids for the proposed construction.

Project Description

The project need, developed jointly by Ballinger/RCG Joint Venture and UMAB, consisted of installing a chilled water loop. This project was part of expansion activities associated with the construction of proposed alterations and additions to the School of Nursing. It would consolidate chilled water needs of several portions of the campus.

A portion of this chilled water loop was planned to be constructed in two vertical shafts and a tunnel to carry the chilled water lines from the new School of Nursing Building under Penn Street to the existing Allied Health Building. The proposed tunnel, as shown in Figure 1, was planned to be approximately 50.3 m (165 ft.) long with a 1.524 m (60 in.) inside diameter. Due to the presence of caisson foundations bearing on rock below the 5-story Allied Health Building, an alignment was chosen to stay within one of the column bays of the building. Two 406 mm (16 in.) diameter welded steel or ductile iron pipes were planned to be installed in the completed tunnel to convey the chilled water. The tunnel was also planned to contain space for manned access and for additional utility installations in the future. This need was determined by UMAB to avoid similar crossings in the future to accommodate other installations such as fiber optic cables.

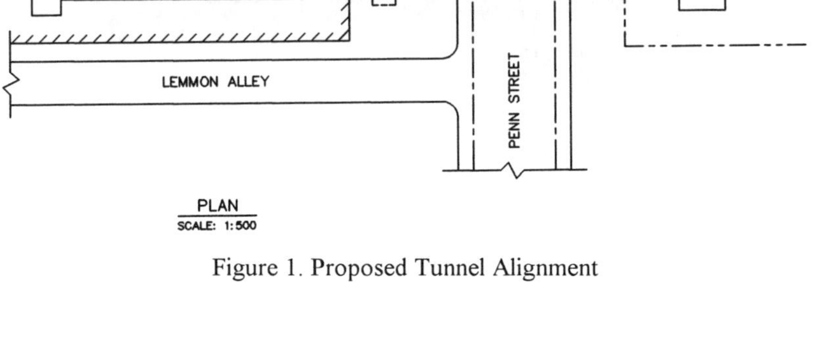

Figure 1. Proposed Tunnel Alignment

A 5.5 m by 1.8 m (18 ft. x 6 ft.) brick and masonry storm sewer exists under Penn Street. The storm sewer is overlain by abandoned trolley tracks on a concrete base. Many below-grade utilities are also present in Penn Street and Lemmon Alley.

Haley & Aldrich, Inc. (H&A) was retained by the Joint Venture to evaluate the subsurface conditions and design and prepare contract documents for the shafts and tunnel. Site surveying and field location of existing subsurface utilities were completed by A. Morton Thomas and Associates, Inc. (AMT). Barton Mallow, construction manager for UMAB, coordinated the development of the final contract documents.

Subsurface Conditions

Based on a review of previous geotechnical reports for facilities near the alignment, a subsurface profile was developed as shown in Figure 2.

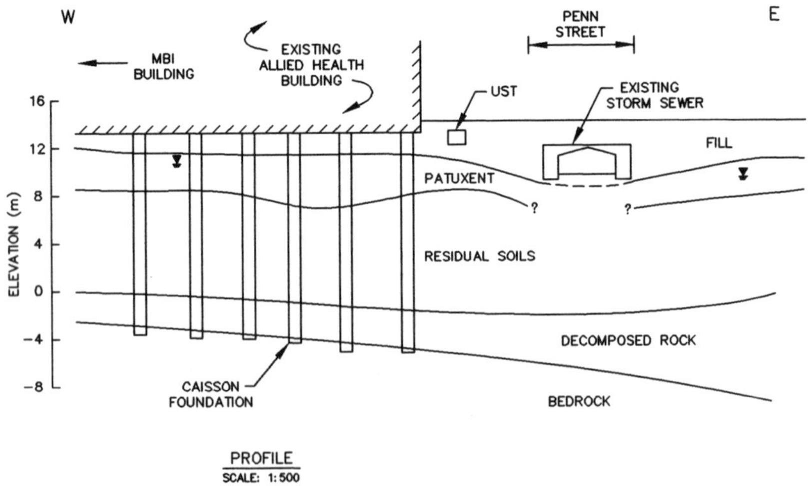

Figure 2. Subsurface Profile

For the tunneling design, the important strata were the Patuxent Formation and the Residual soils. The Patuxent Formation is a very permeable stratum which ranges from a silty sand to a sandy gravel. It is known throughout the Baltimore area as an aquifer and has historically created construction problems at many sites around the city. The Patuxent Formation would have to be penetrated for construction of both shafts and was feared to be present below the Penn Street storm sewer. The Residual soils range from sandy silt to silty clay and are quite competent for conventional tunneling methods and can be readily mined. Historically, the Residual soils have demonstrated good stand up time. Additionally, the Residual soils encountered at this site were generally less permeable than the Patuxent Formation.

Groundwater was observed to range from El. 10.4 m (34 ft.) at the Launch Shaft to El.11.3 m (37 ft.) at the Receiving Shaft.

Design and Construction Considerations

Ideally, the chilled water lines would have to be installed several feet below Penn Street at Lemmon Alley in an open cut. The line would continue under Lemmon Alley, just outside of the Allied Health Building and then enter the Allied Health Building. Two factors impeded design of this preferred alignment. The first was the presence of the large brick and masonry storm sewer conduit under Penn Street. The storm sewer is overlain by abandoned trolley tracks founded on a concrete base leaving no space for the chilled water lines to pass over the sewer. The second was the numerous subsurface utilities located under Lemmon Alley which prevented any consideration of cut-and-cover construction in the alley. Therefore, an alignment was selected which is shown in plan view in Figure 1.

Given the profile noted in Figure 2 and the owner's desire to keep the tunnel as shallow as possible, microtunneling appeared to be the logical solution. However, in order to create a more competitive bidding process, the design team set out to develop a trenchless pipe installation design consisting of either pipe jacking, augering, or microtunneling. Additionally, the designers considered the cost and risk benefits of lowering the vertical alignment relative to potential construction problems in an upper, less favorable, tunneling stratum.

There were several important design and construction considerations evaluated for the proposed alignment. These included the following:

- Prevent potential settlement and disturbance of the existing 80-year old storm sewer below Penn Street during the tunneling operation;
- Prevent potential settlement of the existing slab-on-grade and prevent damage to the caisson foundation system while tunneling below the existing Allied Health Building;

- Maintain full use of the building during construction;
- Design permanent vaults at each end of the tunnel. Specifically, design the receiving shaft vault inside of the existing Allied Health Building to minimize impacts to the existing slab-on-grade and the adjacent MBI Building which is supported on shallow spread footings; and,
- Address the dewatering requirements for construction of the shafts through the permeable Patuxent Formation.

Design Solutions

Microtunneling is a useful and advanced method for boring small to mid-size tunnels through difficult subsurface conditions. It utilizes an unmanned, laser guided tunnel boring machine to bore through difficult ground conditions. In fact, pressurized faced microtunneling machines have successfully penetrated through sandy soils below groundwater level, conditions similar to the Patuxent Formation.

However, microtunneling through the Patuxent Formation was not considered appropriate for this project. One of the drawbacks of using a microtunneling boring machine (MTBM) is that there is no access to the face if problems are encountered. This was of particular concern to the design team while considered for use within the Patuxent Formation. There were uncertainties about the Patuxent Formation and its potential behavior during tunneling which could lead to lost ground and settlement. Although the cost of the vertical shafts would have been significantly reduced due to the shallow alignment, it would have provided minimal cover between the crown of the tunnel and the Penn Street storm sewer and Mechanical Room of the Allied Health Building. Any unforseen problems during tunneling in the highly permeable Patuxent Formation could seriously compromise the success of the project. A rescue shaft would be impossible to install at many locations along the alignment.

Based on the design and construction considerations noted above, the design team evaluated the tunneling, excavation and ground support options and developed the final design for the project. It was decided to lower the vertical alignment to keep the tunnel within the Residual soils, a less permeable and more favorable tunneling stratum than the Patuxent Formation. This reduced some of the Owner's risk and created an opportunity to utilize alternative construction methods. The final design consisted of conventional auger boring or pipe jacking a 1.524 m (60 in.) diameter pipe from the Launch Shaft to the Receiving Shaft. Both of these more conventional methods would allow emergency access to the face if needed. The invert elevation, ranging from El. 4.0 m (13.0 ft.) at the Launch Shaft to 4.4 m (14.5 ft.) at the Receiving Shaft, was selected to lower the crown of the tunnel a depth of two tunnel diameters below the bottom of the storm sewer. This would keep the tunnel within the Residual soils, as shown in Figure 3.

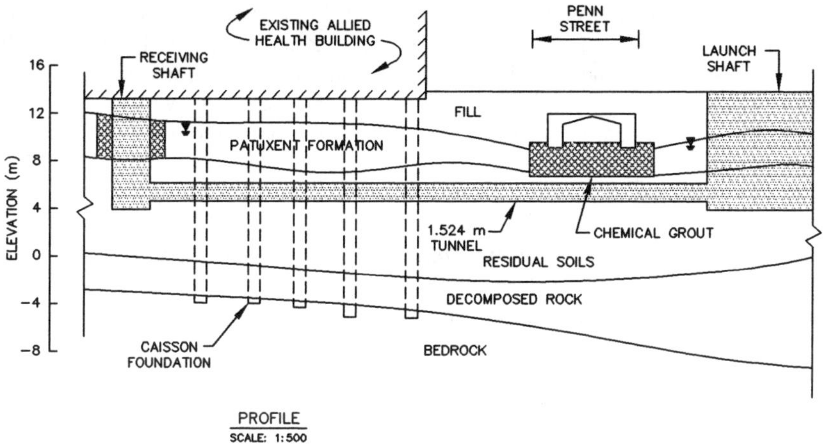

Figure 3. Tunnel Profile

Auger Boring / Pipe Jacking

Auger boring and pipe jacking are simpler tunneling methodologies than microtunneling, because they require less support equipment and do not involve slurry systems. Problems can generally be addressed directly at the face. Both tunneling systems must be conducted in relatively dry soil conditions. Therefore, predraining the soil or a groundwater control system along the alignment is required. Predraining could be accomplished by wellpoints. Alternatively, the pipe jacking or boring system would need a demonstrated capability to control groundwater inflows and ground movements at the cutting head. Borehole stability could therefore be maintained at all times throughout the bore.

Auger boring consists of pushing the pipe into the soil with a boring auger rotating within the pipe to remove spoil. The leading edge of the pipe provides a mechanical means to prevent the auger and cutting head from leading the pipe, eliminating the need for supporting the excavation ahead of the pipe.

Pipe jacking consists of jacking a steel shield which is bolted or welded to the first section of pipe. Spoil removal takes place a short distance behind the shield. Mining ahead of the casing pipe is not required thereby eliminating the need of any type of boring, augering or drilling equipment.

The steel casing pipe to be jacked was specified to be new, 1.524 m (60 in.) nominal diameter, minimum 21.4 mm (0.844 in.) wall thickness, conforming to

ASTM A53, Grade B, with welded joints and having a minimum yield strength of 241 MPa (35,000 psi). This thickness was specified so that the casing pipe would be able to withstand anticipated jacking stresses and to address corrosion concerns. Steel casing was also specified to be painted inside and out with two coats of bitumastic enamel coating paint in accordance with AWWA C209.

Chemical Grouting

Chemical grouting is utilized for groundwater cutoff and ground stabilization. It consists of injecting into the ground liquid sodium silicate, approved reactant and water and, when required, an accelerator. PVC grout pipes are drilled into the ground, as required, to obtain specified minimum grout coverage and strength. Utilizing double packers, chemical grout is injected into the intended grouting zones through properly located grout ports in sleeve pipes. Typical gel times for chemical grout range from 20 to 40 minutes.

Two locations of chemical grout injection were required for this project. These sites were the Receiving Shaft and the Penn Street storm sewer. For the Receiving Shaft, a two-pronged pre-excavation chemical grouting program was developed. The purpose of the chemical grouting was to reduce ground subsidence due to subsequent tunneling and/or excavation operations near the existing MBI Building footings, located within the zone of influence of the Receiving Shaft, defined as a 1 horizontal to 1 vertical envelope up and out from the bottom of the Receiving Shaft. In addition, chemical grouting would be used to minimize groundwater flow into the Receiving Shaft excavation. Grout had to be capable of reducing permeability of grouted sand to at least 10^{-4} cm/sec. Chemical grouting was also permitted to be used at the Contractor's option to control groundwater inflow at the Launch Shaft excavation.

Additionally, chemical grouting was required under the existing Penn Street storm sewer, where it crossed the proposed tunnel alignment. Due to the uncertain ground conditions directly below the Penn Street storm sewer, it was desireable to improve those conditions by grouting rather than risk movements or loss of ground below the storm sewer.

Chemical grouting was planned to be performed from the existing ground floor of the Allied Health Building and through the Penn Street drain. Cement grout also would be used to fill the annular space around the jacked pipe, to minimize subsequent ground movements.

Temporary Excavation Support for Shafts

The type and shape of the excavation support systems were designed to accommodate the Contractor's tunneling preference. For the Launch Shaft and

Receiving Shaft, excavation support wall systems could include, but would not be limited to, soldier piles and lagging, continuously interlocked steel sheet piling, steel liner plate, continuous jet grouted soil columns, or combinations of these. Jet grouting could have been utilized at the Receiving Shaft to provide both excavation support, groundwater cutoff, and ground improvement near the MBI Building footings.

Conclusion

A careful evaluation of subsurface conditions, site constraints, project requirements, and economic and risk considerations led the design team to allow conventional tunneling alternatives in addition to the latest tunneling technology for this project. This allowed more contractors to have an opportunity to bid the work, providing the owner with a better chance to receive a competitive bid price.

While new technology is developed every day providing the opportunity to do things not previously imaginable, it is always prudent to consider traditional approaches. Not every project warrants the use of cutting edge technology to be successful. In this case, by modifying the vertical profile slightly and including minor ground modification requirements, a very difficult crossing was transformed into a very constructable one.

At the time this paper was published, UMAB was evaluating funding priorities and has not yet moved forward with the chilled water loop project.

Acknowledgments

The authors wish to acknowledge the support of the University of Maryland at Baltimore. In particular, Ms. Kyle Bressant and Mr. Terrence Morse were supportive of this project. Additionally, Messrs. Jonathan Friedan and Michael Davidoff of the Ballinger/RCG Joint Venture worked closely with the authors on this project.

References

The Robert B. Balter Company, "Preliminary Geotechnical Engineering Report, School of Nursing Building, University of Maryland at Baltimore, Baltimore, Maryland, 10 April 1995.

University of Maryland at Baltimore, "Alterations and Additions to the School of Nursing," Project No. 95001, Drawing Nos. T1 through T4, Specification Section Nos. 02130, 02150, 02860, and 02945, 2 July 1996.

A NINE PIPE JACKED CROSSING OF I-93
IN DOWNTOWN BOSTON

Ronald C. Strand, P.E.[1]
Russel J. Ross, P.E. , Member ASCE[2]

Abstract

In fulfillment of a major planning element of the Central Artery/Tunnel (CA/T) Roadway Project in downtown Boston, nine pipes were designed for jacking in a corridor across I-93 at Oliver Street. Steel sleeves 1200 mm (48-in) in diameter were used to accommodate electric (including 115 Kv lines), water and telephone utilities. A sanitary sewer and combined sewer of 1700 mm (66-in) and 1800 mm (72-in) diameter respectively were designed for direct jacking. Associated work included underpinning of the existing I-93 retaining walls at each end of the jacking. Detailed sequencing of the construction was required so that existing utilities could be maintained in service until no longer needed, and so that space could be provided for installing the jacked utilities among the myriad of existing utilities. The design included consideration of settlement monitoring of I-93 during construction, support of the jacked pipes (beyond the limits of the crossing proper) during future excavation for construction of the roadway tunnel box and location of the jacked pipelines to minimize intrusion of the pipelines into the roof slab of the roadway tunnel box.

Introduction

At a current estimate of $7.8 billion, the Massachusetts Highway Department's (MHD) Central Artery/Tunnel Project (see Figure 1) is the country's largest transportation project. It has provided a third harbor tunnel (the Ted Williams Tunnel) to Logan International Airport that was opened to commercial traffic on December 15, 1995, and is scheduled to be opened to all traffic in 2002.

1,2 Principal and Vice President, respectively: Camp Dresser & McKee Inc., Ten Cambridge Center, Cambridge, MA 02142

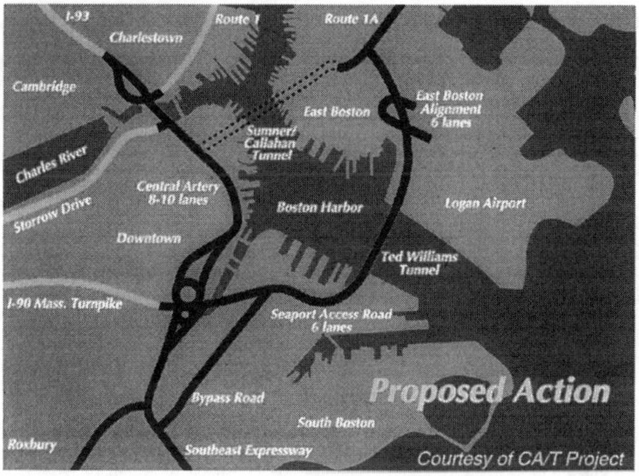

Figure 1. The CA/T Project

By the end of 2004, the project is scheduled to provide a new interchange between I-290 (the Massachusetts Turnpike) at its west end terminus and I-93 at the south end of the city, and a new interchange between I-93 and Route 1 (the Mystic River Bridge) at the north end of the city. Between these two points, which traverses downtown Boston, the existing I-93 mostly overhead viaduct, known locally as the Central Artery, will be depressed into an underground tunnel.

Roadway Construction Methods

This roadway tunnel is designed to be constructed by open cut methods, after the installation of soldier pile tremi concrete slurry walls (slurry walls) which will support the open cut excavation as well as the decking system that will accommodate local street traffic during construction (see Figure 2). Three slurry walls-- east, middle and west-- comprise the walls for the two main line tunnel roadway boxes, one carrying northbound traffic and one carrying southbound traffic. Additional cast-in-place concrete walls are used for the various on and off ramps connecting the new tunnels to the surface streets.

As illustrated in Figure 3, a myriad of existing utilities including water, sewer, drain, combined sewer, electric, gas, steam and signal (telephone, cable TV, telegraph and traffic signal) criss-cross the location of the proposed slurry walls.

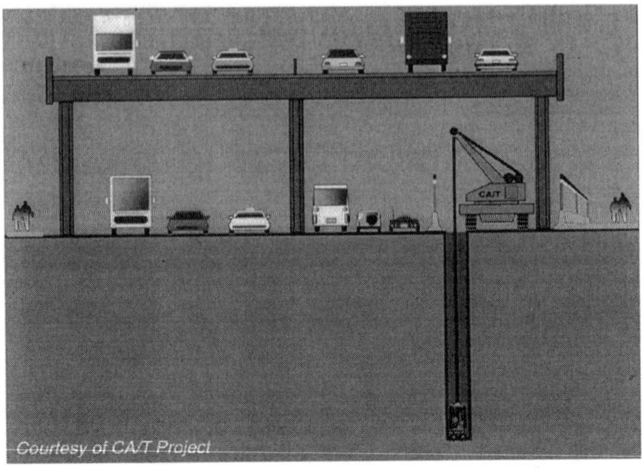

Figure 2. Slurry Wall Construction

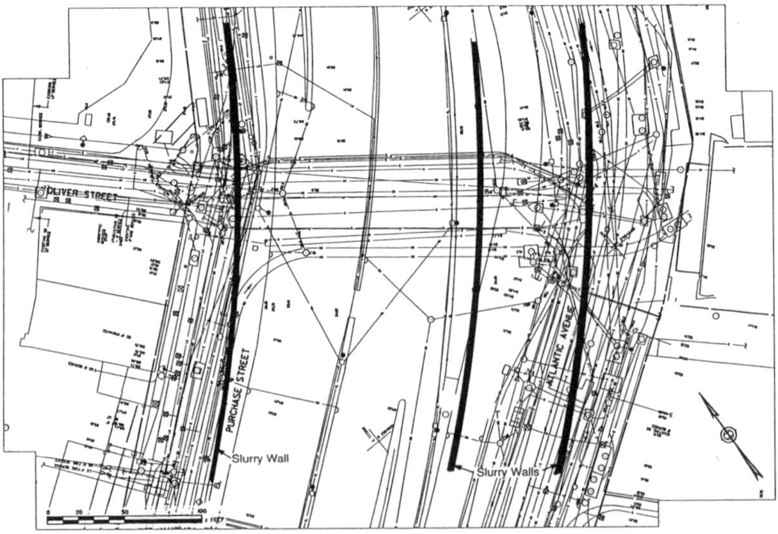

Figure 3. Existing Utilities

Placement of Relocated Utilities

As illustrated in Figure 4, relocated utilities were placed in longitudinal corridors, located, wherever possible, outside the east and west slurry walls so that they would not interfere with future excavation for the roadway tunnel boxes.

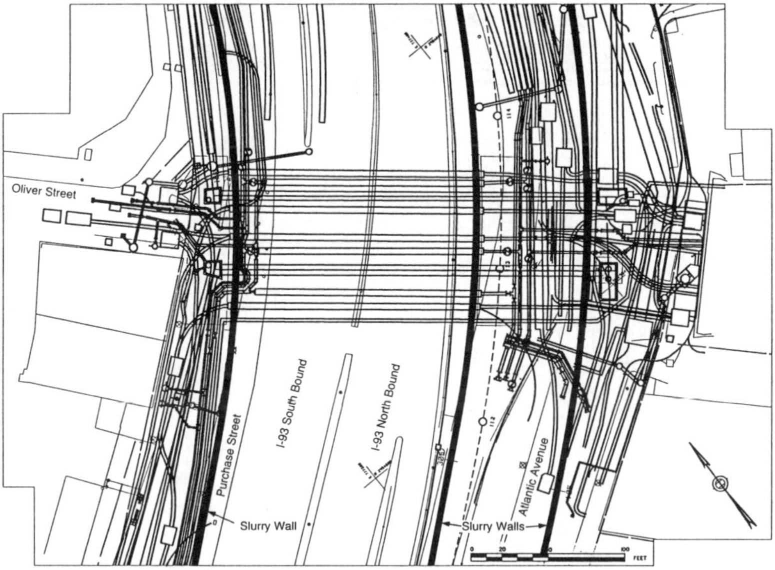

Figure 4. Corridors for Proposed Utilities

Corridors are also intended to minimize temporary relocations and to comply with the primary mitigation requirement of minimizing traffic disruption. As such, open cut excavation of the existing artery in this area was not considered a viable alternative.

Support of Jacked Pipes During Future Excavations

Where, because of space limitations, relocated utilities had to be placed longitudinally inside the slurry walls, or where the relocated utilities were scheduled to cross (generally perpendicularly) the roadway tunnel boxes in open cut, an undercarriage support system as shown in Figure 5 was designed to support the utilities. This support system will be used during future excavations for the roadway tunnel boxes.

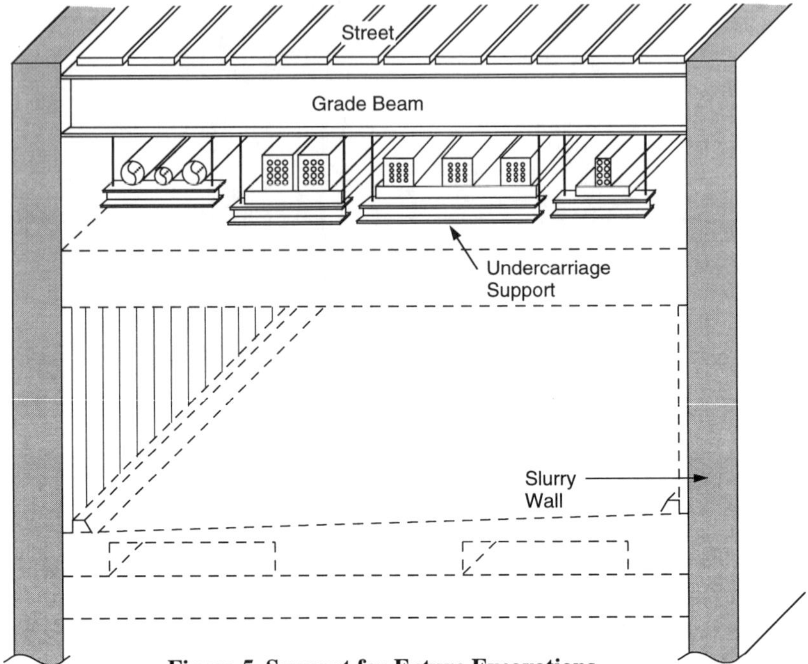

Figure 5. Support for Future Excavations

Oliver Street Crossing

Crossings between the longitudinal corridors are required for distribution of services throughout the city. The Oliver Street crossing is located just north of the existing I-93 South Station Tunnel (see Figure 3). The area at each end of this crossing (Purchase Street on the west and Atlantic Avenue on the east) is especially crowded with existing utilities. Nevertheless, as illustrated in Figure. 4, the following relocated utilities are to be installed in this location (listing is from north to south):

1. Fifteen 130 mm (5-in) electric ducts in a 1200 mm (48-in) diameter steel sleeve.

2. 500 mm (20-in) diameter high service (HS) water main in a 1200 mm (48-in) diameter steel sleeve.

3. 1800 mm (66-in) diameter reinforced concrete sanitary sewer (SS), direct jack.

4. Twenty-four 100 mm (4-in) diameter signal ducts for telephone telegraph, TV and traffic in a 1200 mm (48-in) steel sleeve.

5. Four 200 mm (8.6-in) diameter steel electric ducts for 115 Kv electric line in a 1200 mm (48-in) diameter steel sleeve.

6. 500 mm (20-in) diameter low service (LS) water main in a 1200 mm (48-in) diameter steel sleeve.
7. 1700 mm (72-in) diameter reinforced concrete combined sewer (CS), direct jack.

8. 500 mm (20-in) diameter high fire (HF) water main in a 1200 mm (48-in) diameter steel sleeve.

9. Fifteen 130 mm (5-in) diameter electric ducts in a 1200 mm (48-in) diameter steel sleeve.

Sleeves were specified for most utility crossings to accommodate construction and to conform with utility company requirements. Additionally, the steel sleeves will be used to support and protect the live utilities during the future construction of the tunnel sections. The space between the sleeve and the casing pipe is to be filled with sand (115 Kv, water) or cement grout (electric, signal) per utility standards. Per project requirements, the gravity lines (66-in. SS and 72-in. CS) were direct jack reinforced concrete pipe. Direct jack pipe used a steel joint similar to that of prestressed concrete cylinder pipe. Corrosion protection of the sleeves was not deemed necessary as they are not required once the roadway tunnel is constructed. We anticipate several sleeves will be removed during roadway tunnel construction.

Underpinning of Existing Retaining Walls

In the Oliver Street area, the existing I-93 Central Artery is in a boat section with the retaining walls being earth supported as illustrated in Figure 6. Because excavation for some of the relocated utilities is to extend below the bottom of these retaining walls, they were underpinned prior to the pipes being jacked. This was accomplished as specified by excavating panels along the length of the walls to below the proposed depth of the utility excavation and backfilling with unreinforced concrete. Proposed utilities are to be jacked below the underpinning.

Subsurface Conditions

The material in which the underpinning has been installed and in which the jackings are to be performed consists of fill and till as noted in Figure 6. Consequently, foundation problems are not anticipated for the installed pipes, although some cobbles may be anticipated during the excavation.

Construction Sequence

In order to keep all existing utilities in service during the installation, and to provide space for new pipe installations, the sequence presented below was suggested for installing the proposed utilities.

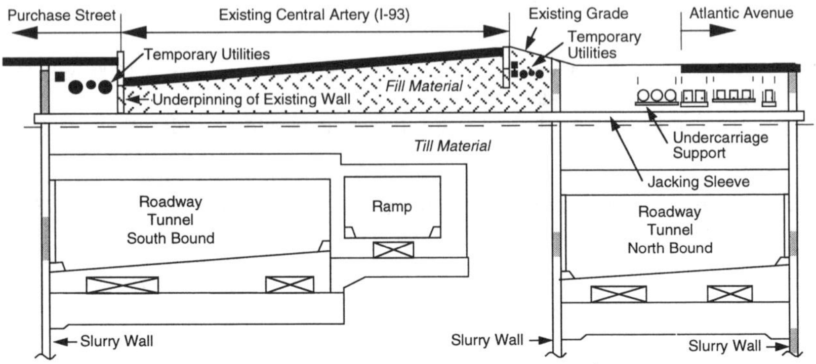

Figure 6. Cross Section at Oliver Street Crossing

First, in order to provide an initial working space, temporary utilities were installed on each side of the crossing. The temporary installations (not show in plan view, but shown in section in Figure 6) included high service (HS), low service (LS) and high service fire (HSF) water mains and Massachusetts Bay Transportation Authority (MBTA) electric ducts. This initial work included underpinning of the existing I-93 retaining walls as noted above.

Once the temporary installations were complete, and activated, the sequence presented below was provided. This sequence is noted on Figure 7, and provides a logical method for installing the new utilities while maintaining existing services and meeting the requirements for slurry wall panel installation. The noted sequence also provides for pulling cables within the installed ducts, making connections to existing utilities at each end of the crossing and activating the newly installed systems:

1. Fifteen 130 mm (5-in) electric ducts in 1200 mm (48 -in) steel sleeve.
2. 500 mm (20-in) diameter HS water in 1200 mm (48-in) steel sleeve.
3. 115 Kv electric line in 1200 mm (48-in) steel sleeve.
4. 1800 mm (72-in) diameter CS direct jack.
5. Telephone and other signal ducts in 1200 mm (48-in) steel sleeve.
6. Fifteen 130 mm (5-in) electric ducts in 1200 mm (48-in) steel sleeve.
7. 1700 mm (66-in) diameter SS direct jack.
8. 500 mm (20-in) diameter LS water in 1200 mm (48-in) steel sleeve.
9. 500 mm (20-in) diameter HF water in 1200 mm (48-in) steel sleeve.

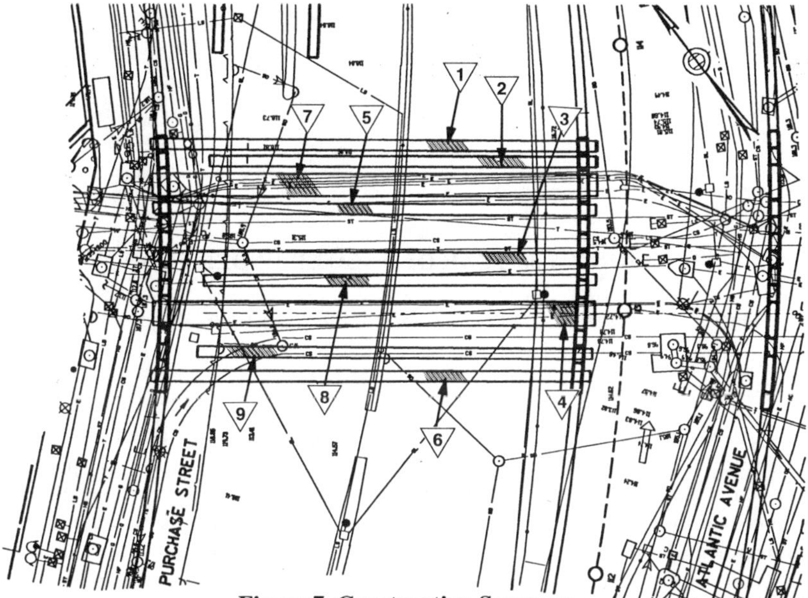

Figure 7. Construction Sequence

Settlement Monitoring

The specifications require that the jackings be performed with no more than 5 cm (0.16 ft) of existing roadway surface settlement. Deflection monitoring points have been installed in the roadway to monitor this requirement. The monitoring system has been established such that readings can be taken from Purchase Street or the pedestrian overpass crossing I-93 without the existing roadway being occupied, thus eliminating the requirement for elaborate traffic control measures to ensure worker safety.

Conclusion

In order to provide an efficient means for constructing a new interstate roadway tunnel in downtown Boston, existing utilities were relocated into corridors outside, for the most part, of the tunnel walls. This corridor concept included a nine pipe jacked crossing at Oliver Street. Design of this crossing included consideration of settlement monitoring during construction, support of installed pipes during future excavation for the roadway tunnel, and the sequence in which the work must be performed.

SLIPLINING A 50 YEAR OLD SEWER UNDER FLOW

Dennis W. Seeman P.E.[1], Lea S. Sanford, P.E.[2], Russell C. Ford, P.E.[3]

Abstract

As part of the city-wide overflow elimination program, the Greater Houston Wastewater Program authorized WSBC Civil Engineers, Inc. to provide construction documents for sliplining 4,877 meters (16,000 feet) of monolithic, reinforced concrete (MRC), 152.4 cm (60-inch) diameter sewer constructed in 1946. Hydrogen sulfide corrosion had severely deteriorated the 1946 trunk sewer to the point where sections had completely collapsed. Five junction structures, constructed in 1973 were also suffering severe deterioration.

Project Description

WSBC Civil Engineers, Inc. was awarded a project to rehabilitate approximately 4,877 meters (16,000 feet) of 152.4-cm (60-inch) MRC sanitary sewer constructed in 1946. This sewer is the main trunk sewer that conveys flow from the North MacGregor Way No. 1 Lift Station (located in south central Houston) 9,754 meters (32,000 feet) east to the Sims Bayou Wastewater Treatment Plant. WSBC's portion of the project alignment is shown on the Project Vicinity Map. Two main design criteria were determined in the modeling and Alternatives Analysis stage. This project, as well as the three projects to rehabilitate the remaining half of the sewer, would be accomplished using sliplining methods and the minimum internal diameter required to convey the required flows would be 137.2 cm (54-inch). The original design was constructed in three sections. Section 3 was located between Telephone Road and Kuhlman Gully (where an aerial crossing was constructed in Brock Street right-of-way). Section 4 continued west and terminated approximately 1,524 meters (5,000 feet) from the west end of the proposed project, and Section 5 continued west past the west end of the project.

[1,2]Project Manager, WSBC Civil Engineers, Inc., 1177 W. Loop South, Suite 310, Houston, TX 77027

[3]Sales Representative, Hobas Pipe USA, Inc., 6620 Long Drive, Houston, TX 77089

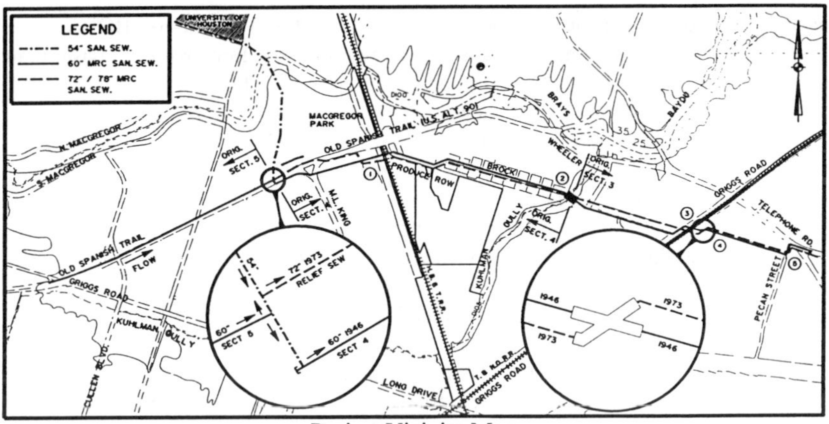

Project Vicinity Map

Design Considerations

The first design issue consisted of a problem located between the original Section 4 and Section 5 of the trunk sewer. Section 4 was constructed on the south side of O.S.T. with a 152.4-cm (60-inch) internal diameter. A 137.2 cm (54-inch) branch sewer crossing O.S.T. to the north was also constructed as a part of Section 4. The Section 4 plans show a 152.4-cm (60-inch) stub-out along the south side of the O.S.T. right-of-way that was to continue with Section 5 in the future. As actually constructed however, Section 5 was built on the north side of O.S.T. and the 152.4-cm (60-inch) diameter Section 5 trunk sewer was connected to the smaller 137.2 cm (54-inch) branch sewer to cross O.S.T. approximately 24.4 meters (80 feet) before re-entering the 152.4-cm (60-inch) diameter Section 4 trunk sewer. In 1973 the City constructed a relief sewer in a dedicated easement north of the O.S.T. right-of-way from this point east. The 182.9 cm (72-inch) diameter relief sewer was tied into the smaller 137.2 cm (54-inch) branch sewer north of the Section 5 tie point. This meant that as flow traveled east along the 152.4-cm (60-inch) diameter Section 5 trunk sewer, it reached the smaller 137.2 cm (54-inch) cross sewer where part of the flow made an abrupt 90° bend, flowed north against grade, conflicted with flow headed south in the branch sewer, and made another abrupt 90° bend east to enter the 182.9 cm (72-inch) diameter relief sewer constructed in 1973. The remaining Section 5 flow made an abrupt 90° bend in the opposite direction, flowed south with the grade through the smaller sewer, and made another abrupt 90° bend east to enter the original 152.4-cm (60-inch) diameter Section 4 trunk sewer. Video tapes showed a great deal of turbulence which produced large quantities of H_2S and showed that the concrete of the junction structures in this area were severely deteriorated.

The second major design issue included the alignment of the original 1946 sewer and the 1973 relief sewer. When the 1973 relief sewer was constructed, it crossed and re-crossed the original 1946 sewer five times within WSBC's project limits. The 1973 project included a series of junction structures with the basic plan section of an "X" constructed of unlined reinforced concrete over the original 1946 sewer. After removal of the 1946 sewer within the junction structures, the flows in both the 1946 sewer and the 1973 relief sewer conflicted at each structure resulting in turbulence and the release of H_2S. Video tapes showed greater H_2S corrosion in the 1973 junction structures than in the original 1946 sewer further downstream. H_2S corrosion had progressed past the first layer of reinforcing steel and WSBC determined that these junction structures would have to be rehabilitated.

A third design issue was the method of construction of the 1946 sewer. The trunk sewer was constructed in an open cut trench as a MRC sewer. This sewer was constructed along an alignment which consisted of a series of straight segments connected by curves of various radii. Since there was no surface evidence of the curves and record drawings showing field dimensions of the curves did not exist, it was not possible to determine the exact field location of the sewer prior to construction. The actual constructed interior diameter of the host sewer was also a cause of concern. The 137.2 cm (54-inch) inside diameter and the wall thickness of the sliplining pipe left a theoretical 7.4 cm (2.9 inch) annular space between the exterior face of the slipline pipe and the interior face of the host pipe. The curved sections of host sewer would have smaller clearance due to the straight pipe joints to be installed in the curved host sewer. This led to concerns as to how to structure the bid documents so that the contractors could receive fair payment for various conditions found in the field.

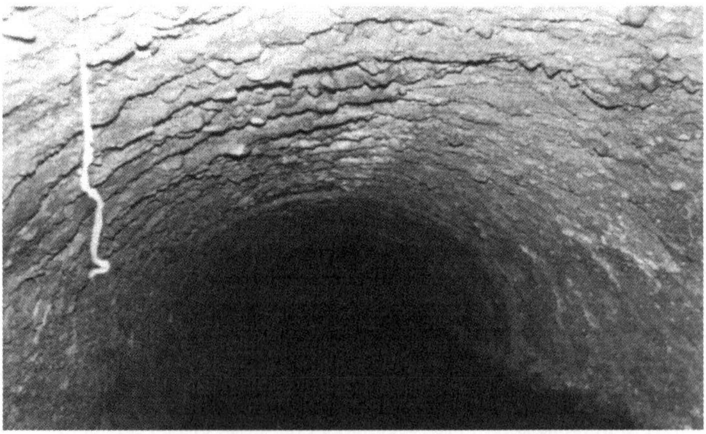

Typical Corrosion Levels in 1946 Host Sewer

The fourth design issue was the structural condition of the host sewer pipe, the junction structures, the manholes, and the pipe bridge over Kuhlman Gully. The entire sewer reach showed severe corrosion and large portions of the project showed corrosion which had progressed completely through the pipe crown. One section under Pecan Street had collapsed with enough debris in the sewer to prohibit access by the video camera. Fortunately this collapsed section was located on the eastern end of the project where the 1973 relief sewer allowed the flow to bypass around the collapsed portion of the 1946 sewer. The bid documents for the 1946 sewer showed that the crown of the MRC sewer was 17.8 cm (7 inches) thick. Corrosion of this magnitude in the sewer led to concerns about the structural integrity of the Kuhlman Gully pipe bridge and whether or not it would be able to safely withstand the construction loads of sewage, sliplining pipe, and annular space grout.

Design Solutions

Since large portions of the sewer were structurally unsound, WSBC recommended that a structural pipe should be specified which could withstand higher grouting pressures and would not depend on the structural integrity of the host sewer for its strength. In addition, it was decided that a stiffer pipe with a minimum pipe stiffness of 302 kPa (46 psi) would hold its shape better during liner insertion. For these reasons, WSBC recommended centrifugally cast reinforced fiberglass (CCRF) pipe to the GHWP, City of Houston, and the other consultants for this project. This material would also allow fabrication of various special fittings to fit the alignment of the MRC sewer.

The smaller 137.2 cm (54-inch) pipe and the poor hydraulics caused by the pipe configuration across O.S.T. was solved by constructing three new junction boxes and a wye section immediately downstream of the original tee intersection. A containment structure was constructed around the original tee in order to retain the sewage while the top of the pipe was removed. This containment structure served as an insertion pit for liner pipe to the west. The east wall of the containment structure contained the single leg of the new fabricated wye pipe which would split the flow between the 1973 relief sewer and Section 4 of the original 1946 trunk sewer. One additional junction structure was constructed at each end of the two downstream legs of the wye over the 1973 relief sewer and the 1946 trunk sewer. After the upstream section of the sewer was lined, the new wye section was constructed, and the lining was inserted into the first downstream section of the 1946 trunk sewer, the existing sewer within the new junction boxes was removed. Specially fabricated 45° bends were installed into the junction structures and connected to the legs of the wye and to the 1973 relief sewer and the new liner in the 1946 trunk sewer. A closure piece of CCRF pipe was installed in the upstream containment structure and all three structures were filled with grout around the new fittings. A fourth junction structure was constructed to direct flow from the 137.2 cm (54-inch) existing branch sewer directly to the 1973 relief sewer without the previous conflicting flow movements. All of this work was accomplished while the sewer was under flow.

Rehabilitation of the six downstream "X" shaped junction structures constructed as part of the 1973 relief sewer was a more difficult problem to solve. These sewers were always more than half full and carried average daily flows in excess of 1.75 m^3/s (40 mgd). Bypass pumps of this capacity would require special wet wells to be constructed so that sufficient submergence and space could be obtained to operate properly. Temporary force mains large enough to handle these flows would also be difficult to install and anchor to resist the thrust developed at the various bends. In addition, these temporary force mains would almost certainly have to be buried so that traffic could be maintained. For these reasons it was determined that the design would allow rehabilitation of the junction structures under flow. Since the sewer and the junction structures were always at least half full, the structure below the low water level was not deteriorated. The design selected to accomplish these parameters consisted of excavating around the outside of the existing junction structure, removing the top of the structure, and constructing a new structure around the exterior of the existing structure. The existing structure was used as the interior forms for the new structure and a layer of T-Loc pvc liner was installed on the interior face of the new wall. The walls of the new structure were doweled into the base of the original structure and a new precast reinforced concrete top with stainless steel panels protecting the bottom side from H_2S corrosion were set in place on top. Over the years, the remains of the original walls will serve as sacrificial concrete until the new pvc liner is exposed. The City has plans to eventually line the 1973 relief sewer and the junction structures were designed with removable tops so that the structures may be used for insertion pits. After the liner is installed, the top of the junction structure can be replaced and the pvc liner installed as a part of this project can be tied to the new liner pipe in the 1973 relief sewer to provide continuous corrosion protection. This design allowed the rehabilitation of these structures under flow.

The remainder of the project was divided into three separate categories for bidding purposes. The contract documents required that straight sections of host sewer be tested with a mandrel to ascertain that the liner pipe would fit into the host sewer without binding. Insertion pits were constructed approximately every 122 meters (400 feet) in the straight sections and the liner pipe was pushed into the host sewer one joint at a time. Curved sections of the host sewer which had a large enough radius to receive shorter standard length pipe, as short as 2 meters (6.67 feet), without binding were lined as normal straight pipe. Curved sections of host sewer with radii small enough that the shorter sections of liner pipe would bind were open cut down to the spring line of the host sewer pipe, the host pipe widened at the liner joint by hand if necessary, the new liner installed in the half pipe, and the pipe was backfilled using standard open cut construction methods. The radii of some curved sections of the host sewer were so irregular that the CCRF manufacturer fabricated special fittings to custom fit the host sewer half pipe section. The last method of construction occurred where the host sewer had collapsed completely in Pecan Street. Flow was bypassed through the 1973 relief sewer while a new section

Typical Insertion Pit with Liner Ready for Grouting

Typical Open Cut Construction Used In Small Radius Curves

of pipe was constructed using standard open cut construction methods to replace the collapsed section. The quantities of each type of construction were estimated and bid as unit price construction. The contractor was then paid based on the actual length of each type of pipe constructed.

Rehabilitation of the sewer and the junction structures left two major structural components to be rehabilitated. The first component was the existing brick and mortar manholes which were in an extremely weakened condition due to H_2S corrosion of the mortar between the bricks and the corrosion of the host sewer. It was determined that these structures would be abandoned and backfilled. New manholes would be constructed at the insertion pits. This provided closer manhole spacings and, because the insertion pits had to be constructed at the beginning and end of every curve, allowed the placement of manholes at these points for easier maintenance. The new manholes were designed to be CCRF fabricated tees placed in the vertical position. These tees served the dual purpose of providing new corrosion resistant manholes and acting as closure pieces between sliplined sections.

The last remaining component was the Kuhlman Gully pipe bridge. This structure was designed as a series of simply supported reinforced concrete beams resting on bents. The gully slopes were covered with pillow type concrete slope protection. The interior of the beams consisted of a round 152.4 cm (60-inch) diameter cross section carrying the sewage and a rectangular exterior cross section. An expansion joint at each end consisted of a thin copper membrane. The interior of each beam was unprotected and had suffered considerable corrosion. A sonic survey showed that sufficient sound concrete remained to enable removal of the unsound concrete by water blasting, insertion of the new liner, and filling of the annular space with new structural grout.

To insert the liner, the top half of the last 6.1 meters (20 feet) of the east beam had to be removed. This was required because the alignment contained a deflection large enough so that the joints of liner pipe could not be pushed past the deflection. The original plans showed that the beam was supported by the east embankment. However, field excavation revealed that the buried joint at the end of the bridge beam/host sewer connection had no expansion joint, was leaking, and had, over the years, piped out the supporting soil under the beam. The existing slope protection concealed this condition and when the contractor removed the top half of the beam, it failed and dropped into the concealed cavity beneath it.

The City asked WSBC to design a replacement beam and prepare emergency bid documents to replace it. The time required to construct the beam in place, let it cure, and line it would have been significant. The City was bypassing the flow through the 1973 relief sewer but was concerned about its ability to handle peak wet weather flows. For this reason, WSBC designed the new beam with a steel pipe interior serving as the structural member. The concrete cast around the exterior to bring it back to the original cross section served as exterior corrosion protection and kept the stream hydraulics the same as they had been previously. The steel interior pipe meant that if necessary, flow could be diverted back through the bridge at any time during repair construction. The new beam was sliplined with 137.2 cm (54-inch) CCRF pipe as was the remainder of the structure. However, since the main structural element in the new beam is the steel pipe, the contractor was allowed to fill the annular space with the cellular grout in this portion of the bridge in lieu of the structural grout required in the remainder of the bridge.

The contractor for the project was Jalco, Inc. who also repaired the damaged pipe bridge. WSBC's estimate for the project was within 1% of the low bid. The finished cost of $7,572,745, based on installed quantities, was within 5% of the original bid amount. The contractor reported an average of 46 cm (18 inches) of debris in the host sewer. 4,870.4 meters (15,979 feet) of sewer was lined with CCRF pipe manufactured by Hobas, USA. Of the total length, 487.7 meters (1,600 feet) was open cut with shortened joint lengths. The contractor surveyed the mitered joints used in the tight radius curves and the pipe manufacturer fabricated them to the contractor's requirements and delivered them to the jobsite within two days. There were four new junction structures and five rehabilitated junction structures. The contractor installed 30 new manholes fabricated as tees by the pipe manufacturer. The annular space was grouted with cellular grout provided by Gulf Coast Grout. The 137.2 cm (54 inch) diameter crossing of O.S.T. was installed by hand mining and used 1,448 kPa (210 psi) pipe stiffness CCRF pipe.

View of Interior of Finished Sliplined Sewer

Design and Installation of Large Diameter Slipliner Pipe in Lakeland, Florida

Joseph E. Hickle, Jr., P.E.[1]
Kenneth L. Glasgow, ASCE affiliate[2]

Abstract
The application of practical design criteria to determine the suitability of large diameter sewer pipe for rehabilitation by sliplining is presented in light of a real project installed in Lakeland, Florida during 1994. The project consisted of rehabilitating 6.71 km (22,000 ft) of gravity sanitary sewer mains ranging in size from 61 cm (24 in) to 122 cm (48 in) diameters. The determination of pipe rehabilitation techniques with consideration to flow monitoring and televised inspection is presented.

The appropriate pipe stiffness to resist long term hydrostatic buckling and short term grouting pressure is noted. The application of field controls to insure proper installation and acceptance testing is discussed. Practical cleaning processes, pit preparation, pushing techniques, and field fabrication techniques are reviewed. An innovative method of making closures is described. Methods of grouting a small annular space are described along with methods of building bulkheads, grout pressure control and grout delivery.

Introduction

The City of Lakeland, Florida desired to perform a Sanitary Sewer Evaluation Study (SSES) of its Eastern Trunk Sewer in 1992. This program was intended to determine the conditions of the existing sewer and to recommend how the sewer system could be rehabilitated by currently available methods.

----

1 Project Engineer, Camp Dresser & McKee, 1715 Westshore Blvd, Suite 875, Tampa, FL 33607

2 Regional Manager, Lamson Vylon Pipe, 25701 Science Park Drive, Cleveland, OH 44122

The system comprised of approximately 83 manholes and 6.7 km (22,000 ft) of gravity sanitary sewers ranging from 61 cm (24 in) to 122 cm (48 in) in diameter. The majority of the sewers were reinforced concrete pipe (RCP) or vitrified clay pipe (VCP) which were installed approximately 50 years ago. Several system problems were recorded and the City chose to take a proactive approach to investigating the condition of its sewer system and planning for rehabilitation prior to system failures. The City retained Camp Dresser & McKee, Inc. as the engineer for the system evaluation and subsequent rehabilitation design project. Kimmins Contracting Corporation was the rehabilitation contractor for the project.

System inspection.

A program to observe rainfall induced flows in the sewer was conducted for 30 consecutive days. Rainfall was measured during this period to develop some correlation of the system response time to precipitation.

All manholes were entered and inspected for structural integrity and leaks. Each sewer run between manholes was televised and videotaped to record the actual condition of the sewer. Numerous problems were recorded that lead to design decisions of how certain sections of the sewer line should be considered for rehabilitation. Manholes were to be either lined with fiberglass inserts or coated with cementitious material by a spraying process.

Video inspection of the sewer revealed corrosion, offset or dropped joints, broken joints, and protruding laterals. Many of these pipe defects often call for point repairs. In addition to defect identification, service laterals were also noted for future reference. Proper consideration of existing pipe conditions and surface inspections enabled the contractor to locate the optimum position for insertion pit placement.

Hydraulic capacity analysis

An existing condition hydraulic analysis was performed to determine the flow capacity of the sewer. The methodology used was to determine flow characteristics according to the Manning Equation and to calculate the hydraulic grade line to investigate any sewer surcharging or partial flow for design flow rates planned by the City. Observed flows during a thirty day period were measured as well as precipitation to determine if any rainfall induced correlation could be established. Rainfall data only supported the belief that the sewers did receive infiltration and needed to be rehabilitated so that the City could reduce treatment volumes at its wastewater treatment facility.

Each run of existing sewer was noted for the type of pipe material, roughness, diameter, and slope between manholes. The Manning Equation was then calculated for each run of sewer to compute the hydraulic grade line for design flows planned in the system. This would determine if adequate flow capacity would exist after sewer rehabilitation.

A post rehabilitation capacity analysis was performed in the same manner as the existing condition analysis with the exception of modifications to pipe diameter and roughness. Generally, pipe diameter reduction with smoother slipliner materials retains adequate system capacity. Observed peak flows as well as planned design flows were induced on the rehabilitated sewer system and a resulting hydraulic grade line calculated. In areas where chemical grout sealing was proposed, the pipe diameter would remain the same, but the roughness factor would be similar to that of normal concrete with n=0.013. Other segments proposed to be repaired by sliplining were estimated to be improved to a relatively smooth roughness factor between 0.009 to 0.011.

Rehabilitation methods consideration

The SSES considered the methods of chemical grouting leaking pipe joints, sewer line rehabilitation by a sliplining method, or complete pipe run replacement. Unit costs for sewer rehabilitation by one of the four conventional methods of cured-in-place, sliplining, spiral wound, or complete excavation and replacement were considered.

Excavation and replacement was rejected for further design consideration due to excessive costs and disruption to traffic and local residents. Cured-in-place technology presented the highest unit costs for sewer rehabilitation, but had to be accepted in areas of the sewer system where hydraulic capacities were marginal and the design team wanted to retain as much of the original pipe diameter as possible. In other areas of the sewer system, diameter reduction by sliplining rigid pipe into the host pipe was acceptable when hydraulic concerns were not as critical.

Project design and specification

Construction plan and profile drawings were prepared by the design engineer to control the installation of the slipliner process. The construction documents addressed where each type of sewer system rehab method should be employed to construct the project. The contractor chose the final locations of the insertion pits from suggested locations provided by the engineer. Occasionally, planned insertions pits would be abandoned due to conflicts observed underground once a pit had been started.

Various types of leak-free, corrosion resistant pipe materials were considered as suitable slipliners, such as polyvinyl chloride (PVC), high density polyethylene (HDPE), fiberglass and cured-in-place liners. Sufficient pipe stiffness to withstand buckling pressures and acceptable flow capacities were considered prior to inclusion into the specifications.

Initial technical specifications for the project were assembled, in part, from similar sliplining projects, listing acceptable pipeline material types and construction requirements to accomplish a successful rehabilitation of the existing sewer system. A variety of sewer system rehabilitation methods were considered and included in the technical specifications for the project. Inclusion of segmented, closed-profile PVC slipliner pipe and various manhole lining materials were late entries in the technical specifications. These materials were specified only after exhaustive investigations ensured that they were suitable for sanitary sewers and met long-term criteria established by the engineer.

Due to the critical nature of the existing sewer, the most stringent of these investigations involved the suitability of closed-profile PVC slipliner pipe. Prior to this project, the engineer was unaware of closed-profile PVC slipliner, although it had seen limited use in the southwestern United States. This PVC slipliner met, and in many areas, exceeded the required design criteria. The following are the criteria under which the slipliner pipe materials were reviewed and accepted.

- Joint tightness - All slipliner products specified must pass the ASTM D3212 joint test (where applicable).
- Pipe stiffness - All slipliner products must pass a minimum 248 kPa (36 psi) pipe stiffness test in accordance with ASTM D2412. The PVC slipliner exceeded this requirement by 28%.
- Chemical resistance - All slipliner products must be suitable for long-term exposure in a variety of chemical concentrations, pH levels and temperature ranges. Chemical resistance for PVC was tested in accordance with ASTM D1784; the PVC slipliner was found to provide grade "A" chemical resistance.
- Axial strength - All slipliner products were required to provide sufficient wall and joint structure to withstand high jacking loads which may be encountered during insertion. Closed-profile PVC slipliner demonstrated sufficient axial strength to allow for jacking distances in excess of 610 m (2000 ft).
- Long-term strength characteristics - Deflection and buckling calculations ensured that all products could provide long-term performance in both sliplining and direct bury environments. Existing pipe and grout were assumed not to provide any structural support to the slipliner pipes in these calculations.
- Project references - All slipliner products were required to have been used on similar projects. Follow-up inquiries were conducted with the engineers, owners and contractors of these prior projects.

Sliplining methodology

Trenchless technology was chosen to rehabilitate the sewer system primarily due to cost factors and minimal impact to traffic, businesses and local residents. Sections of existing sewers planned to be sliplined were investigated for practical locations to construct insertion pits.

System rehabilitation would be accomplished with the use of cured-in-place pipe, pushing rigid slipliner pipe, and pulling flexible HDPE pipe where appropriate. Since cured-in-place pipe could conform to the walls of the host pipe, it would be used where a reduction in the diameter of the existing pipe needed to be minimized to preserve as much hydraulic capacity as possible.

In areas of the sewer system where hydraulic capacity was not critical, the next smaller diameter of slipliner pipe would be pushed into the host pipe from insertion pits. In some sewer runs where 7.62 cm (3 in) downsizing was desired, the existing pipe had settled and misaligned pipe joints would not allow the pushing of a rigid slipliner pipe. In these rare situations, the contractor would have to adapt his construction method to use the more flexible HDPE pipe that would be pulled from manhole to manhole rather than being pushed. This method would require that a bypass pumping system be established. Since using slipliner pipe would create an annular space, the annulus between the host and liner pipes needed to be grouted. Grouting pressures would be limited so that the slipliner pipe would not buckle during grouting of the annulus.

Contractor's material selection
The low bidder selected to use segmented, closed-profile PVC slipliner, as provided by Lamson Vylon Pipe, as the primary slipliner material for this project. Although the material cost of the PVC slipliner exceeded the material cost of the butt-fused HDPE, installed cost for the PVC slipliner was much less. Among the deciding factors involved in the contractor's cost analysis were:

- Bypass pumping would not be necessary for segmented PVC slipliner, whereas, the continuous nature of butt-fused HDPE pipe required bypass pumping for the flow conditions in the existing line;
- Rental of a fusion machine and technician was needed for continuous HDPE pipe;
- Continuous HDPE required an insertion pit much longer than for segmented PVC;
- Storage and fusion of the HDPE created additional traffic control concerns.

Vylon Slipliner is a segmented, closed-profile PVC slipliner pipe meeting the requirements of ASTM F794. The joint consists of a gasketed fiberglass coupling which is designed to join the pipe by insertion into an internal channel machined into the ends of the pipe (see figure). The groove created by this routing process enables the joint to articulate 2° . The pipe's unique joint geometry creates a flush exterior and interior allowing for greater annular clearance than conventional joining methods.

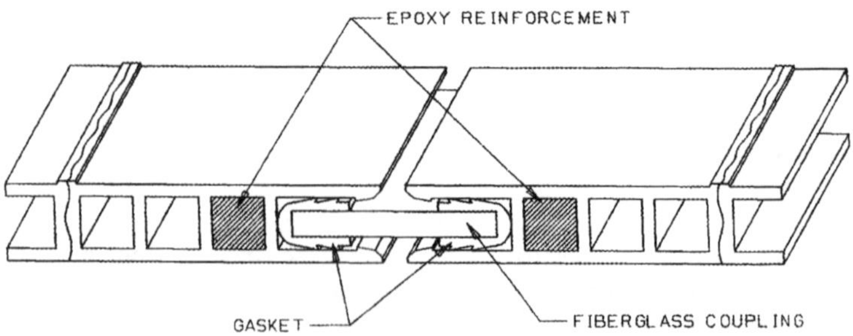

Cross Section of Vylon Slipliner Pipe Wall and Joint

The internal I-beam structure between the dual walls receives the axial loading transferred across the coupling and transmits the loading throughout the entire wall section. Epoxy fills the void between the first and second rib, thus, providing additional support and raising allowable jacking loads.

Pit preparation

Prior to any construction activity, cleaning and video inspection were required. For initial cleaning, a bucket machine (similar to a drag line) was employed to remove large build-ups of solids and debris. A final cleaning using a standard jetting nozzle and vacuum truck removed most residual solids.

Proper consideration of existing pipe conditions, necessary point repairs and surface geography enabled the contractor to locate the optimum position for insertion pit placement. Since the PVC slipliner could be pushed equally as well upstream or downstream, flow direction was not a consideration when selecting pit location.

Construction and preparation procedures remained similar throughout the entire project, regardless of pipe diameter. Due to shallow ground water conditions, a well pointing system was normally established around the proposed pit area. Typical length of an insertion pit was approximately 6 m (20 ft). The width of each pit varied by diameter, but was typically 2.4 m (8 ft). Excavation depth was ordinarily limited to the springline of the existing pipe (1/2 OD). A trench box and steel sheeting were placed and a concrete floor was poured at this point to stabilize the work area.

The crown of the existing pipe was removed by saw cutting along the springline. Crown sections were stored intact to serve as a cover when construction procedures were interrupted. With the old sewer pipe interior exposed, a piece of PVC slipliner was lowered into a cradle formed by the invert of the host pipe. This test piece would be pulled through the existing pipe in both direction to ensure that the slipliner could fit through the entire section to be lined.

Slipliner insertion

When the PVC slipliner was inserted, the gasketed coupling was oriented at the leading end. This enabled the bullnose of the rubber gasket to absorb any impact received on the leading edge during insertion. To distribute the jacking loads evenly throughout the slipliner, a push ring was supplied by the pipe manufacturer.

The sewers that were sliplined were typically flowing between 1/3 and 1/2 full, depending on the time of day and recent rainfall. Once the slipliner pipe was placed in the cradle and submerged, it was pushed into the host pipe. The slipliner pipe was pushed into place with the assistance of a backhoe (Caterpillar 225). The contractor found that placing the bucket on the push plate and crawling the backhoe forward or backward provided the most even and consistent jacking loads.

The trailing two feet of each slipliner joint was left extending into the pit to allow for joining the next piece of slipliner. To prevent the slipliner from moving while being joined, the contractor used air bags to anchor the slipliner pipe. These bags were placed in the annulus at the 10 o'clock and 2 o'clock positions and inflated, thereby creating a wedge effect that anchored the slipliner during joining.

With this system, the contractor was able to achieve an installation rate of one 4.6 m (15 ft) joint every three minutes or 90 m/hour (300 ft/hr), even though the sewers were flowing as much as half-full during installation. Sliplining continued through manholes where the host pipe was properly aligned. The greatest jacking distance encountered on this project was approximately 610 m (2000 ft) and required only seven hours of actual installation time. A 45° elbow at a manhole prevented pushing any further.

When pushing slipliner in two directions from a single installation pit, a closure piece was needed to join the slipliner together. Lamson Vylon Pipe provided a field routing tool in order to create a factory-style joint on the jobsite. This portable router allowed the contractor to cut pipe to length in the field, create a new joint, and close the slipliner pipe at the pit without creating a hydraulic disturbance at the closure.

Annular grouting

When a push was completed, bulkheads were built at each manhole interface. Bulkhead construction was completed first at the upstream end and continued downstream as the annulus drained. Since the annulus was never entirely dry due to groundwater infiltration, bulkheads were constructed of virgin oakum and chemical grout (3M #5600) below the spring line. This not only created an initial seal, but also aided in constructing a 20.3 cm (8 in) thick structural bulkhead with hydraulic cement. Two tubes were built into each bulkhead to allow for grout injection or venting. After constructing bulkheads, service laterals were excavated and reconnected. Prior to reconnection, waste from the service laterals had been allowed to flow into the annulus, thereby, not interrupting local sewer service.

A subcontractor performed annular grouting services. The slipliner pipe was grouted from manhole to manhole. In areas where a four inch annulus was present, the grout was pumped in multiple lifts in order to prevent floating the slipliner. In areas where a one inch annulus was present, a monolithic pour was performed. The annular grout consisted primarily of Type II Portland cement, pozzolanic fly ash, and water. To improve flowability, sand content was kept to a minimum. Grout pump pressures were normally limited to 34.5 kPa (5 psi); however, in areas where long grout runs or tight annular spaces were encountered, pump pressures in excess of 68.9 kPa (10 psi) were necessary.

The grout was pumped into the annulus through the two tubes built into the bulkhead. As grout was pumped, annular sewage was monitored as it flowed out of the vent tubes at the other end of the annulus. The annular sewage flowed through the vent tubes as it was being displaced by the heavier grout. As the volumes of grout approached the theoretical annular volume, pump speed was slowed, pressures were observed and the color of the displaced sewage was noted. When grout began venting from the annulus, pumping ceased. As a quality control measure, a post-grouting video was performed to ensure that all joints remained sealed and no buckling of the pipe wall had occurred during grouting.

Summary

The Eastern Trunk Sanitary Sewer of the City of Lakeland, Florida was successfully rehabilitated using cured-in-place pipe and PVC slipliner pipe as the primary trenchless technology methods. This rehabilitation has reduced infiltration and inflow to the wastewater treatment facility and has restored the structural integrity of the system for extended service life without major disruption to local businesses, traffic, schools, and local residents. This project is noteworthy since new sliplining materials were used with great success.

SPECIFICATIONS THAT ENSURE SLIPLINING SUCCESS

John Goodwin[1], Dave Gellings[2], Douglas McMillan, Jr.[3], Grace Chow[4], Hamid Sarvi[5], Rod Stude[6]

Abstract

The variable conditions inevitably encountered in underground pipeline rehabilitation give rise to a significant question: Can construction specifications be written with such accuracy that they ensure performance of the finished project? This paper relates the authors' experience in creating detailed specification requirements for trenchless rehabilitation of a sanitary sewer pipeline in Pleasanton, California. Bringing together the expertise of the system owner, engineering consultants, pipe manufacturers, and grouting contractors, an effort was made to write, issue and execute construction contract specifications sufficiently detailed to describe the intended performance requirements for each key operation. This paper describes detailed specifications for 1) preliminary examination of the pipeline, 2) preparation of the existing line for rehabilitation, 3) requirements for pipe strength, durability, and joint tightness, 4) requirements for grouting the annular space, and 5) acceptance testing of the final product. Cooperation within the project team, among the various disciplines involved, produced performance-based, descriptive specifications that resulted in a successful installation.

Introduction

The East Amador Trunk Sewer, commonly known as EATS, was constructed in 1968 within the City of Pleasanton, California. The EATS collection system serves an area of approximately 2,000 acres consisting of

[1] Principal Engineer, Brown and Caldwell, 3480 Buskirk Avenue, Pleasant Hill, CA 94523-4342
[2] Principal, Trenchless Resources International, 2923 Alvarado Drive, Bellingham, WA 98226
[3] Associate Civil Engineer, City of Pleasanton, P.O. Box 520, Pleasanton, CA 94566-0802
[4] Project Manager, Brown and Caldwell, 3480 Buskirk Avenue, Pleasant Hill, CA 94523-4342
[5] Const. Man.,G.S. Dodson & Associates, 165 Lennon Lane, Suite 105, Walnut Creek, CA 94598
[6] Regional Manager, Lamson Vylon Pipe, 25701 Science Park Drive, Cleveland, OH

residential, commercial, and office land use. The trunk sewer consists of approximately 11,000 feet of 18- to 33-inch diameter reinforced concrete pipe (RCP) and 6,500 feet of 15-inch diameter vitrified clay pipe (VCP).

Need To Rehabilitate The Pipeline. In 1989, the City's sewer collection maintenance crews noticed that the RCP portion of the EATS was deteriorating in the downstream reach between the DSRSD wastewater treatment plant and the Chabot Canal in the Hacienda Business Park. The length of this reach is approximately 4,300 feet. The concrete of both the pipe and manhole was corroding and thinning, resulting in the exposure of reinforcement bars and the formation of some holes. Figure 1 illustrates the pipeline corrosion and depicts a hole in the crown of the existing pipe. The City concluded that hydrogen sulfide gas, released during turbulence in the manholes, was corroding the unprotected concrete. The City's concern was that the structural integrity of the RCP portion of the EATS would be adversely affected.

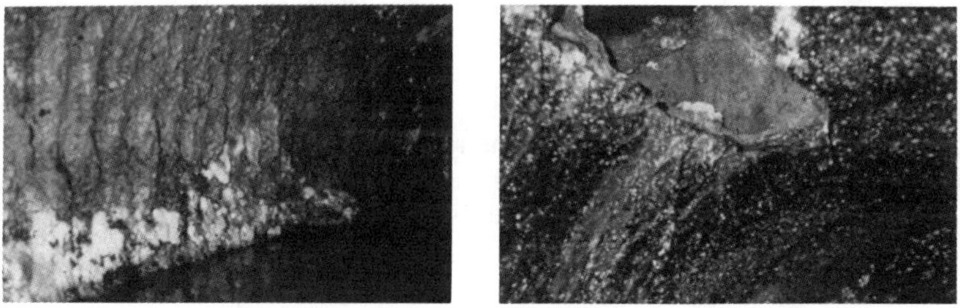

Figure 1. Existing Pipe Condition

Short-term remedial repairs at the worst locations were implemented by the City. Next, Brown and Caldwell was contracted to inspect and evaluate the problem, identify long-term alternative solutions, and develop plans and specifications for a rehabilitation project. From the City's perspective, the project had to maintain hydraulic capacity to the maximum possible extent and preserve the structural integrity of the EATS while minimizing negative impacts on the residential neighborhood, the high traffic of Hopyard Road, and the businesses within Hacienda Business Park.

Rehabilitation Options Considered

Pipeline rehabilitation options considered included cured-in-place-pipe (CIPP); sliplining using fusion-welded high-density polyethylene pipe (HDPE) or gasketed HDPE; polyvinyl chloride (PVC) pipe; and reinforced plastic mortar pipe (RPMP). A rehabilitation technique was selected after evaluating criteria such as the hydraulic capacity of the completed project, the need for bypass pumping during construction, the impacts upon residents and traffic, competitive bidding, and construction cost.

Hydraulic Capacity. Each of the rehabilitation alternatives reduced the hydraulic capacity of the original EATS pipeline. Land use information and occupancy and build-out projections provided by the City was evaluated to project the design flows for the lower 33-inch diameter reach and the upper 30-inch diameter reach of the EATS. Figure 2 illustrates the projected flows for both reaches between the years 1991 and 2011. Projections indicated that the existing trunk sewer's capacity (with an inside diameter of 33 inches) would be exceeded by the year 2010. The rehabilitation alternatives, each offering a different finished inside diameter, varied in their projections of capacity and year capacity would be reached.

Bypass Pumping. The CIPP and sliplining using fusion-bonded HDPE rehabilitation alternatives required complete bypass pumping of the trunk sewer for installation. Sliplining using gasketed pipe could be completed without any bypass pumping relatively easily, as long as the flow could be maintained below the springline of the existing pipe. This would probably have required installing flow-through plugs to limit flow, or performing the installation during the off-peak-flow hours.

Impacts Upon Residents and Traffic. The CIPP could be installed through existing manholes and usually did not require excavation. Some manholes with small openings required that manhole cones and frames be removed and replaced. All of the sliplining alternatives considered required excavation of insertion pits approximately 10 feet wide and 24-36 feet long.

Bypass pumping would have required aboveground pumps, generators, and temporary piping placed along the length of the project, potentially causing odor and noise problems for nearby residents and businesses. The piping would have restricted access to driveways, parking, and traffic. Bypass piping could have been recessed below ground to minimize access, parking, and traffic impacts, but that would have entailed added cost and temporary disruption.

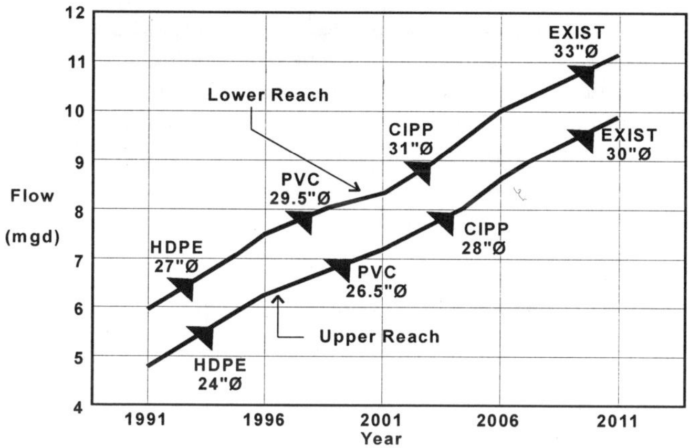

Figure 2. Projected Capacity for Upper Reach of EATS

Competitive Bidding and Rehabilitation Cost. It is a good idea to specify at least two rehabilitation alternatives to insure competitive bid prices. Considering the added cost of bypass pumping, CIPP is usually not competitive compared to sliplining with gasketed-joint pipe. Considering the costs, the City elected to specify sliplining gasketed-joint pipe with alternative pipe materials.

Selected Alternative. For the existing 33- and 30-inch diameter trunk sewers, sliplining with gasketed-joint PVC pipe provided the maximum hydraulic capacity and the lowest cost. Figure 3 illustrates sliplining pipe alternatives and dimensions for the upper reach of the EATS. The results were similar for the lower reach of the EATS.

Sliplining Alternative	Host I.D. (in.)	Liner I.D. (in.)	Liner O.D. (in.)	Annular Space (in.)
HDPE (fusion)	30.0	25.8	28.0	2.0
HDPE (gasketed)	30.0	24.4	27.8	2.2
RPMP	30.0	24.8	25.8	4.2
PVC	30.0	26.5	28.2	1.8

Figure 3. Sliplining Pipe Alternatives for Upper Reach of EATS

Sliplining Specifications

Following are a number of recommendations for performance-based sliplining specification.

With the emergence of segmented slipline products to supplement the existing technology of butt-fused polyethylene sliplining, there has been a continual struggle to address annular space grouting (grouting the space between the host pipe and the new liner pipe). There have been many documented slipline grouting failures throughout the country due to a poor understanding of proper grout mixes and procedures. These failures generally have taken the form of unwanted buoyant uplift, over deflecting, or catastrophic failure (collapse) of the new liner pipe.

In the 1980s, the City of Pleasanton had such a failure. With this in mind, particular attention was paid to the annular space grouting portion of the specifications and the subsequent grouting operation. A strong, performance-based grouting specification was developed incorporating the Annular Space Grouting portion of the Standard Specifications for Public Works Construction (Greenbook), Section 500-3, 1994 Edition and the 1995 and 1996 supplements. Items in the submittals and required calculations sections of the specification were strictly followed.

Qualifications. The sliplining contractor and foreman should have completed a minimum of three similar sliplining projects (type and size) within the past three years. The pipe manufacturer should have a successful record of operation and manufacturing for a minimum of five years. Finally, the grouting subcontractor should specialize in the placement of annular grout and have a minimum of five years of experience, having completed at least three similar projects within the last year (including experience with a similar pipe diameter, strict quality control, size of annular space, and length of run).

Submittals. Following is a partial list of submittals that should be requested.

- Contractor, pipe manufacturer, and grouting subcontractor qualifications.
- Installation details and liner dimensions.
- Proposed grouting mix design. Specifically, request the proposed grout density and viscosity, initial grout set time, the 24-hour and 28-day compressive strengths, and the grout working time before a 15 percent change in density or viscosity occurs. The submittal data should be derived from trial batches of grout tested by an approved, independent testing laboratory.
- Description of grouting method, including maximum injection pressures, proposed grout stage volumes, buoyant force calculations, pipe anchor plan to allow the pipe to set on the invert of the host pipe until the grout is set, vent location plans, and acceptance letter from pipe manufacturer indicating complete review and approval of the grouting method and procedures.

- Pressure gauge certification. Pressure gauges should have a maximum range not exceeding 200 percent of the maximum injection pressure.
- Pressure monitoring recorder details.

Our experience has shown that the pressure monitoring recorder is critical to assure that the maximum grouting pressure has not been exceeded. It is impossible to have one person monitor and record pressure accurately. (This is probably the item that the Contractor will resist most.). Some grouting contractors are willing to exceed recommended maximum grouting pressures because they know there are some safety factors included. By requiring this equipment, the specifier is sending a strong signal to the contractor that quality control is important.

Pipe Material and Dimensions. Specify acceptable pipe materials such as PVC, HDPE, and RPMP. For each reach of sewer to be sliplined, indicate the existing pipe material and actual inside diameter, the minimum acceptable inside diameter of the liner pipe, and the minimum pipe stiffness or wall thickness required to handle anticipated jacking loads or prevent buckling during grouting.

Sequence of work. The following sequence of work was specified for the EATS Project.

1. Clean the Sewer
2. Conduct internal television inspection
3. Mandrel test the existing pipe
4. Perform spot repairs as required
5. Clean the Sewer
6. Proof the Sewer with liner pipe section
7. Install the liner pipe
8. Leak test liner pipe
9. Conduct internal television inspection
10. Grout annular space
11. Mandrel test the liner pipe
12. Conduct internal television inspection

Because of the very limited annular space available for the EATS project, mandrel testing of the existing pipe (item 3) was required prior to purchase and shipping of the liner pipe. This procedure assured that the new slipliner pipe would fit prior to ordering several thousand feet of liner pipe. Television inspection of the liner pipe prior to grouting (item 9) provides a good visual check of the liner pipe joints.

Testing. The specifications should specify pre-television inspection to determine if spot repairs are required and post-television inspection to verify quality of the finished product. For this project, we required television inspection after insertion of the liner and before grouting to verify joint integrity of the liner pipe. A load cell to measure the jacking load for the installation of the liner pipe should be provided by the Contractor to verify that total jacking loads are not exceeded during installation. Mandrel testing of the host pipe prior to installation is required to verify that the liner can be installed successfully. The liner pipe should be mandrel tested after grouting to verify that the liner roundness is maintained.

Testing of the grout installation is critical. Field testing the density and viscosity of each batch of grout is required. Injection pressures are monitored by a chart recorder to provide positive verification that recommended pressures are not exceeded. Grout specimens shall be tested after 24 hours and 28 days to verify specified strengths.

Installation

To allow installation of the liner, several insertion pits were excavated and shored during the course of the project. Insertion pits were approximately 10 feet wide by 30 feet long by 21 feet deep, and were shored in a manner to allow the lowering of liner segments into proper position. The top of the RCP trunk sewer was sawcut and removed to a depth just above the flowline to allow wastewater to pass through, while allowing access to lower the liner down to pipe invert prior to pushing it into the host pipe. Pulleys were welded to the sheet piling at the bottom of the excavation to allow the backhoe excavator to pull up on the cable and thereby exert a horizontal force on the end of the liner through the pulley system.

Despite the project's successful completion, there were some tense moments when the 27-inch diameter PVC liner stuck approximately halfway down the first reach in the 30-inch diameter portion of the trunk sewer. We determined that excessive amounts of grit and small rocks in the invert of the host pipe were being pushed and wedged under the liner, causing it to lift and rub against the crown of the host pipe. The City, contractor, design engineer, construction manager, and liner pipe supplier convened a meeting. A couple of participants suggested the unconventional idea of using a steel can to clean and proof the host pipe prior to installing the liner. The steel can could, it was conjectured, scour the invert of the RCP pipe and loosen any grit deposits and remove any minor offset joints. If the steel can could reach its destination, it was the general consensus that the liner had a good chance of making it too. The steel can was 5 feet in length, with a wall thickness of ¼ inch.

Because of the close tolerances between the host pipe and new liner pipe (1.5 to 1.8 inches) and grout runs of approximately 400 linear feet, maintaining strict QC/QA procedures on the grout mixes and grouting operations was critical. Grouting procedures were followed, grout mixes were tested per batch for density (ASTM C138) and viscosity (ASTM C939), and all grout runs were grouted under the pipe manufacturers recommended grouting pressure of 8 psi. Pressures were monitored and recorded with calibrated, certified pressure gauges and the pressure recorder.

Lessons Learned During Installation

After implementing the above modifications, the installation of the liner and the grouting operations went smoothly and the project was completed on time. In summary, the project taught us the following:

- A 1-1/2-inch annular space was achievable; however, it required extra care in cleaning the host pipe, especially the invert.
- The use of a steel can approximately one inch smaller in diameter than the host pipe proved quite effective in proofing the line while removing obstructions such as grit.
- The grouting of the annular space was completed without major difficulties because of the strict adherence to specified procedures. The only problem was that grout leaked through one of the bulkheads and hardened in the liner in one of the reaches. This grout was later broken up using a small jack hammer and removed.
- Again, several things need careful checking to ensure a successful grouting operation: The bulkheads must be completely sealed to prevent grout from leaking; grout viscosity and consistency must be continuously monitored; and grout pressure must be checked and recorded on a strip chart to ensure that the liner does not collapse or bulge from excessive grout pressures.

Project Team. In addition to Brown and Caldwell's services as design engineer, the City contracted the services of G.S. Dodson and Associates, in conjunction with Montgomery Watson, to perform a biddability/constructability review. G.S. Dodson and Associates also served as construction manager, and Montgomery Watson provided inspection services. During construction, technical assistance was provided by Lamson Vylon Pipe and Trenchless Resources International. The general construction contractor was McNamara & Smallman, and the grouting subcontractor was Pacific International Grout Corporation. The engineering section of the City's Public Works Department managed the project through the design and bidding stages while the City's inspection section oversaw the project during construction.

The Application of Continuous Reflection Profiling Methods to Trenchless Pipeline Projects

Richard E. Sylwester[1]

Abstract

Directional drilling in the terrestrial and offshore environment often takes place with only a limited amount of information regarding subsurface conditions. The presence of unknown buried utilities or tanks, shallow groundwater, possible localized sources of contaminants and anomalous soil conditions along the corridor can cause project delays resulting in the need to reroute the boring or to repair equipment or utilities damaged by impact. A number of geophysical techniques, including ground penetrating radar (GPR), electromagnetics (EM), magnetometry and offshore seismic reflection profiling have been used to map the subsurface along pipeline corridors to help reduce the risk to drilling associated with unpredictable conditions. On river and lake crossings ground penetrating radar was used to map the subsurface stratigraphy to a depth of 25 feet clearly identifying the top of the bedrock and variation in sediments including the presence of till, sand, gravels and soft mud. On marine crossings where GPR is ineffective continuous seismic reflection profiling can be used to map the thickness and lateral variation in sediments to a depth of several hundred feet. On terrestrial crossings ground penetrating radar and electromagnetic methods clearly located subsurface utilities in addition to mapping sedimentary units, enabling the drilling contractor to work under and over the obstructions. These geophysical techniques have not only been very successful for evaluating complex subsurface conditions but are rapid to employ, produce results in the field and are relatively inexpensive. It is recommended that the use of one or more of these geophysical methods be evaluated or considered during the planning and costing of a trenchless pipeline and cable projects.

[1] Associate, Golder Associates Inc., 4104 148th Avenue N.E., Redmond, WA 98052.

Introduction

Trenchless technology is a generic term that includes a number of techniques for conducting underground installations without disrupting the surface (Kramer, McDonald & Thomson, 1992). The costs associated with these methods are often substantially less than those associated with surface excavation methods which require relatively expensive surface restoration. The methods include pipejacking, microtunneling, and horizontal directional drilling; categories that are based on the size of the bore and or the equipment utilized. Application of these various techniques include the installation of pipes and cables from several inches to ten or more feet in diameter under streets and highways; subsurface crossing of rivers and streams with fuel lines and power cables; and the installation of horizontal wells for in-situ soil and groundwater remediation (Cooper, 1994).

During trenchless excavation activities, it is usually possible to determine the lateral location, depth and pitch of the drillhead. This position information is used to direct or guide the drillhead in the vertical and horizontal plane to close tolerances. The degree of directional control or steerage is dependent on the sophistication of the system being used. However, in spite of the ability to control the drillhead, unknown subsurface conditions can obstruct advancement of the drill, prevent the operator from following the desired route, cause damage to the equipment or create hazardous or unsafe conditions. This then suggests that trenchless technologies require, in addition to careful planning and traditional soils investigations, the use of investigative techniques to obtain knowledge of subsurface site conditions preferably along the entire proposed alignment (Wylock, 1994). The use of geophysics can be a viable means to obtain this knowledge. Geophysical surveys are usually cost-effective, provide non-intrusive subsurface information for initial site reconnaissance investigations, provide detailed subsurface information for planning the final route, and can be used to monitor the location of the subsurface installation following completion of the project.

A number of geophysical techniques are presently being used to help to reduce the possible risks and increase in costs associated with encountering unknown subsurface conditions during trenchless technologies activities. The simplest of these non-intrusive methods include standard utility and metal locators and electromagnetics. These instruments can detect and determine the lateral position of many subsurface objects. However, they cannot reliably determine the depth of the objects, detect changes in subsurface soil conditions, locate natural (rocks, voids) or man-made (foundations) obstructions. This paper describes the continuous reflection method, using either seismic or ground penetrating radar technology, which is a geophysical technique that can overcome the limitations of the other methods.

Continuous Reflection Profiling

Continuous reflection profiling is a geophysical method for obtaining a profile image of the subsurface along a selected route or transect. Continuous reflection equipment can be rapidly mobilized to a site, the data can be acquired rapidly (5 to 500 feet/ minute typical) and the subsurface images are produced in real-time. Subsurface utilities, natural and man-made obstructions, lateral changes in soils, the presence of groundwater and subsurface voids and anomalous conditions are often detected with this method. Depending on the speed at which the system is moved along the traverse it is possible to obtain a subsurface image at an interval of several inches to several feet, hence the term continuous reflection profiling (CRP).

Continuous reflection data can be acquired with instruments that use either acoustic (sound) or electromagnetic energy. The first method is commonly referred to as continuous seismic reflection profiling or CSRP (Figure 1) and the second as ground penetrating radar or GPR (Figure 2). Although the way in which the data are gathered by the two methods are almost identical the underlying physical principals and the conditions or environment in which the two can operate can be quite different.

Continuous Seismic Reflection Profiling (CSRP)

This geophysical method uses acoustic or sound energy, and is used for studies in water covered areas including lakes, rivers, and oceans (Sylwester, 1989). The energy source for the CSRP system produce an acoustic pulse or sound wave with a device called a transducer. The acoustic pulse travels through the water and upon contact with the water-sediment interface several things can occur. Either most of the energy is reflected back toward the surface or some portion of the energy is reflected back to the surface and the remaining energy travels through the water-sediment interface into the underlying sediment (energy is also scattered by the sediment at the interface). At the next interface or boundary, which will be between two sediment layers with different physical properties, the process is repeated. The amount of energy that is reflected, and conversely the amount transmitted through the interface, is dependent upon the relative physical properties of the two mediums on either side of the interface. The physical properties are the density and compressional wave velocity (for this discussion we are ignoring the cause for other losses such as scattering). If there is a large contrast in density and velocity (water to bedrock) at the interface nearly all of the energy is reflected back to the surface. Therefore, very little subsurface information can be obtained. For a small relative contrast (water to silt) very little energy is reflected and most of the energy is transmitted through the interface to be reflected at the next relative change in velocity and density.

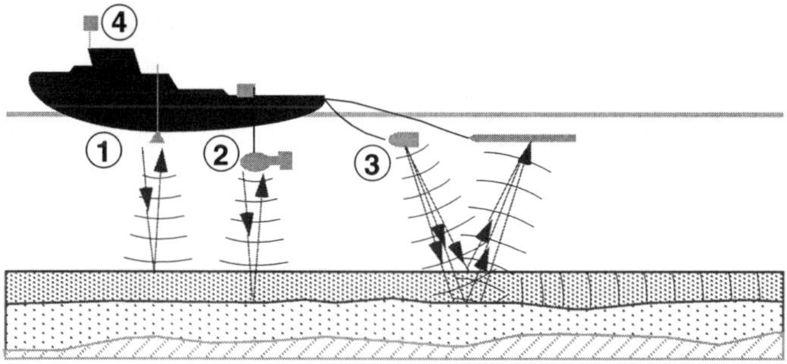

	SYSTEM	INFORMATION	DEPTH/RANGE
1	Echo-Sounder	Water Depth	2-10,000 ft below water surface
2	High-Resolution Sub-Bottom Profiler	Shallow Subsurface Stratigraphy	1-50 feet below sea floor
3	Seismic Reflection Profiling System	Medium to Deep Stratigraphy	5-1000 ft below sea floor
4	Navigation System	Horizontal Position	±6 feet horizontal accuracy

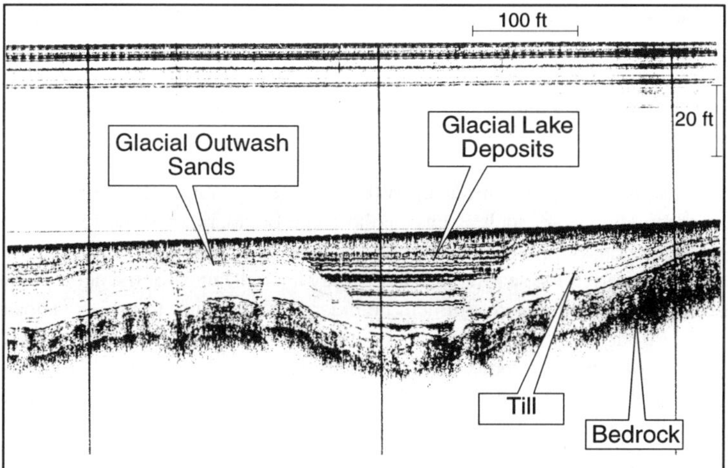

Example of acoustic subbottom profile

Figure 1: Offshore continuous reflection profiling techniques for mapping subsurface features.

The reflected energy is received at the water surface by either the transmitting transducer or a separate hydrophone, which convert the acoustic energy to an electrical signal. The electrical signal is then processed by an amplifier and printed on a graphic recorder (Figure 1, lower panel).

The simplest acoustic system is the common echosounder which operates at a relatively high frequency of 100 to 200 kilohertz, (kHz). These high frequencies are needed to produce a detailed profile or map of the floor of the water covered environment. However, the high frequency energy is unable to penetrate through the bottom sediment since most of the energy is scattered or attenuated at the water-sediment boundary. To achieve subbottom penetration requires lower frequencies and increased acoustic energy. This is done with acoustic systems that have the generic term of sub-bottom profilers (SBP). These systems operate at frequencies ranging from 2 to 12 kHz and can achieve subsurface penetration of up to 50 feet in fine-grained sediment such as silt and clay. However, if the sediments are coarse-grained or semiconsolidated, it is necessary to use systems capable of producing acoustic signals with even lower frequencies and more energy to achieve subsurface penetration. These acoustic instruments, referred to as seismic reflection systems, operate in the frequency range of 400Hz to 2kHz. They also differ from the first two acoustic systems in that they use a separate device for generating the acoustic signal (acoustic source) and for receiving the reflected acoustic energy (hydrophone). Seismic reflection systems can achieve subsurface penetration to depths ranging from several hundred to several thousand feet.

However, as the subsurface penetration capabilities of an acoustic system increases (lower frequency and more energy) the resolution, or ability to measure small vertical variations, decreases. For instance, a subbottom profiler system can resolve vertical strata that are 6 inches thick. A seismic reflection system, however, may only be able to resolve layers that are from 1 to 5 feet in thickness, depending on the acoustic characteristics of the energy source.

The subsurface penetration capability of CSRP systems is limited in sediments that are extremely coarse, such as large cobble and boulders, or if the sediments contain organic debris. Organic debris produces gas when it degrades and if these gas bubbles are trapped in the sediment (usually clay and silt) most of the acoustic energy is reflected at the gas-charged sediment boundary. This failure to achieve subsurface penetration through gas-charged sediments results from the large contrast in density and compressional wave velocity that exists between water, or non gassy sediment and the gas-charged sediments. Although the contrast in physical properties are quite opposite from the contrast at a water-bedrock interface (low density and velocity to high density and velocity), it is the relative change that is the controlling factor.

Ground Penetrating Radar (GPR)

Ground penetrating radar uses high-frequency radio waves (electromagnetic energy) to image the subsurface. An antenna (analogous to the acoustic transducers) contained in a fiberglass box, is pulled along the ground or floated on the water surface (Figure 2). The antenna converts an electrical signal into an electromagnetic pulse and then reverses the process when it receives the reflected electromagnetic energy. The reflected data is processed and displayed on a graphic recorder in the same fashion as the acoustic systems (Figure 2, lower panel).

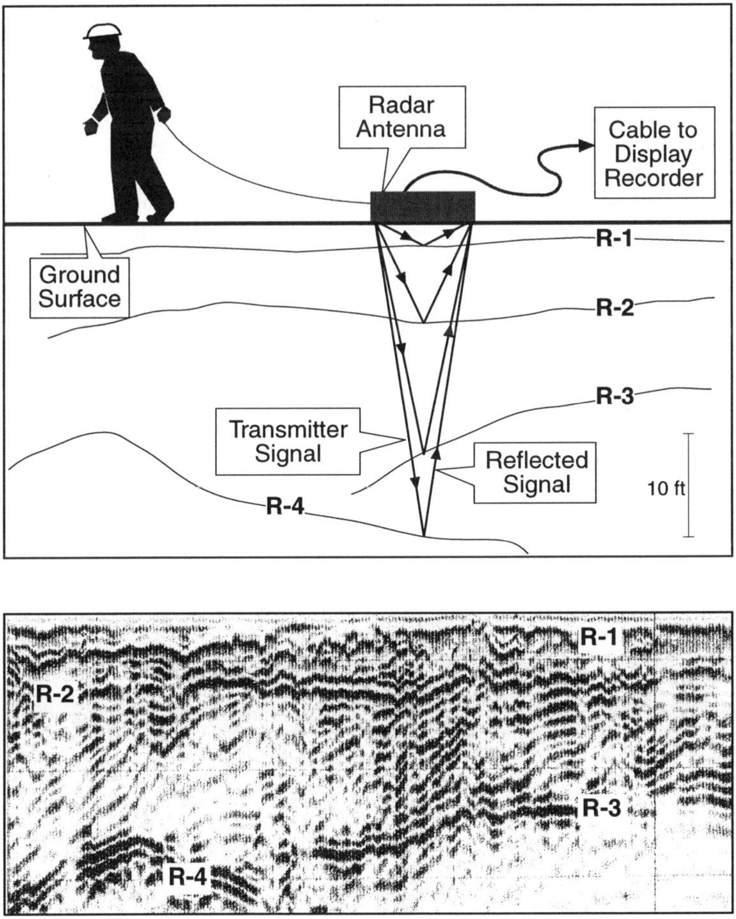

Figure 2: Ground Penetrating Radar (GPR), a geophysical technique for mapping subsurface features on land and over fresh water.

The electromagnetic energy is reflected when it contacts a surface or object where a change in electrical properties occurs (conductivity and dielectric property). These changes in electrical properties are related to the type of material, the amount of water, the presence of dissolved ions, or the electrical properties of a discrete object such as a buried cable, pipe or conduit. It is not necessary that a discrete object, such as a pipe, be conductive as is the case with a utility or metal locator. In the case of soils, for example, if a dry sand unit overlays a water saturated sand unit, a distinct image or reflection will appear on the display that represents the water saturated interface. Lateral changes in soil, such as till to sand and gravel, can also be detected. And finally, objects such as utilities, boulders, miscellaneous debris and voids are generally very apparent on the GPR record, as will be shown later.

The radar system can be interfaced with different frequency antennas (15 Mhz to 1GHz) to optimize either the resolution or subsurface penetration. This is similar to selecting different frequency acoustic transducers to optimize penetration or resolution. The high frequency antenna are usually used to detect small features, such as pipes, cable and other utilities at subsurface depths of several centimeters to several meters. The medium and low frequency antennas are more applicable to mapping subsurface soils, and detecting large tanks and voids to depths of 5 to 30 or 40 feet. Maximum subsurface penetration is achieved in coarse-grained, dry sediments and minimum penetration in fine-grained, saturated sediment. In fresh water, a low frequency antenna can often produce much better results than an acoustic system, particularly if the sediments are gas charged or if the sediments are extremely coarse or dense. However, because of various energy loss mechanisms, the GPR system is usually limited to water depths less than 30 feet and subsurface penetration of 10 to 30 feet.

Table 1 summarizes the capabilities, characteristics, limitations and advantages of the two continuous reflection methods.

Table 1: Summary of Continuous Profiling Techniques

	GPR	CSRP
On Land	Ideal	Not applicable
Fresh Water	Very good results	Excellent results
Salt Water	Very poor results	Excellent results
Ice Covered	Very good, fresh water	Poor results
Penetration	1 to 35 feet	1 to 1000 feet
Resolution	3 inches	6 inches to 3 feet
Application	Utility, soils, ground water	Map stratigraphy, locate large pipes

Examples of Continuous Reflection Profiling Data

A number of examples are presented in this section showing subsurface data from the two continuous reflection methods described above. Most of these examples are from geophysical surveys conducted on projects associated with trenchless technology activities. Several examples simply illustrate how these methods could have prevented possible difficulties if the geologic conditions shown were encountered during trenchless technology activities. For examples the acoustic image in the lower portion of Figure 1 shows a basin filled with soft sediments. This basin cannot be identified by a bathymetric survey since the sea floor is featureless. Furthermore, it is entirely possible that this basin or channel would not be identified during a soils reconnaissance survey using conventional boring techniques. Figure 2 is an example of a GPR profile, the other continuous reflection technique. This example illustrates how this method can be used to map lateral changes in soil and determine the depth to various soil horizons.

GPR-Profile of Subsurface Utilities

Knowing the lateral location and depth of subsurface utilities is usually extremely important for planning a trenchless technology activity. Figure 3 is a profile along a proposed directional drill route located beneath a concrete road surface. The reinforcement bars and the utilities were clearly imaged and it was possibly to plan a safe route for installing a subsurface cable beneath the road surface and above the utilities.

GPR-Profile of Water Table

Knowledge about the presence of saturated sediments or the depth to the water table may be important to determine if drilling is possible or what techniques and procedure should be recommended under these conditions. On the other hand, if borehole information has identified contaminated groundwater, the GPR profile could provide important information for designing a horizontal drilling program for installation of a horizontal remediation well.

GPR-Profile of Contaminated Groundwater

The hydrocarbon, resulting from a leak in a jet fuel storage tank, is clearly evident in the subsurface in Figure 5. The GPR was used to map the lateral extent of the hydrocarbon plume prior to installing extraction wells. This work was done prior to the development of drilling and installing horizontal wells which would have been appropriate techniques for remediating this site; for example by extraction or in-situ remediation.

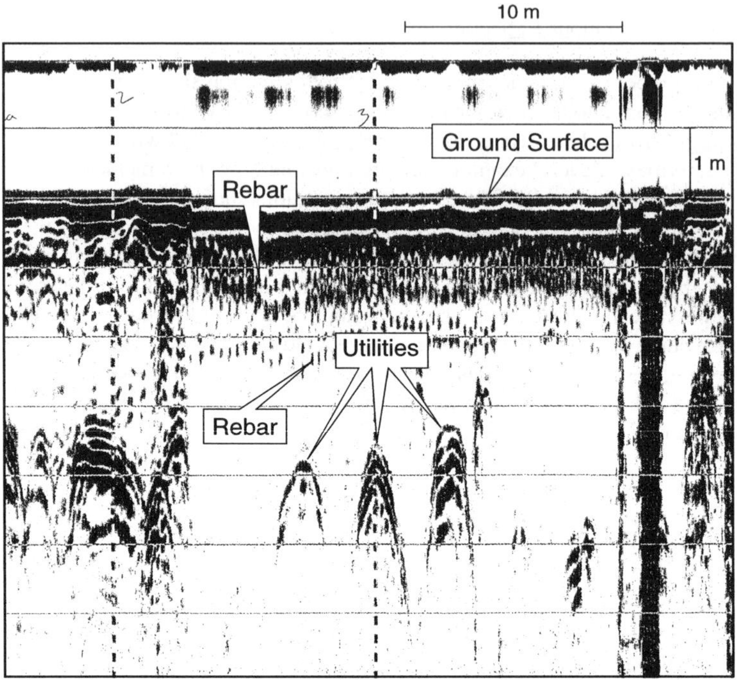

Figure 3: Ground penetrating radar (GPR) profile along a proposed route for directional drilling, clearly shows subsurface utilities and reinforcement bars.

GPR-Profile of Top of Bedrock

Prior to installation of a directionally drilled gas pipe line in a lake, a CSRP survey was conducted to map the depth to the top of bedrock along the proposed alignment. The results were poor because of the presence of organic material and gas-charged sediment. In addition the sediments were quite thin and the limited resolution of the acoustic reflection system was unable to distinguish the sediment-bedrock contact in many places. A follow-up survey, done with the GPR, was able to clearly identify top of bedrock along the entire alignment (Figure 6).

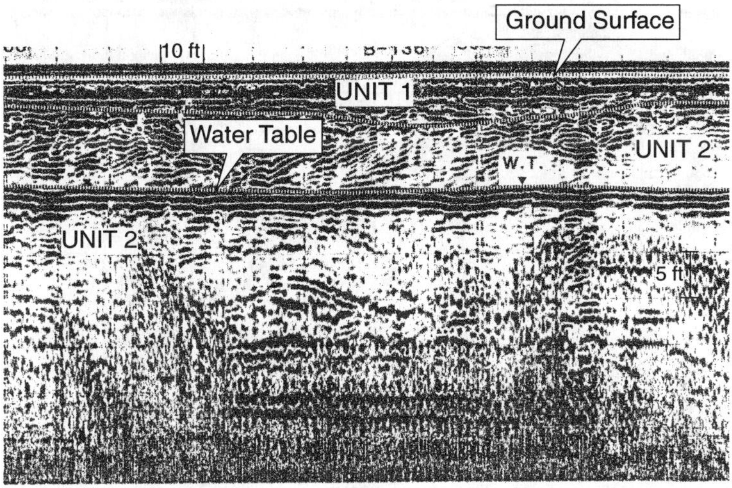

UNIT 1 Sand and fill
UNIT 2 Gravel

Figure 4: Ground penetrating radar (GPR) image along a proposed route for directional drilling that clearly shows the water table.

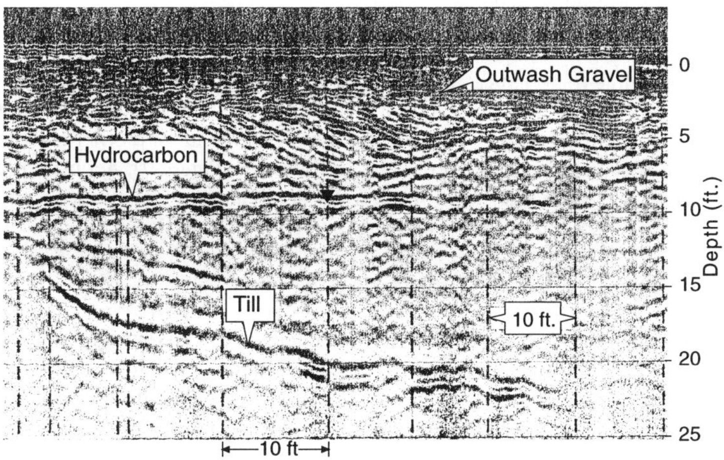

Figure 5: Ground penetrating radar profile showing the presence of contaminants on the water table. The presence of hydrocarbons, lithology and depths were confirmed with borings.

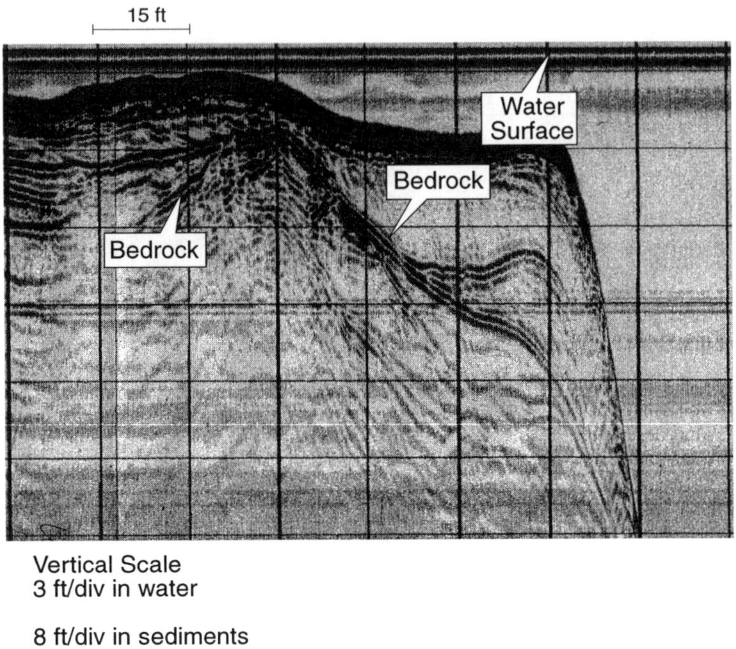

Figure 6: Ground penetrating radar profile in fresh water lake showing the top of bedrock.

Acoustic SBP-Profile of Bedrock

Sampling soil boring done during a reconnaissance investigations along a proposed gas pipeline route indicated that a silt/clay layer rested on a very irregular bedrock surface. The use of additional borings to define the bedrock surface would have been prohibitively expensive. A CSRP, using both a subbottom profiler and a reflection system was conducted to map the top of bedrock (Figure 7). Acoustic methods were used instead of GPR because of the water depth and the fine-grained sediments, conductive sediment (clay) which are opaque to electromagnetic energy.

GPR-Profile Showing Lateral Variability in Soil

Although trenchless activities were not conducted at this site the example illustrates how test pit data may not provide information to properly characterize soil conditions at a site. The two test pits indicate that the soils at this site consists predominantly of 20 to 25 feet of sand and gravel. However, a GPR profile between the two test pits indicates the presence of dense clay that could impact the success of a trenchless technology activity were it planned for this area (Figure 8). This data is a copy of the actual field record that was acquired in less than ten minutes of time.

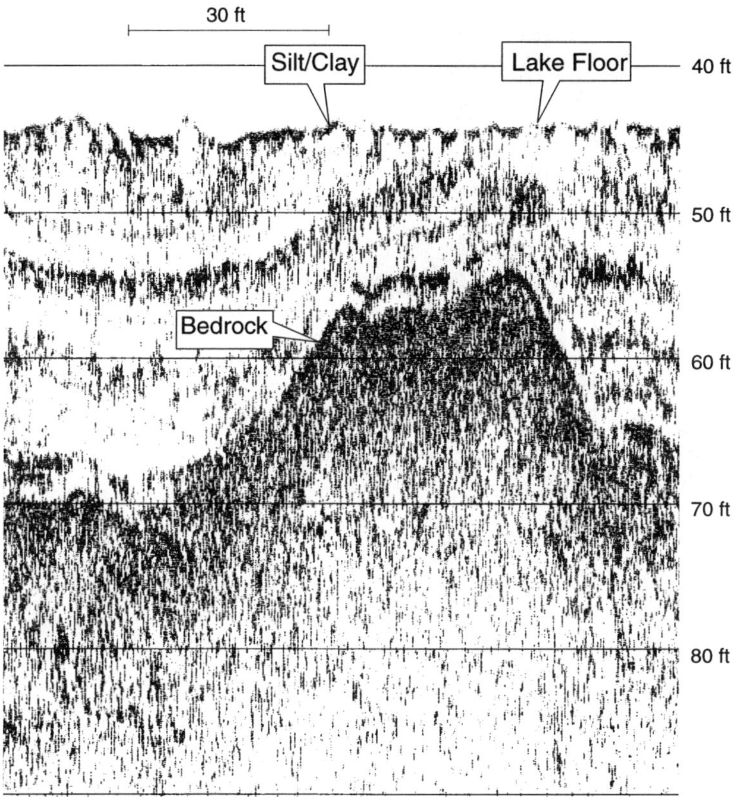

Figure 7: A 3.5kHz subbottom profile record showing the variation in depth to the top of bedrock.

GPR-Profile of River Crossing

An onshore boring had identified a relatively shallow bedrock surface near the entry for a proposed directionally drilled pipeline. The river floor was expected to be covered with coarse grained sediment since it has high flows during most of the year. Therefore it was decided to use GPR instead of acoustic methods to map the top of bedrock (Figure 9). Interpretation of the sediments suggest that the sediments consist of sand and gravel that overlay the bedrock. One or two selected cores or boring could be obtained to confirm the interpretation or to make any necessary changes.

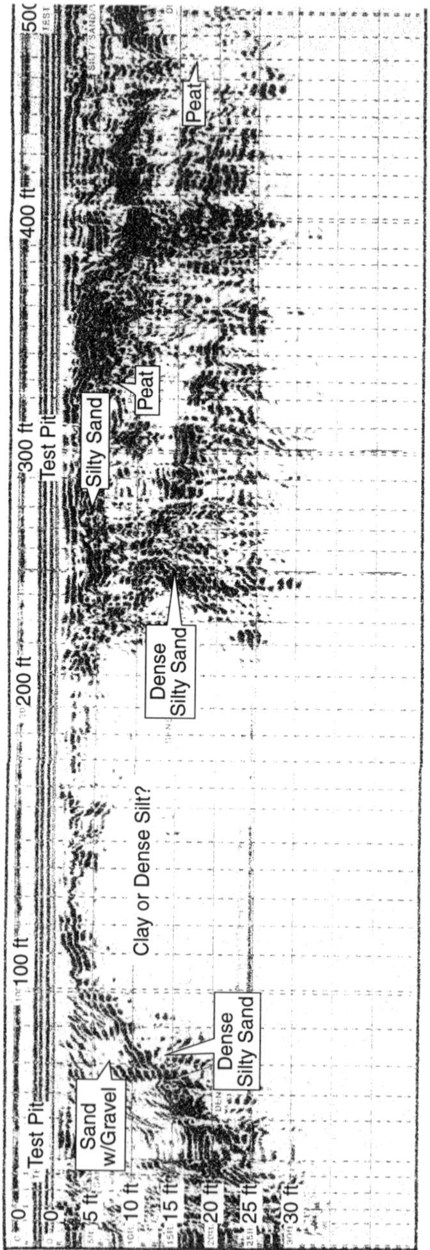

Figure 8: Ground penetrating radar record showing the lateral variability in subsurface soil type that could pose a problem to a planned directional drill.

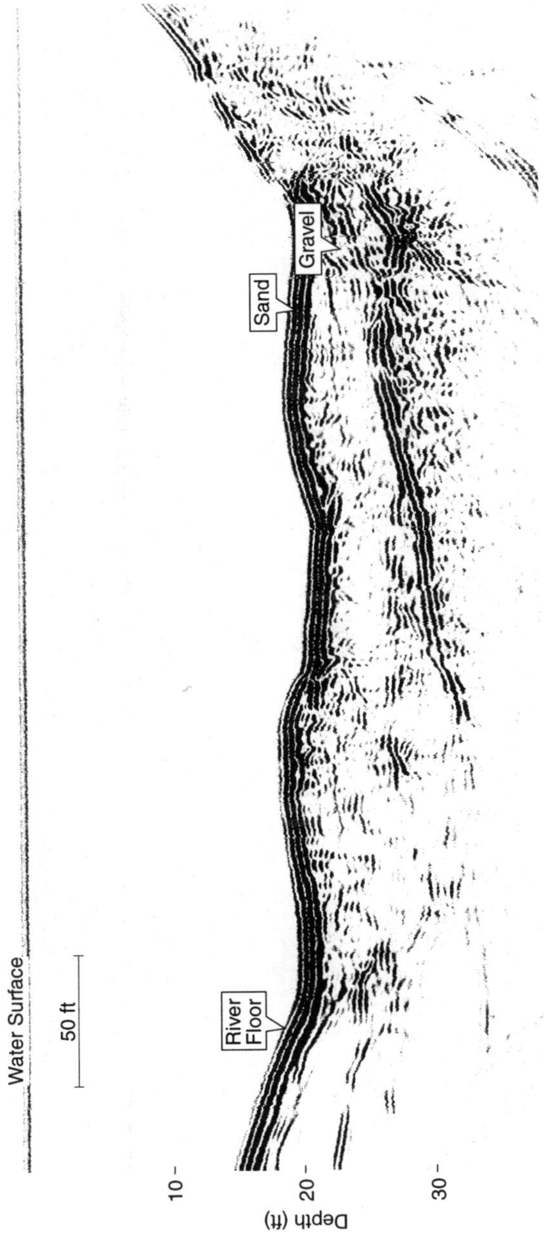

Figure 9: Ground penetrating radar profile on a proposed directional drill river crossing.

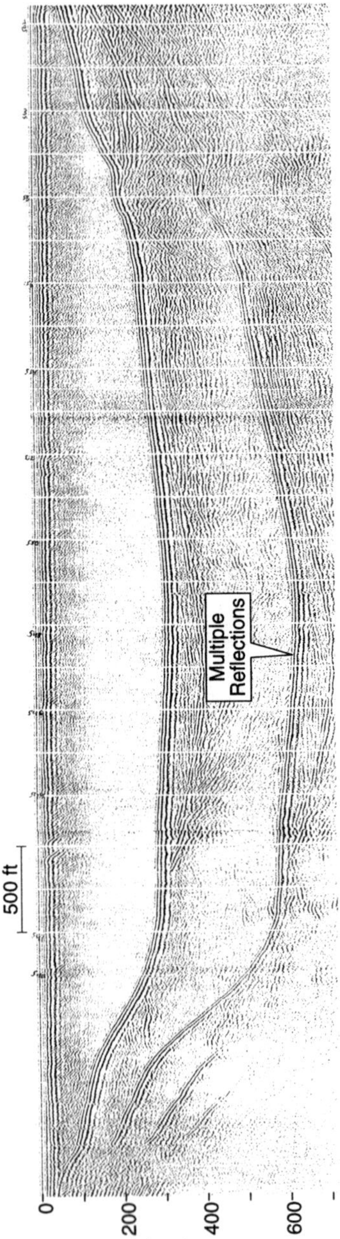

Figure 10: Subbottom profile record along alignment for proposed crossing of deep-water estuary.

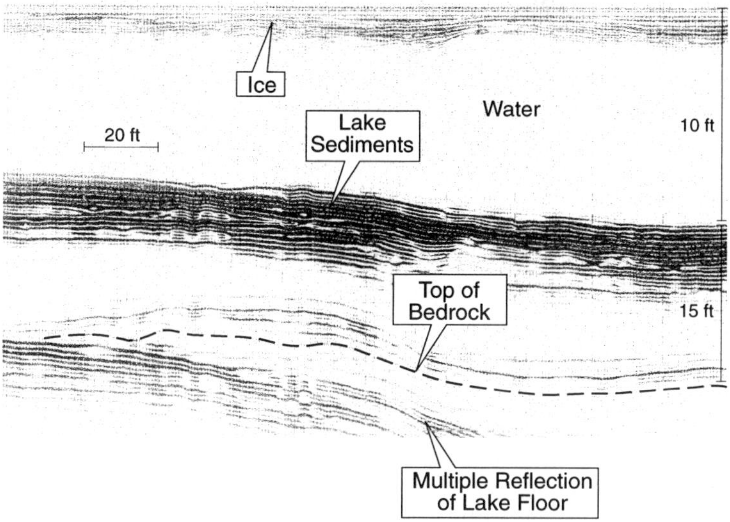

Figure 11: Ground penetrating radar profile made through ice on a fresh-water arctic lake.

Continuous Seismic Reflection-Profile for Deep-Water Pipeline Crossing

A directionally drilled crossing for a gas pipeline in a deep-water, high current marine environment required information on the subsurface sediment. The subsurface geology was expected to be complex based on the known glacial history of the area. Using conventional drilling methods, from a barge, would expensive due to the deep-water and high current (5 to 7 knots) conditions. Furthermore, it was felt that a large number of borings would be needed to characterize the complex subsurface conditions. A continuous seismic reflection survey was conducted in the area to help select a preferred route for the directional drill and to select locations for soil borings (Figure 10).

GPR-Profile On Arctic Lake

Information on the nature of the subsurface soils and the depth to top of bedrock were needed to install a subsurface water pipeline beneath an arctic lake that was to become a reservoir for a small village. Attempts were made to obtain sediment samples during the summer months but high winds made the task nearly impossible and dangerous. It was decided to conduct a geophysical survey on the lake in the early spring when the lake was still ice covered (Figure 11). GPR was selected as the geophysical method since it is well known that radar signals can readily penetrate ice (In Antarctica GPR has achieved over 5,000 feet of penetration). Following the survey an ice auger was used to bore through the ice and hand held coring devices were used to obtain soil samples and calibrate the GPR data.

SUMMARY AND CONCLUSIONS

Thorough investigation of subsurface conditions is extremely important if trenchless technology is to be applied successfully and cost effectively. A number of geophysical instruments including magnetometers, utility locators and conductivity meters (EM) are traditionally used to locate buried utilities, buried drums and tanks, miscellaneous debris, and soil conductivity. These methods however do not provide reliable depth information to discrete anomalies; cannot identify information or vertical changes in soils; cannot locate boulders or voids and cannot determine the presence of groundwater. The most reliable and cost effective geophysical method for obtaining this information is with continuous reflection profiling.

On land, ground penetrating radar is the most effective geophysical tool for imaging the subsurface. The data can identify lateral changes in soil conditions, detect discrete objects such as utilities, rocks and voids, and map the top of the water table or bedrock. These systems can be used on land and on fresh water, or frozen ground or water. They typically achieve subsurface penetration from several feet to 30 or 40 feet. However, the GPR method is generally ineffective if the sediment are conductive, for instance near the shoreline where there is salt water intrusion or in wet clay environment, In addition, miscellaneous subsurface debris, such as in land fills, often limits the effectiveness of this method.

The continuous seismic reflection profiling method uses acoustics or sound waves to image the subsurface. These systems can only be used in water covered areas and are used primarily to map subsurface stratigraphy. However, buried utilities and miscellaneous debris can also be detected. This method is not limited by the depth of water however dense, coarse-grained sediment, or sediment that contain organic debris and gas are generally acoustically opaque. Under most conditions an acoustic system can be selected that will provide subsurface penetration ranging from ten feet to several hundreds feet and still maintain good vertical resolution.

Although either of these systems can provide an image of the subsurface they should not be considered as a substitute for conventional methods used to obtain geotechnical or geologic information. These geophysical methods should be used to:

- Optimize the placement of borings or other intrusive site investigation techniques
- Provide preliminary reconnaissance information to assist in rapidly evaluating proposed routes for trenchless technology activities
- Assist in selecting or designing an appropriate trenchless technology program

REFERENCES

Cooper, George A., 1994. "Directional drilling", Scientific American, May, 1994, p82-87.

Hair, J, & Hair, C. "Considerations in the design and installation horizontally drilled river crossings. ASCE Pipeline Infrastructure Conference, Boston, 1988.

Kramer, McDonald & Thomson, 1992. "An introduction to trenchless technology". Chapman-Hall, 1992.

Sylwester, Richard E., 1983. "Single-channel, high-resolution, seismic reflection profiling: A review of the fundamental and instrumentation". Geophysical Exploration at Sea. CRC Press, Boca Raton, FL.

Wylock, John D. "Integrated surface geophysical methods-a cost effective approach to site characterization. The Professional Geologist, December, 1994.

Parametric Study for Buckling of Liners: Effect of Liner Geometry and Imperfections

Khaled El-Sawy[1] & Ian D. Moore[2], Member, ASCE

Abstract

Non-linear Finite Element analysis is used to investigate liner stability under external fluid pressure. Liners are assumed to be circular or elliptical with initial local or global geometrical imperfections. The typical local imperfections examined are defined as a wavy intrusion into the liner. Global imperfections are also investigated in the form of an ovality in the liner or a gap between the liner and the host pipe. A parametric study is undertaken to determine the effect of the initial liner geometry. It is shown that critical external fluid pressure depends heavily on the ratio of liner thickness to radius. Local geometrical imperfections and the size of the gap between the liner and the host pipe have very substantial effects on the liner buckling strength particularly for thin liners. The results of the parametric study are used to develop a new design method for liners used in rigid pipe rehabilitation.

Introduction

The trenchless rehabilitation of damaged pipelines is developing rapidly as a competitive alternative to conventional methods of pipeline replacement. Most of the damaged pipelines are structurally safe, but due to hydraulic requirements (e.g. leakage) renovation is required. In this case, any loads due to the soil surrounding the pipeline are carried by the rigid pipeline while the liner carries only the pressure from any internal fluid or the ground water pressure which is transmitted through the cracked pipeline to the exterior liner surface. The following sections report on a parametric study examining the effect of liner geometry and imperfections on the critical pressure, P_{cr}.

Problem Definition

The buckling of an unsupported circular ring subjected to an external pressure was studied in the last century by Levy (1884). The buckling pressure for an unsupported pipe of radius, R, cross sectional moment of inertia, I, Poisson's ratio, ν, and modulus of elasticity, E, is defined by

$$\bar{P}_{cr} = \frac{3\,EI}{(1-\nu^2)\,R^3} \tag{1}$$

---

[1] Postdoctoral Fellow, Geotechnical Research Centre, University of Western Ontario

[2] Professor, Geotechnical Research Centre, University of Western Ontario, Canada

where the elastic solution for the pipe case is adopted from the ring case by simply adjusting the ring elastic Young's modulus, to be $E/(1-\nu^2)$.

The concept of comparing the buckling pressures for unsupported and encased liners has been employed by both researchers and developers of design codes. The idea has been to determine a quantity called *enhancement factor*, *K*, which is used to define the critical pressure for the encased liner as

$$P_{cr} = K \times \bar{P}_{cr} \tag{2}$$

where P_{cr}, and $\bar{P}_{cr}$ are the buckling pressure for encased and free liners respectively. The same concept is used herein in addition to other reduction factors due to liner ovality and local imperfections.

Literature Review

The encased ring (or liner) stability problem appeared in the literature in many different forms according to the engineering application being examined. The interested reader may refer to El-Sawy, 1996, for a comprehensive review of the literature.

Glock, 1977, analysed the liner stability problem. A non-linear deformation theory and the principle of minimum potential energy were used to develop Glock's solution. Glock's solution is based on a pre-assumption of the deformed shape of the detached part of the ring and for an encased liner it can be written as

$$P_{cr} = 1.0025 \frac{E}{1-\nu^2}\left(\frac{t}{D}\right)^{\frac{11}{5}} \quad \text{or simply} \quad P_{cr} = \frac{E}{1-\nu^2}\left(\frac{t}{D}\right)^{\frac{11}{5}} \tag{3}$$

Later, a theoretical study of a liner with a single local imperfection was reported by Kyriakides and Youn, 1984. The imperfection used earlier by Bucciarelli and Pian (1964) for the problem of a ring encased in an imperfect cavity and subjected to parallel distributed load, is considered again but for the ring itself. The imperfection parameters are its amplitude Δ_o and its arc length, $2s_o$ (Fig. 1). They concluded that the buckling pressure is a function of the wavy imperfection amplitude, Δ_o.

Lo and Zhang, 1994, studied the effect of the loose fitting imperfection on the critical pressure for Cured-In-Place liners. An initial radial gap of size Δ_1 exists between the liner and the host pipe due to the thermal contraction of the liner after being installed. The size of this initial gap is calculated by simply calculating the thermal contraction of the liner material due to drop in temperature equal to the difference between the curing process temperature and the liner temperature at the time of experiment. They also assumed that during the application of the external pressure, extra circumferential deformations occur in the liner which add an amount of Δ_2 to the initial radial gap size (i.e. the radial gap size Δ at buckling is defined by $\Delta = \Delta_1 + \Delta_2$). They further assumed that no more liner circumferential deformations occur within the buckling process. In their study, an approximate solution for the critical buckling pressure was obtained as a function of the gap size Δ. The solution is based on the analogy between the buckling of the detached part of the liner and the buckling of a shallow arch. The symmetrical one buckle case and asymmetrical two buckle case were both studied and a range for the expected critical pressures was calculated.

Li and Guice, 1995, studied the reduction factor of the hoop thrust at buckling of an elliptical ring compared to the circular ring of a radius R_o equal to the average of both the ellipse radii a and b (i.e., $R_o=(a+b)/2$) where a is the long radius. The ring is assumed to be subjected to a thermal expansion. They used Chicurel's approach (Chicurel, 1968) where the buckled portion of the ellipse is assumed to be an axially compressed beam with a given initial shape. They obtained an approximate closed form solution for the ovality reduction factor α given by

$$\alpha = \left(1 + \frac{q}{100}\right)^{-\frac{8}{5}} \left(1 - \frac{q}{100}\right)^{-\frac{1}{5}} \quad \textbf{(4)}$$

where q is percentage ovality defined by $100*(R_o-b)/R_o$ or $100*(a-R_o)/R_o$. They also assumed the influence of ellipticity on the thermal expansion solution to be identical to ovality effects for the case of ring under pressure.

Various ASTM standards (F1216-93) involve the calculation of the critical pressure of the elliptical (oval) pipe through application of a reduction factor α to the value of the critical pressure for the perfect circular pipe of radius R_o equal to $(a+b)/2$, where $2a$ and $2b$ are the length of the ellipse axes. The following equation is shown in ASTM specifications F1216-93.

$$\alpha = \left[\left(1 - \frac{q}{100}\right) / \left(1 + \frac{q}{100}\right)^2\right]^3 \quad \textbf{(5)}$$

The ASTM reduction factor α is calculated for the unsupported oval pipe case and is assumed to be applicable on the encased liner.

Method of Analysis

The Finite Element Method (FEM) has been used successfully to analyse many problems in various fields. A general purpose computer program is developed and is used throughout the study. The details of the FE model and are given in El-Sawy, 1996. Many key issues should be considered in the Finite Element Analysis used for the liner problem. These are summarized as follows:

Plane Strain Assumption Versus Three Dimensional Analysis.

It is assumed that plane strain conditions will prevail for the liner problem. It is recognized that some real liner configurations involve variations along the liner axis (where localized imperfections exist in the liner longitudinal direction, for example) but such considerations are beyond the scope of this work. Plane strain analysis involves, in most of the cases, higher stresses in the liner for a given external fluid pressure and reductions in stability relative to the three dimensional problem.

Geometrical Non-linearities

As large liner deformations are expected at the critical pressure, the FEM analysis should allow for geometrical non-linearity (in a small strain scheme). Therefore, the Updated Lagrangian approach is used.

Interaction Between The Liner And The Rigid Cavity

The modelling of the interaction between the liner and the rigid cavity strongly influences the behaviour of the encased liner. A conservative assumption of a smooth interface is used herein. A new interaction method is developed to rigorously model the smooth or rough interface between the liner and the rigid host pipe (El-Sawy and Moore, 1996).

Time-dependant Behaviour of Polymers

The behaviour of polymer materials is significantly time-dependent and ultimately the

Finite Element Analysis should accurately model the rheology of the material. The work in the present study is directed towards an evaluation of geometrical effects on liner buckling using elastic approximations for polymer behaviour.

For design purposes, a suitable approximation is to use the long-term Young's modulus (i.e. the value at the end of the lifetime of the liner) with any elastic solution for the critical pressure to represent the effect of time on the reduction of the pressure. A better approximation, is to consider a time dependant Young's modulus of the

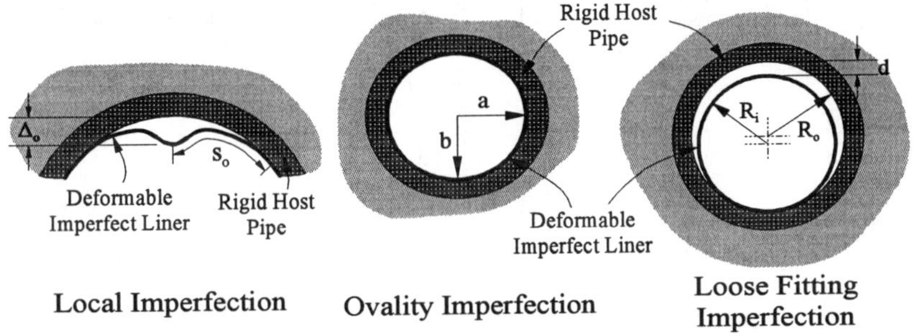

Figure 1. Different Types of Liner Imperfection

polymer material (Boot and Welch, 1996) and perform a time-dependant analysis. Although the previously mentioned approximations may look appropriate, they suffer from serious problems and there is no measure of their conservativeness. The major problem is that the time-dependant behaviour of polymers is more complicated than just assuming a time-dependent Young's modulus. The behaviour is visco-elasto-plastic and is strongly stress dependent. A more rigorous model is required to deal with such strong time-dependent behaviour. This is the target for a future research.

Geometry of Imperfection

Three types of imperfections are considered. The first type is local and takes the shape of a wavy intrusion into the liner (proposed by Pian and Bucciarelli 1967). The second is due to the possible ovality (global imperfection) of the damaged pipelines (referred to as *ovality imperfection*). The third type is due to loose fitting of a smaller diameter liner within the host pipeline (referred to as *loose fitting imperfection*). Figure 1 shows the three types of imperfections considered.

Parametric Study

The results for the parametric study are described for the three types of imperfect liners in the following sections.

Results for Circular Liners

To overcome the sudden change in the behaviour (bifurcation) at buckling, a very small imperfection is imposed on the system to obtain the critical external pressure. Different types of small imperfections are introduced and found to have little impact on the liner stability (El-Sawy 1996). The small imperfection used throughout the parametric study for circular liners is ovality imperfection. The ratio between the ellipse axis lengths, *a* and *b*, is chosen such that any decrease in this ratio should not change the analysis results. The ratio a/b=1.005 is found satisfactory. Figure 2 shows the normalized critical pressure, $P_{cr}\,(1-\nu^2)\,/\,E$, and the enhancement factor, K, versus the liner thickness to radius ratio (t/R). It also shows a comparison between the numerical FE solution and Glock's semi-analytical solution (Glock, 1977). There is an excellent agreement between the two solutions although Glock's solution was based

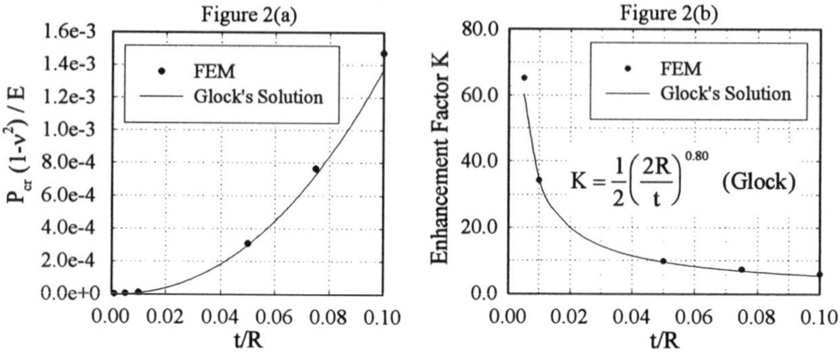

Figure 2. Finite Element Results versus Glock's Solution

on an *a priori* assumption of the mathematical function of the buckled portion of the liner. The buckling pressure and the enhancement factor, K, of the encased circular liner are simplified as

$$P_{cr} = \frac{E}{1-\nu^2}\left(\frac{t}{2R}\right)^{2.2} \quad \text{or} \quad K = \frac{1}{2}\left(\frac{2R}{t}\right)^{0.8} \qquad \textbf{(6)}$$

It is obvious from Fig. 2(b) that the enhancement factor K increases as t/R decreases. This is mainly due to the smaller circumferential deformations that occur for small values of t/R (i.e. for the cases of higher hoop stiffness) compared to those for the higher values of t/R.

Results for Liners with a Local Imperfection

In practice, local imperfections in the liner are always present due either to the liner installation technique or to the uneven shape of the rehabilitated pipe. The same imperfection used by Bucciarelli and Pian (1964) and Kyriakides and Youn (1984) is used herein. The imperfection is a single wave of length $2s_o$ and amplitude Δ_o (Fig. 1).

The normalized critical pressure P_{cr} is plotted versus the normalized imperfection size, Δ_o/R, for a liner with t/R=0.002 and different normalized imperfection wave lengths, $S_o = s_o / \pi R$. A conservative envelope for a series of these results for different imperfection wave lengths is also shown in Fig. 3(a). Envelopes for different liner t/R ratios are constructed and shown in Fig. 3(b).

From Fig. 3(a), it is clear that for any given imperfection length s_o, the critical pressure P_{cr} decreases with increases in the imperfection amplitude Δ_o and that the effect of the imperfection amplitude Δ_o on the critical pressure P_{cr} is more important than the imperfection length s_o. For practical cases of thin liners with large local imperfection amplitudes, the enhancement factor K=3.5 to 4.0 seems appropriate (Fig. 3(b)). The critical pressure, for small values of t/R, drops rapidly with the existence of a very small imperfection. The sensitivity of the critical pressure to changes in the imperfection amplitude decreases as the imperfection amplitude becomes bigger.

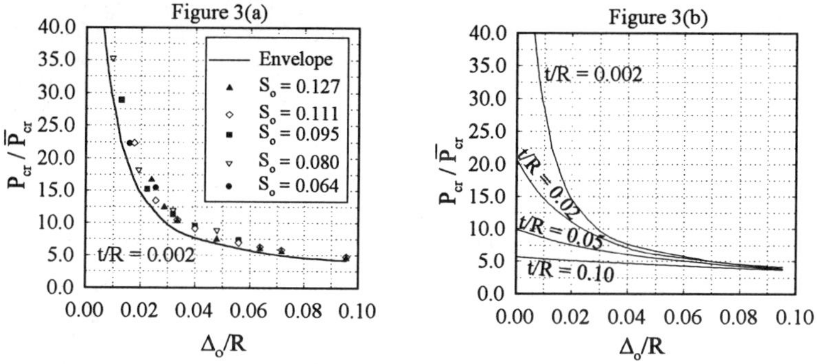

Figure 3. Effect of Local Imperfection Size on the Critical Pressure

Results for Oval Liners

A comparison between the available reduction factors proposed by Li and Guice and the ASTM standards is shown (Fig. 4(a)). The reduction factor proposed by Li and Guice is not conservative and differs by more than 55% from the FEM solutions. This may be attributed to the fact that their solution was derived from a solution to the thermal buckling problem.

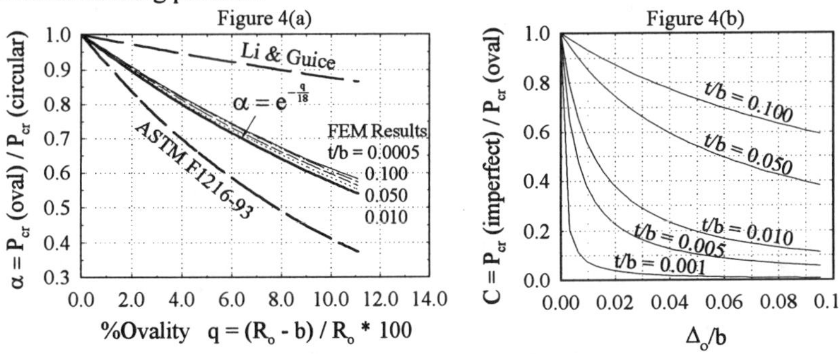

Figure 4. Reduction factors due to Ovality and Local Imperfection

The FEM solution is more rigorous than the ASTM reduction factor and differs by up to 40%. This is because of the inappropriate mathematical model used to develop this ASTM reduction factor. An equation for the reduction factor, based on an envelope developed from the FEM results, for a practical range of t/b, is given by

$$\alpha = e^{-\frac{q}{18}} \tag{7}$$

This equation is a lower bound for all the practical t/b ratios considered. Using Eqn. 7 and the equation discussed before for the critical pressure for an encased circular liner (Eqn. 6), the critical pressure for elliptical liners is given by

$$P_{cr} = \alpha \frac{E}{1-\nu^2} \left(\frac{t}{2R_o}\right)^{2.2} \tag{8}$$

Figure 4(b) also shows a series of conservative envelopes for the reduction factor C due to a single local imperfection in an oval liner (a/b<1.25). From Fig. 4(a), it is shown that the effect of liner thickness to radius ratio, t/b, on the reduction factor for critical buckling pressure, α, for ovality q<11.11% is not significant. The increase in ovality q (or a/b ratio) reduces the critical pressure. This behaviour is due to the increase in the radius of curvature of the oval liner as the buckling occurs in the flatter part of the liner.

Results for Loose Liners

A loosely fitted circular liner of radius R_i in a rigid circular cavity of radius R_o, is considered. A very small ovality imperfection (a/b=1.005) is introduced to the liner. Figure 5 shows the enhancement factor K versus the change in d/R_o ratios for different t/R_o ratios. The critical pressure drops significantly as the gap size increases. An increase in the gap size from d/R_o=0.0005 to 0.05 leads to a drop in the critical pressure by more than 80% for t/R_o=0.01 and a drop of about 40% for t/R_o=0.10. Figure 5 shows also that the larger the t/R_o ratio the less the enhancement factor K.

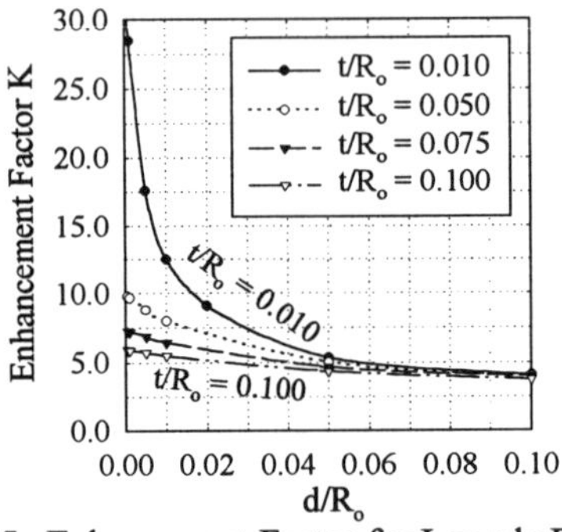

Figure 5. Enhancement Factor for Loosely Fitted Liner

Design of Liners for Rigid Pipe Rehabilitation

The parametric study has examined geometrical effects on liner stability, and can be used to develop a new design method to overcome the shortcomings of the current ASTM design specifications.

Buckling Pressure for Tightly Fitted Liners

The critical buckling pressure of a tightly fitted liner used for rigid pipe rehabilitation (i.e. partially deteriorated pipe) can be written as

$$P_{cr} = \frac{2E}{1-\nu^2}\left(\frac{t}{2R_o}\right)^3 K\, C\, \alpha \qquad \textbf{(9)}$$

The first part in the expression for the critical pressure, represents the critical pressure for the unsupported circular liner of radius R_o (Levy, 1884). The enhancement factor K quantifies the increase in critical pressure for the circular liner due to encasement and is defined by

$$K = \frac{1}{2}\left(\frac{2R_o}{t}\right)^{0.8} \qquad \textbf{(10)}$$

The factor α, is a reduction factor due to ovality. The reduction factor α, is defined by the exponential form $\alpha = e^{-q/18}$. The factor C accounts for the reduction in the critical pressure due to a single local imperfection. The reduction factor C is obtained from Fig. 4(a).

Buckling Pressure for Loosely Fitted Liners

The second type of liner considered is a loosely fitted liner with an outer radius R_i. The rigid host pipeline has an inside radius R_o. The gap between the liner and the circular cavity is defined by d, where d=2(R_o-R_i). The elastic buckling pressure is defined by

$$P_{cr} = \frac{2E}{1-\nu^2}\left(\frac{t}{2R_o}\right)^3 K \tag{11}$$

where K is the enhancement factor due to encasement (Fig. 5)

References

1. Boot, J. C. and Welch, A. J., *Creep Buckling of Thin-walled Polymeric Pipe Linings Subject to External Groundwater Pressure*, Thin-Walled Structures, Vol. 24, pp. 191-210, 1996.
2. Bucciarelli, L. L. Jr., and Pian, T. H. H., *Effect of Initial Imperfection on the Instability of a Ring Confined in an Imperfect Rigid Boundary*, J. Appl. Mech, Trans. ASME, pp. 979-984, 1964.
3. Chicurel, R., *Shrink Buckling of Thin Circular Rings*, ASME J. Appl. Mech., Vol. 35, No. 3, pp. 608-610, 1968.
4. El-Sawy, K., *Non-Linear Analysis of Liners for Rigid Pipe Rehabilitation*, Ph.D. Thesis, University of Western Ontario, July 1996.
5. El-Sawy, K. and Moore, I. D., *A Two-Level Iterative FEM Technique for Rigorous Solution of Non-Linear Interaction Problems Under Large Deformations*, Computers and Structures, Vol. 61, No. 1, pp. 43-54, 1996.
6. Glock, D., English Translation is "*Behaviour of Liners for Rigid Pipeline Under External Water Pressure and Thermal Expansion*", Der Stahlbau, 7, 212-217, 1977.
7. Kyriakides, S., and Youn, S.-K., *On the Collapse of Circular Confined Rings Under External Pressure*, Int. J. Solids Structures Vol. 20, No. 7, pp. 699-713, 1984.
8. Levy, M., Me'memoire sur un nouveau cas inte'grable du proble'm de l'elastique et l'une de ses applications, [Memoir on a New Integrable Case of the Problem of Elasticity and One of its Applications], J. Math. Pure et Appl., (Lioville), Series 3, Vol. 10, pp. 5-42, 1884.
9. Li, J.Y., and Guice, L. K., *Buckling of Encased Elliptic Thin Ring*, ASCE, J. Engng. Mech., Vol. 121, No. 12, pp. 1325-1329, 1995.
10. Lo, K. H. and Zhang, J. Q., *Collapse Resistance Modelling of Encased Pipes*, Buried Plastic Pipe Technology: 2nd volume, ASTM STP1222, Dave Eckstein, Ed., American Society for Testing and Materials, Philadelphia, 1994.
11. Moore, I. D. and El-Sawy, K., *Buckling Strength of Polymer Liners Used in Sewer Rehabilitation*, Transportation Research Record 1541, pp. 127-132. 1996.
12. *Standard Practice for Rehabilitation of Existing Pipelines and Conduits by the Inversion and Curing of a Resin-Impregnated Tube*, F1216-93, ASTM, Philadelphia, Pa, 1993.

Trenchless Technologies Used To Rehabilitate Pipeline Under Landfill

Michael Miller, ASCE Member[1] and William Lecuyer, ASCE Member[2]

Abstract

This paper is a case study illustrating a practical application for trenchless technologies. Due to the imminent structural failure of a 2.13 meter (84 inch) diameter corrugated metal pipe, Clough, Harbour & Associates LLP (CHA) was commissioned to identify and evaluate feasible design alternatives for the rehabilitation of the existing pipe. Due to characteristics of the site, neither standard open-trench or sliplining techniques were practical alternatives. In order to provide a solution, trenchless techniques were evaluated and eventually employed to provide an innovative, cost-effective solution. This paper describes the analysis, design and construction processes performed to complete the rehabilitation of the pipeline.

Introduction

The 487 meter (1,600 feet) long pipeline is located under the Cheltingham Avenue Landfill in the City of Schenectady, New York. The system is approximately 25 years old and serves to convey storm water runoff under the landfill. Over the period in which the landfill was in operation, a depth of approximately 18 to 24 meters (60 to 80 feet) of waste material had accumulated over the pipeline. The pipeline was experiencing a structural failure along a 139 meter (455 foot) reach. Portions of the pipe had experienced a deflection of 30 percent in the rise of the pipe; which had previously required the installation of temporary steel bracing. The rehabilitation problem was further compounded by the fact that the pipe was installed with a series of 20 to 30 degree horizontal bends along its alignment (see Figure 1).

----

[1]Project Manager, Clough, Harbour & Associates LLP, 3 Winners Circle, Albany, New York, 12205

[2]Project Engineer, Clough, Harbour & Associates LLP, 3 Winners Circle, Albany, New York, 12205

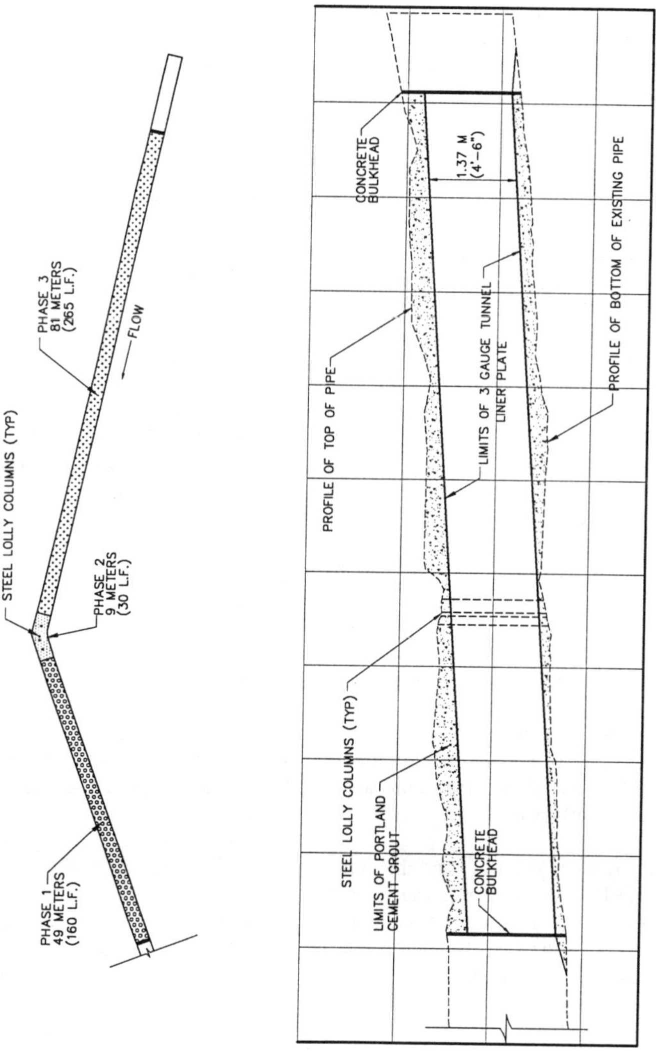

FIGURE 1 PLAN AND PROFILE OF PIPELINE REHABILITATION

This paper includes, but is not limited to, a discussion of the following items:

- Investigations performed to assess the condition of the existing system;
- Quantifying the impacts of potential mitigation measures on the upstream watershed;
- Identification of various methods to repair the damaged pipe, as well as preventive measures to preserve the remaining portions of the pipe;
- Overview of the construction process, along with some good practical "lessons learned" and advice on typical problems to consider and avoid.

Assessing the Condition of the Existing System

The last known inspection of the system was conducted by the City of Schenectady in 1985. As a result of this inspection, the steel bracing mentioned previously had been installed along the centerline of a 4.27 meter (14 foot) section of deformed pipe to provide additional structural support. This deformation occurred at the first horizontal bend in the system (estimated at approximately 25 degrees). These supports appeared to have partially stabilized the pipe. However, based on a current inspection conducted as part of our work, it was observed that the bottom of the columns had begun to rotate. The columns were rotated 5 to 10 degrees from their assumed vertical position. The columns appeared to have been properly sized to withstand the load, however, no structural connection had been made to the existing culvert. Due to the fact that no inspections were recorded between the time the columns were installed and the current inspection, it was unclear as to whether the columns were still moving or at what rate movement was taking place.

In order to evaluate the severity of the deformation that had occurred, the inside of the pipe was surveyed to establish horizontal and vertical control. This survey not only served to identify portions of the pipe in need of immediate repair, but also serves as a baseline to monitor any potential future movement of the pipe. The rise and span of the pipe were measured at approximately 15.24 meter (50 foot) intervals. In those areas where structural deformation was observed, the interval was reduced to 1.52 meter (5 foot) intervals. The results of the survey confirmed that several sections of the pipe were in need of immediate repair; with the severity of the deformation varying substantially with no uniform pattern. The worst deformation was found to be in the location of the steel bracing. The dimensions of the pipe in this area consisted of a 1.50 meter (4.9 feet) rise by 2.50 meter (8.2 feet) span.

The culvert is 25 years old and theoretically should have had an additional 15 to 25 years of service life. A number of scenarios were developed as to why the culvert had failed prematurely, and to the possible mode of failure. The first scenario theorized that due to ground water infiltration through the joints, which was prevalent throughout the culvert, piping had occurred reducing the lateral support around the pipe (EPA, 1991). A second theory was that the impact of the water on the bend had damaged the pipe due to an absence of a properly designed thrust block. Finally, there was speculation that due to the nature of the landfills' organic materials, the surrounding soil had decomposed causing the loss of lateral support (WRC, 1990). Unfortunately, there was no cost effective way to verify the mode of failure. Borings or test pits would have been costly and would have required extensive permitting due to the unknown nature of the overburden.

Additionally, inquiries to non-destructive testing subconsultants revealed that due to the non-uniform nature of the materials surrounding the pipe, non-destructive testing methods would have more than likely proven inconclusive.

Quantifying Potential Impacts on the Upstream Watershed

The watershed that conveys surface runoff to the Cheltingham Avenue Landfill System is comprised of approximately 576 hectares (1,423 acres). The watershed is generally comprised of highly developed urbanized and residential areas. The channel upstream of the landfill is contained within a rather deep ravine. Upstream crossings serve to reduce flows conveyed to the landfill by detaining water upstream of their embankments, thereby utilizing a substantial amount of available flood plain storage within the ravine.

Potential storm water impacts were evaluated for the 10-, 25-, 50- and 100-year 24 hour storm events. Peak flow rates and volumes of runoff were generated for the upstream watershed using the SCS USDA TR-20 computer program (USDA SCS, 1983). This program was chosen due to the fact that it could generate the peak flows and runoff volumes for the watershed, while at the same time evaluate the impact of potential alternatives on the water surface elevations upstream of the site.

In order to evaluate the impact on the water surface elevations upstream of the site, it was first necessary to determine the stage storage and discharge rating tables for the landfill system and upstream crossings. Stage storage tables were created using the USGS quadrangles, while stage discharge tables were generated using methodologies presented in HEC-5 (USDC, 1965). These rating tables were then input directly into the TR-20 program.

The impacts of potential alternatives were evaluated by incrementally reducing the size of the Cheltingham Avenue Landfill culvert until an unacceptable increase in the upstream water surface elevations was predicted. Based upon the analysis, it was determined that the waterway opening of the pipe could be reduced to the equivalent opening of a 1.68 meter (66 inch) diameter pipe without adversely affecting upstream property owners.

Identification and Evaluation of the Rehabilitation Alternatives

Several alternative methods were evaluated in regard to the needed repairs and rehabilitation of the system. The purpose of this section is to provide a brief overview of these alternatives, and to show why these alternatives were deemed unfeasible or cost prohibitive.

The "do nothing" alternative was not an option due to the imminent failure of the pipe. In general, the majority of the alternatives identified involved the abandonment of the existing system, or the portions of the system containing the deformed pipe sections, in conjunction with the installation of a replacement pipe. The following alternatives were considered as possible mitigation measures:

- Open-trench excavation: As mentioned previously, the existing pipeline was buried beneath 18 to 24 meters (60 to 80 feet) of waste material. As such, standard open-trench excavation would have been very difficult due to the complicated excavation stabilization methods. This alternative would have been costly and would have required extensive permitting efforts due to the excavation of the waste material. The estimated construction cost for this alternative was $2.5 million.

- Microtunneling and Pipe Jacking: These options would have been very difficult due to the unknown nature of the waste material. Once again, these alternatives would have been costly and difficult to permit due to the removal of the waste material. The estimated construction cost associated with this option was $2.05 million.

- Relocation of the system: This alternative was eliminated due to the fact that it was cost prohibitive. The cost associated with this option was estimated at over $5.0 million.

As a result of the above, various trenchless technologies were evaluated to repair the failed portions of the system. A summary of the techniques evaluated are identified below:

- Spray-on linings: Due to the nature of the failure, it was determined that this type of rehabilitation measure would not provide the structural strength required to reinforce the damaged sections of pipe. As such, this alternative was deemed unfeasible.

- Standard sliplining techniques: This process normally involves inserting a smaller diameter pipe inside of the existing pipe. However, this option was not physically possible due to existing bends in the pipe. Furthermore, this option would have been further complicated by the bracing system that had been previously installed. As such, this option was eliminated from further consideration.

Detailing the Solution to the Problem

The solution to the problem required a lining system that possessed the required strength to withstand the vertical load, but was also flexible enough to navigate through the existing pipe. It was decided to utilize standard sliplining techniques in conjunction with an innovative use for tunnel liner plate. Tunnel liner plate, as the name suggests, is typically used for the installation of tunnels. Due to the fact that the tunnel liner plate could be assembled in segments inside of the pipe, the problem of navigating a rigid liner through the existing pipe was eliminated. In addition, tunnel liner plate could be fabricated in the shape of the existing bend.

For this application, 3 gauge 2 flanged galvanized tunnel liner plate was selected with an oval shaped cross section. The oval section had a 1.37 meter (54 inch) rise with a 2.06 meter (81 inch) span, fitting within the collapsing section of pipe while at the same time providing the required flow opening (see Figure 2). It was proposed to install 139 meters (455 feet) of the tunnel liner plate and then fill the annular space created

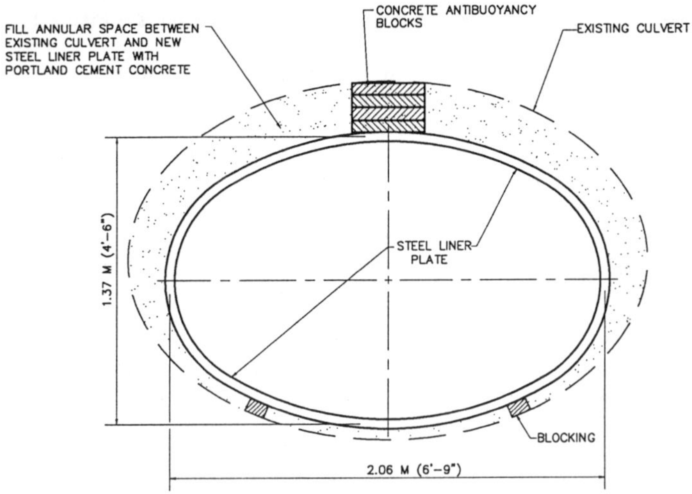

FIGURE 2 TYPICAL INSTALLATION SECTION

between the existing culvert and the new liner plate with portland cement concrete. Calculations showed that the tunnel liner plate could withstand both the vertical dead load and the horizontal forces imposed at the bends. In addition, pressure grouting the remaining culvert pipe joints with a hydrophobic polyurethane chemical grout was proposed as a preventive measure. Hydrophobic grout expands 15 to 20 times its original volume when coming in contact with water, forming a tough, flexible, compound that serves to prevent infiltration. The grout serves to effectively seal the existing joints, thereby, surrounding the pipe joints with an impermeable grout "envelope". This approach is effective for infiltration control due to the fact that the grout fills existing soil voids and consolidates the surrounding soil helping to prevent future leaks that may occur as water migrates.

Construction Phase

The project was publicly bid and awarded to Delaney Construction Corporation, of Mayfield New York, in the amount of $450,000. The construction began in January 1995. Construction of the tunnel liner plate consisted of the installation of over 1,600 pieces of liner plate, connected by over 21,000 nuts and bolts. In order to accomplish this task, a number of logistical problems had to be overcome throughout the construction process.

It was decided that the installation of the liner plate would be completed in three phases, starting at the downstream end of the pipeline and proceeding upstream. The first phase of the project included the installation of approximately 49 meters (160 feet) of the liner plate in the most distant downstream portion of the culvert; the second phase consisted of the installation of the fabricated bend; and lastly, the third phase consisted of the installation of approximately 81 meters (265 feet) upstream of the bend (see Figure 1).

Initially, two problems faced the construction team. First, how to move the 1,600 pieces of tunnel liner plate into the tunnel in an efficient manner; and second, how to control the flow of water through the tunnel. Each piece of liner plate weighed about 34 Kg, (75 pounds) and had a curved surface. It was very cumbersome to move them about the site, much less 139 meters (455 feet) into the pipe. In order to move individual pieces into the pipe, small heavy duty pull carts were utilized to transport the plates. Although the flow through the tunnel during the time of construction was relatively low, it was necessary to reduce the flow during the liner installation. Due to the existing topography and the volume of flow, bypass pumping was not possible. In order to reduce the flow, a restrictor plate was installed at the inlet of an upstream culvert. As mentioned previously, the upstream ravine is very deep and provides a substantial amount of storage. The restrictor plate was installed in a manner that allowed the plate to be periodically removed to allow any built-up water to pass downstream in a safe controlled manner.

Prior to the commencement of the liner plate installation, the chemical grouting procedure was performed along portions of the pipeline where the liner plate would not be installed. Prior to installation, a walk through was conducted to determine which joints would be grouted. However, fearing future ground water migration, nearly all the joints within the culvert received pressure grouting to some degree. The procedure required the drilling of a series of holes offset six to twelve inches on either side of the joint. Water was then injected in the joint to act as a catalyst for the grout. The grout was injected into the holes until it was observed to be coming out of adjacent holes. As expected, the volume of grout used varied from joint to joint. The majority of the joints used 11.3 liters (3 gallons) of grout. However, some of the joints took as much as 22.6 liters (6 gallons) of grout.

As the liner plate installation began, the major hindrance was the water inside the culvert. The water was just above freezing and collected in low spots in the culvert. The workers were outfitted in hip boots to attempt to shield feet and legs from the icy water. However, due to the nature of the assembly, it was impossible for the workers to avoid contact with the water. Tools, nuts, bolts and equipment were wet and cold, making working conditions inside the culvert less than desirable. Additionally, a blower system was required to ventilate the pipe to provide fresh air within the working environment.

As construction progressed, it was necessary to remove the existing steel bracing to install the liner plate around the bend. Safety of the workers was the immediate concern of all parties during construction. It was decided that the best course of action would involve a two step construction sequence. First, 15 foot sections of liner plate would be installed upstream and downstream of the bend to serve as a safe areas in the event of a catastrophic collapse. Secondly, a 50 ton bridge jack would be brought into the system to provide for temporary structural support as the columns were removed. The jack would be moved along accordingly to allow for the installation of the liner plate as work progressed.

When the time came to remove the first column, it was decided that the screw jack at the top of the column would be turned very slowly. It was assumed that if the jack turned easily then the load would not be directly transferred to the jack. A four foot pipe wrench was used to turn the jack, and the column was successfully removed. As was anticipated, the culvert had the structural integrity to withstand the temporary load. The remaining columns were removed in a similar manner, and the plate assembly at the bend was completed in two days.

The final phase of liner installation was completed quickly. Upon completion of the liner plate installation, two bulkheads were constructed at the ends of the reconstructed section and the annular space between the existing culvert and the tunnel liner plate was filled with portland cement grout. The tunnel liner plates came with prefabricated grouting holes and plugs that were installed along the rehabilitated section at approximately 6.10 meter (20 foot) intervals. Concrete antibuoyancy blocks were also installed between the crown of the tunnel liner plate and the existing culvert to counteract the buoyancy forces experienced during the grouting process. The grouting process was completed in four lifts using predetermined volumes of concrete grout per section. This process was employed to minimize the buoyancy forces created during the process. Upon completion of this work in the Spring of 1996, the culvert rehabilitation was complete and the rehabilitation of the pipeline was deemed a success.

Summary

This project utilized a conventional form of trenchless technology, applied in an innovative manner, to provide a cost-effective solution. During the construction phase, the project was treated as a team effort by the contractor, the engineer and the City of Schenectady with all sharing the common goal of producing a quality project. This project illustrates how trenchless technologies can be used to combat adverse field conditions that preclude the use of open-trench excavation or standard sliplining practices, and is a good example of how trenchless technologies can provide solutions to the challenge of rehabilitating aging infrastructure.

References

USDC (1965), "Hydraulic Charts for the Selection of Highway Culverts, Hydraulic Engineering Circular No. 5", US Department of Commerce, Bureau of Public Roads, Washington, DC.

USDA SCS (1983), "Technical Release No. 20, Project Formulation Hydrology, Second Edition", Soil Conservation Service, Washington, DC.

EPA (1991), "Sewer System Infrastructure Analysis and Rehabilitation Handbook", US Environmental Protection Agency, Office of Research and Development, Cincinnati, OH.

WRC (1990), "Sewerage Rehabilitation Manual, Second Edition", Water Research Center, Huntingdon Valley, PA.

Credits

Bernard Sisson and Milton Mitchell, City of Schenectady
Richard Bovee, Frank Griggs and Paul VanGelder, CHA
Delaney Construction Corporation and Grout Tech Inc.

DESIGN GUIDELINES FOR DIRECTIONAL DRILLED PE PIPE

Larry J. Petroff, M. ASCE[1]

Abstract

New construction technologies are often developed solely by the ingenuity of the contractor. As the technology gains acceptance, it is often applied to increasingly more difficult projects, causing the emphasis in development to shift from construction techniques to design issues. This is true of horizontal directional drilling (HDD) of polyethylene (PE) pipes. The major development of HDD has been due to construction innovations. Initially developed for steel pipes, HDD was then applied to small diameter PE pipes. Now, PE pipes are installed in increasingly larger diameters, longer lengths, and deeper crossings. Applications have become more critical, making proper, thoughtful design more relevant. To be effective, the designer has to understand the loads acting on the pipe and the reactions of the pipe both during installation and during operational service. This paper addresses external loading and reaction of PE pipe during pull-back and during its service life. The equations presented herein allow the designer to compare the forces applied to the pipe against pipe performance limits. Performance limits for directional drilled pipe include tensile strength, external collapse, bending, and ring deformation. Methods of estimating earth pressures acting on the pipe are discussed. Guidelines are given to help designers determine the suitability of the equations for their applications.

Pipe Design Process

The forces applied to HDD pipe are difficult to determine as they depend on complex soil-pipe interaction during pull-back as well as during the pipe's service life. Any methodology for this process is approximate. As more data is collected and more experience obtained, the accuracy of the calculations will improve. Today, the engineer must exercise careful judgment in determining the applied loads and in selecting the design parameters. Thought should be given to the applicability of the equations themselves to any specific installation.

Engineering Supervisor, PLEXCO, 1050 IL State Route 83, Suite 200, Bensenville, Il 60106

Load Analysis
Directional drilled applications differ somewhat from direct burial applications in that often the highest loads occur during pipe installation. But, this may not always be the case. Each step of the installation procedure and all operational conditions should be carefully examined to identify (1) all ring directed loads from earth and live-load pressure, groundwater and annular space fluid pressure, and internal pressure (vacuum) and (2) all tensile loads including pulling and frictional forces. The duration of all loads must also be determined.

One of the key parameters in this process is the fluid pressure differential between inside and outside of the pipe, also referred to as the net external pressure;

$$P_N = P_E - P_I \tag{1}$$

where P_N = net external fluid pressure (kPa), P_E = external hydrostatic pressure due to annular space fluid (groundwater and/or slurry) (kPa), and P_I = internal pressure (kPa). External pressure can create significant pipe ring stresses when the line is empty, under vacuum, or filled with a fluid having a unit weight less than the drilling slurry. P_N along with the earth pressure (given in a following section) should be calculated at the maximum depth for each identified load.

Time-dependent Behavior
Most performance limits for PE pipe in directional drilled applications are dependent on the modulus of elasticity and the tensile strength. Both of these properties are time-dependent. PE pipe's resistance to a newly applied load increment decreases with time as the molecular structure rearranges due to viscoelasticity. This results in a higher resistance to short-term loading than to long-term loading. Careful consideration must be given to the duration and frequency of each load, so that the performance limit associated with that load can be calculated using PE material properties representative of that time period. For instance, during pull-back, the pipe's tensile strength decreases with pulling time, so its safe pulling stress is a function of time. See Table 1 for typical values for high density PE (HDPE) and medium density PE (MDPE).

For viscoelastic materials, the ratio of the applied stress to strain is called the apparent modulus of elasticity, because the ratio varies with load rate. Typical values for the apparent modulus of elasticity at 23°C (73°F) are presented in Table 1. Consult the manufacturer for specific applications.

PE pipe is normally specified by material, diameter, and dimension ratio (DR). (DR is the ratio of the pipe mean outside diameter (OD) to the pipe minimum wall thickness.)

Table 1. Apparent Modulus of Elasticity and Safe Pull Tensile Stress @ 23°C

Typical Apparent Modulus of Elasticity			Typical Safe Pull Stress		
Duration	HDPE	MDPE	Duration	HDPE	MDPE
Short-term	800 MPa	600 MPa	30 min	9.0 MPa	7.3 MPa
10 hours	400 MPa	300 MPa	60 min	8.3 MPa	6.7 MPa
100 hours	350 MPa	250 MPa	12 hours	7.9 MPa	6.4 MPa
50 years	200 MPa	150 MPa	24 hours	7.6 MPa	6.2 MPa

Installation Design

During pull-back the pipe is subjected to axial tensile forces caused by the frictional drag within the hole, the frictional drag on the ground surface, the capstan effect around drill-path bends, and the hydrokinetic drag. Simultaneously, the pipe may be subjected to external hoop pressures due to the net external fluid head. The pipe's collapse resistance to external hoop pressure may be reduced by the axial pulling force. Furthermore, the drill path curvature may be limited by the pipe's bending radius. (Torsional forces occur but are usually negligible when back-reamer swivels are properly designed.) Design consists of determining the maximum installation loads and ensuring that they do not exceed the permitted stress level in the pipe.

Bore Path Curvature

Pipe curvature in the bore path induces bending stresses and increases the pull-back load due to the capstan effect. The design curvature is limited by the bending radii of the drill rod and the pipe. For small pipes, the drill rod usually controls. Typically, the minimum radii of curvature for PE pipes are 20 diameters for DR 13.5 or less and 25 diameters for DRs between 15.5 and 21. However, during pull-back, curvature is further limited by the tendency of the pipe to oval around bends. This effect reduces the pipe's resistance to crimping (bend buckling), and thus some manufacturers limit the minimum bending radius to 40 diameters during pull-back.

Pull-Back Force

Considerable judgment is required to predict the pull-back force because of the complex interaction between pipe and soil. Good sources for information include experienced drillers and engineers, programs such as DRILLPATH (1) and publications. Typically, pull-back force calculations, including those given below, are approximations. Pipe resistance to pull-back in the borehole depends primarily on the frictional force created between the pipe and the borehole or pipe and the ground in the entry area, the frictional drag between pipe and drilling slurry, the capstan effect at bends, and the weight of the pipe. Equation 2 gives the frictional resistance or required pulling force (kPa), F_P, for pipe pulled in straight, level bores or across level ground.

$$F_P = \mu w_B L \tag{2}$$

Where μ = coefficient of friction between pipe and slurry (typically 0.25) or between pipe and ground (typically 0.40), w_B = net downward (or upward) force on pipe (kN/m^3), and L = length (m). When a slurry is present, w_B, is the upward buoyant force of the pipe and its contents. Filling the pipe with fluid significantly reduces this force. For curves in the borehole, the force can be factored into horizontal and vertical components as in the mathematical expression in Fig. 1 or its resultant can be used as given by Svetlik (2). Huey et al. (3) shows an additional frictional force that occurs in steel pipe due to the pressure required by the borehole to keep the steel pipe curved. For bores with a radius of curvature similar to that used for steel pipe, these forces are likely insignificant for PE pipe. For very, small radii of curvature, it may be prudent to consider them. The frictional resistance during a pull is compounded by the capstan effect. As the pipe is pulled around a curve or bend creating an angle θ, there is a compounding of the forces due to the direction of the pulling vectors. The pulling force, F_C, due to the capstan effect is given in Eq. 3. Equations 2 and 3 are applied recursively to the pipe for each section along the pull-back distance. See Fig. 1. This method is credited to Larry Slavin of Bellcore (Middletown, NJ).

$$F_C = e^{\mu\theta}(\mu w_B L) \tag{3}$$

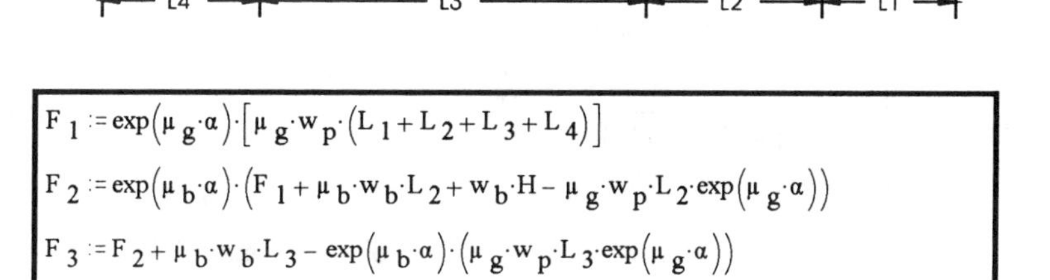

$$F_1 := \exp(\mu_g \cdot \alpha)\cdot\left[\mu_g \cdot w_p \cdot (L_1 + L_2 + L_3 + L_4)\right]$$

$$F_2 := \exp(\mu_b \cdot \alpha)\cdot\left(F_1 + \mu_b \cdot w_b \cdot L_2 + w_b \cdot H - \mu_g \cdot w_p \cdot L_2 \cdot \exp(\mu_g \cdot \alpha)\right)$$

$$F_3 := F_2 + \mu_b \cdot w_b \cdot L_3 - \exp(\mu_b \cdot \alpha)\cdot\left(\mu_g \cdot w_p \cdot L_3 \cdot \exp(\mu_g \cdot \alpha)\right)$$

$$F_4 := \exp(\mu_b \cdot \beta)\cdot\left[F_3 + \mu_b \cdot w_b \cdot L_4 - w_b \cdot H - \exp(\mu_b \cdot \alpha)\cdot\left(\mu_g \cdot w_p \cdot L_4 \cdot \exp(\mu_g \cdot \alpha)\right)\right]$$

Where H = Depth of bore
Fi = Pull Force on pipe at Point i
Li = Horizontal distance of Pull from point to point
u = Coeff. of friction (ground (g) and borehole (b))
w = Pipe weight (p) and Buoyant pipe weight (b)
α,β = Entry and Exit angles

Figure 1. Estimated Pull-back Force Calculation

Hydrokinetic Force

During pulling, pipe movement is resisted by the drag force of the drilling fluid. This hydrokinetic force is difficult to estimate and depends on the drilling slurry, slurry flow rate, and borehole and pipe sizes. Typically, the hydrokinetic pressure is in the 30 to 60 kPa (4 to 8 psi) range.

$$F_{HK} = p\frac{\pi}{8}(D_H{}^2 - OD^2) \tag{4}$$

Where F_{HK} = hydrokinetic force (kN), p = hydrokinetic pressure (kPa), D_H = borehole diameter (m), OD = pipe outside diameter (m).

Tensile stress

The tensile stress in the pipe due to the pull-back force and hydrokinetic pulling force should not exceed the safe pull stress. Bending stress due to pipe curvature is often ignored but should be subtracted from the safe pull stress when the bend radius is less than 40 diameters. During pull-back it is advisable to monitor the pulling force and to use a "weak link" (such as a pipe of higher DR) or other failsafe method to prevent over-stressing the pipe. After pull back, pipe may take several hours (typically up to 24 hours) to recover from the axial strain. It is recommended to over pull the bore by 4 to 5 percent.

Hydrostatic Collapse

External hydrostatic pressure, such as from the drilling slurry, can cause a sudden, large inward deflection resulting in collapse of the pipe. The pipe's resistance to collapse should be checked for the net external fluid load, P_N, under each loading scenario as discussed previously in "Load Analysis". In unstable bores, the earth pressure is treated as fluid pressure. The net external pressure that the pipe can safely resist is called the allowable collapse strength (kPa), P_{AC}, as given below.

$$P_{AC} = \frac{2\,E}{(1-\mu^2)}\left(\frac{1}{DR-1}\right)^3 \frac{f_o\, f_R}{N} \tag{5}$$

Where E = apparent modulus of elasticity (kPa), μ= Poisson's Ratio (0.35, short-term and 0.45, long-term), f_O = ovality compensation factor (see fig. 2), f_R = tensile pull reduction factor, and N = safety factor (typically 2). In choosing the modulus (E) value, the designer must consider the duration of loading as discussed previously.

The collapse resistance is reduced when the pipe ovals or deforms. Most pipes will encounter some ovalization from handling and bending during pull-back. Typically, three percent is assumed. One of the most critical conditions is pull-back of empty pipe. In this case, ovalization due to buoyant uplift should be calculated. See Watkins (4).

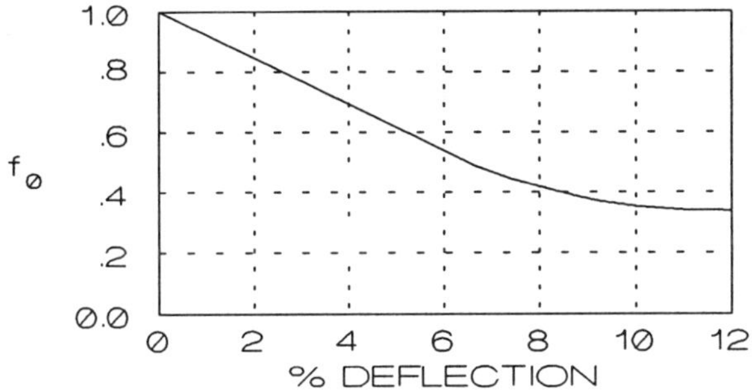

Figure 2. Ovality Reduction Factor

The pulling load in the pipe creates a hoop strain as described by Poisson ratio. The hoop strain reduces the allowable external pressure which is accounted for with the reduction factor, f_R.

$$f_R = \sqrt{(5.57-(r+1.09)^2)}-1.09 \quad \textbf{(6)}$$

$$r = \frac{\sigma_T}{2\sigma} \quad \textbf{(7)}$$

Where σ_T = calculated tensile stress during pull-back (kPa) and σ = safe pull stress (kPa) from Table 1.

Post-Installation Design Assumptions

External post-installation earth and groundwater loads occur as a result of groundwater and slurry in the annular space, deformation or collapse of the borehole, vacuum in the pipe, live loads, etc. Determination of the resulting loads is a complex problem requiring considerable engineering judgment. Since the borehole is typically 25% to 50% larger than the pipe, uniform soil support as seen with entrenched pipes is unlikely. The drilling slurry is thixotropic but has a viscosity typically less than a soft clay. Therefore, prudent design ignores any pipe support from the drilling slurry.

Stable Borehole

Where the hole remains stable, either due to the insitu soil characteristics or due to the drilling mud slurry, hole deformations and soil pressure on the pipe are negligible. In this case, the primary loads are due to the groundwater and the drilling slurry. The net pressure applied to the pipe during post-installation events (vacuum, line shut-down, etc.) should be calculated and checked against the pipe's resistance to collapse using Eq. 5 with an apparent modulus value based on the anticipated loading time.

Plastic Deformation of Borehole

If the crown of the borehole deforms sufficiently to create a plastic state in the soil, arching is mobilized. Although deformation may not be sufficient to transmit load to the pipe, the assumption is made that it does. The earth load (effective stress) applied to the pipe can be calculated using Terzhaghi's equation with the 50% reduction in wall friction angle per Stein et al.[5]. This analysis should only be considered where the depth of cover exceeds 5 borehole diameters or 1 meter of cover, whichever is larger.

$$P_{EV} = \kappa \gamma H \tag{8}$$

$$\kappa = \frac{1 - \exp\left(-2\dfrac{KH}{B}\tan\left(\dfrac{\phi}{2}\right)\right)}{2\dfrac{KH}{B}\tan\left(\dfrac{\phi}{2}\right)} \tag{9}$$

Where P_{EV} = external earth pressure (kPa), γ = effective soil weight (kN/m^3), H = depth of cover (m), K = earth pressure coefficient (typically, 0.5), B = “silo” width, and ϕ = Angle of internal friction (deg). The "silo" width should be estimated based on the application. The width varies between the pipe diameter and the borehole diameter.

P_{EV} will cause diametrical deflection in the pipe. Since the pipe has little or no side support, Spangler's formula is not applicable. Deflection may be estimated by using a ring deflection equation for a line load such as given in Eq. 10.

$$\Delta V = \frac{0.0145\ P_{EV}}{\dfrac{E}{12(DR)^3}} \tag{10}$$

Where ΔV = diametrical deformation as a fraction of the diameter, and E = apparent modulus of elasticity (kPa). In determining load duration, consideration should be given to the fact that while external water pressure does not contribute to the deflection, positive internal pressure resists it and can reround deformed pipe.

Pipe deformation has two consequences: (1) Ring bending stresses occur. Current Industry practice limits them by controlling deflection to typically 5% for DR 11, 6% for DR 17, and 7.5% for DR 21. (Janson [6] has shown that for stable pipe installations, higher limits may be used.) (2) The collapse resistance is reduced, as accounted for by the ovality factor in Fig. 2.

The pipe must have sufficient collapse resistance as given by Eq. 5 to withstand the "total" pressure applied to it. The "total" pressure is the sum of the external earth pressure, P_{EV}, and the net external fluid pressure, P_N.

Unstable borehole (Arching collapses)

If soil arching breaks down, the borehole will likely collapse. Hole collapse applies considerable load to the pipe. The upper limit on the load is the effective weight of the soil prism above the pipe plus external fluid loads. Hole collapse is more likely to occur in shallow applications (H<5 borehole diameters) subjected to live loads, boreholes in unconsolidated sediments such as river crossings, holes subjected to dynamic loads, or holes where the drilling slurry is lost.

When hole collapse occurs, the pipe is subjected to the same analysis as discussed in the preceding section except P_{EV} equals the prism load ($P_{EV} = \gamma H$) plus live load. Live load can be calculated according to the same method used for direct buried pipes. The "total" pressure is the net external fluid pressure plus the effective soil pressure from the prism load and the live load pressure.

Conclusions

HDD is a new and evolving field. The most conservative design approach is to review available literature, seek advice from drillers, and make prudent assumptions regarding the applied forces and pull-back resistances. It is particularly important to analyze the post-installed pipe to assure a sufficient safety factor for long-term performance.

References

1. Kirby, M.J., Kramer, S.R., Pittard, G.T., Mamoun, M. (1996). "Design Guidelines and Procedures for Guided Horizontal Drilling", No-Dig '96, New Orleans, LA.

2. Svetlik, H. (1995). "Design Considerations for HDPE Pipe Used in Directional Drilling", No-Dig Engineering, Vol.2, No.3, March.

3. Huey, D.P., Hair, J.D., McLeod, K.B. (1996). "Installation Loading and Stress Analysis Involved with Pipelines Installed by Horizontal Directional Drilling", No-Dig '96, New Orleans, LA.

4. Watkins, R.K. (1994). "Structural Mechanics of Buried Pipes", Buried Pipes Seminar, University of Colorado at Denver.

5. Stein, D., Mollers , K., Bielecki, R. (1989). Microtunnelling, Ernst & Sohn, Berlin.

6. Janson, L.E. (1991). "Long-Term Studies of PVC and PE Pipes Subjected to Forced Constant Deflection", Report No. 3. KP-Council, Stockholm, Sweden.

Planning The Use of Horizontal Directional Drill To Minimize Site Impacts

F. Ray Bailey, P.E., Member, ASCE
Beth Drylie, P.E.[1]

Abstract

In the past several years the use of Horizontal Directional Drilling (HDD) has gone from being an experimental construction technique to being preferred by regulatory agencies tasked with conservation and preservation. The technique meets the secondary criteria of many of these agencies, which is to minimize any potential impacts to regulated lands, wetlands, or historical sites, when total elimination of impacts is not possible.

When addressed during the planning stages, agencies will often encourage the use of HDD to minimize or eliminate site impacts. By passing beneath river bottoms or wetland areas, the need to disturb sensitive areas can be completely eliminated. When working around historical sites, the amount of excavation required for installation is greatly reduced, thus preserving a greater quantity of the historical record.

By planning for the use of HDD on recent river crossings in eastern Virginia, the City of Norfolk was able to eliminate impacts to wetland areas and minimize the potential impacts to two historical sites. Making use of the HDD technique and planning the use of other complimentary construction techniques, the City was able to reduce the excavation required for the installation of the two crossings to less than one percent of the total historical site area.

Introduction

The geography of the Hampton Roads area provides an excellent opportunity for the use of Horizontal Directionally Drilled (HDD) crossings and other types of

---

[1]Manager - Civil Engineering, Michael Baker Jr., Inc., 770 Lynnhaven Parkway #240, Virginia Beach, VA 23452
Assistant Engineering Manager, City of Norfolk, Department of Utilities, P.O. Box 1080, Norfolk, VA 23501

trenchless pipeline installation techniques. The area is located in the flat, coastal plain of southeastern Virginia. It is drained by many, slow-moving streams, most of which find their way into the Chesapeake Bay. Many of the streams are estuarine in nature,

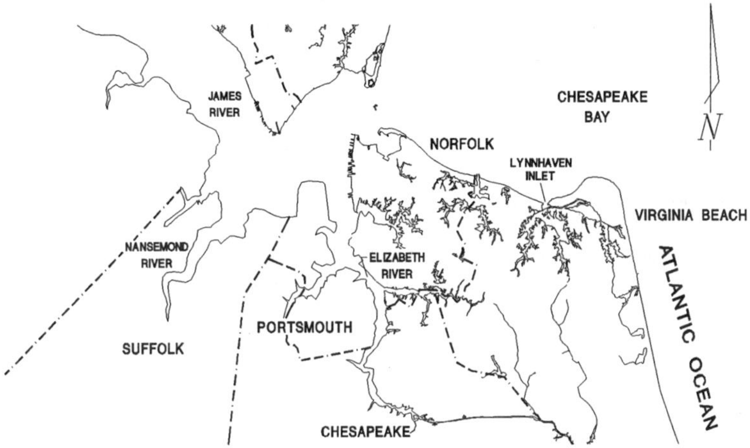

containing brackish water. These waters tend to be highly productive for marine species and to border large areas of tidal or adjacent native wetlands. Figure 1 is a map of a portion of South Hampton Roads, illustrating the nature of the waterways in this area.

Figure 1. South Hampton Roads

The many streams in the area segment the land masses and create peninsular forms that must still be served with water, sewers, and other utilities. Utility departments have historically provided service by crossing the streams to reduce the total length and cost of their facilities. Crossings have been made using the traditional cut-and-cover method, by spanning smaller streams, and by attaching facilities to other bridges.

The impacts of these installation techniques to wetlands and historic resources can vary from minor, temporary construction impacts to significant impacts which must be mitigated. The U. S. Army Corps of Engineer's responsibility in reviewing applications for utility crossings includes the consideration of impacts to wetlands and historic resources. Their primary objective in terms of wetlands is to eliminate any impacts, or to minimize any unavoidable impacts. For historic resources the Corps must request the review of the State Historic Preservation Office (SHPO). In Virginia the Department of Historic Resources is designated as the SHPO.

Much of the early history of English settlement in the new world is written around the Chesapeake Bay and the surrounding rivers and waterways. Prior to the arrival of John Smith the English settlers were known to have made contact with the Indians who lived along the major rivers and estuaries. These sites are highly valued by archeologists as providing a picture of the way of life that existed before contact with the English. The chances of impacting a known or unknown historical site along any of the waterways in the area are very high. In many cases the only question is whether the site has been previously disturbed, thereby reducing or eliminating its value in adding to the historical record of the area.

The Utility

The City of Norfolk, Department of Utilities has for many years been the primary supplier of potable water to the residents of South Hampton Roads. Today the Department treats and delivers nearly 318 million L/day (84 mgd) of water on an average day basis to the area. Most of that water is supplied from freshwater lakes developed by the City in the 1920's, 40's and 50's. The lakes are located within the City and within neighboring cities, like Suffolk. Raw water is pumped from the lakes to the City's treatment plants, where it is treated and pumped to the customers. The pipelines that bring the water to the plants were built at about the same time. As the potable water needs of the area have increased over the years, the City has replaced and strengthened parts of its raw water transmission system to accommodate the higher working pressures required to deliver increasing amounts of raw water.

After years of debate and legal questions, the City of Virginia Beach has recently won the permitting necessary to bring the raw water resources it needs from Lake Gaston on the Virginia-North Carolina border. As part of an intercity agreement Norfolk is expanding its Moores Bridges Water Treatment Plant to be able to treat almost 462 million L/day (122 mgd) on a maximum day basis to accommodate the increased treatment needs of the City of Virginia Beach. Virginia Beach will deliver the water from Lake Gaston to Norfolk's reservoirs, and Norfolk will pump the increased flow from the reservoirs to the treatment plant.

To assure the City's resources are adequate to transport the additional flow, Norfolk evaluated the raw water transmission system with the newly anticipated flows added. Their evaluation concluded that a portion of the system would have to be replaced to accommodate the higher pressures required to move the raw water to the plant. The City then began to consider replacement methods for the various line segments, including a major crossing of the Nansemond River within the City of Suffolk. While considering this crossing the City recognized that not far downstream was another crossing that was considered a weak link in their system.

The Project

The Baker project consisted of the design of the two crossings of the Nansemond River. One of the crossings was to be 1067 mm (42 in) in diameter and was to parallel an existing single line crossing of the same size and provide the backup capability that was missing. The crossing would connect on each side of the river to a

762-mm (30-in) and a 914-mm (36-in) pipeline. The crossing was estimated to be about 630 m (2060 ft) long, including about 170 m (550 ft) of primary marsh along the side of the river. At this point the river is about 6 m (20 ft) deep at its deepest point.

The second crossing was located about 3.6 km (2.2 miles) upstream (south) from the first crossing and is part of a complete replacement of the existing 1219-mm (48-in) diameter pipeline. This crossing was estimated to be about 900 m (2950 ft) long and to cross about 460 m (1500 ft) of designated wetlands. The river has a maximum depth at this point of about 6 m (20 ft).

Consideration was also given to the soil conditions in the area. Borings were made on each side of the river and across the river following as closely as possible the probable alignment of the crossings. The results of the soil borings indicated that within the area of the crossings the soils were consistent below about 23 m (75 ft). From this depth there extends a deep layer of dense to very dense dark grey fine sand with clay throughout the Hampton Roads areas. This layer provides an excellent medium for HDD construction and tends to hold its shape following the installation of the pilot hole and reaming prior to the pull back of the pipeline.

All four upland sites, the entry sites on the east side of the river and the exit sites on the west side of the river, were suspected of being near designated historical sites. In the case of the larger of the two crossings, Civil War fortifications were known to be located very close to the existing pipeline. The historical record was reviewed, and maps of the area were drawn which showed the locations of designated historical sites on the west side of the river. Figure 2 shows the historical site designation for the smaller of the two pipelines, the 1067-mm (42-in).

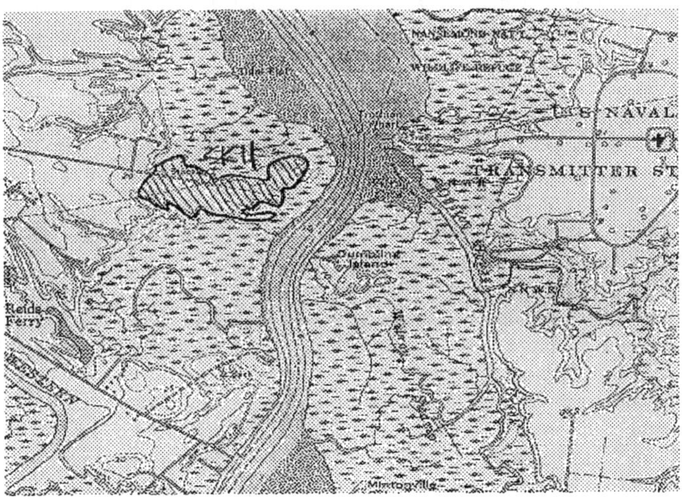

Figure 2. 1067 mm Site Designation (44SK11)

This site covers the entire point of land occupied by the existing pipeline. There was no opportunity to avoid some impacts to this site. Dumpling Island, which lies south of the 44SK11 in the marshes across the main channel of the river, is reported to have been the center of the Nansemond Indian culture.

Figure 3 shows the site of the larger crossing. This site exhibited several, smaller site designations.

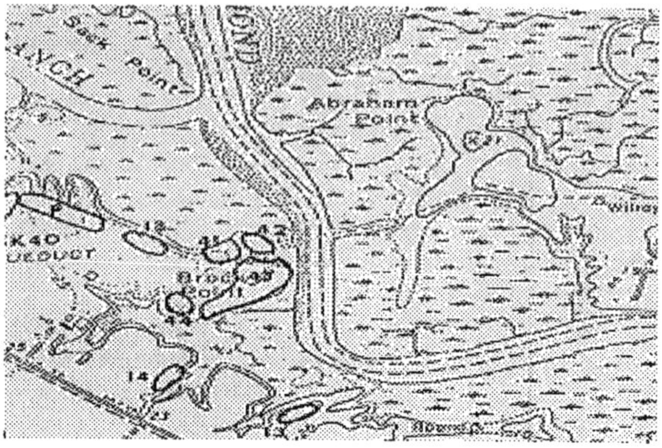

Figure 3. 1219 mm Site Designation

The challenge for the alignment of the crossings was the consideration of wetland impacts and potential mitigation and threading the needle between known historic sites for the 1219-mm (48-in) crossing. The use of HDD technology was seriously considered at this point, particularly in light of the Corp's mandate to minimize wetland impacts. The alternative to the City would have been some mitigation effort for the more than 630 m (2050 ft) of primary marsh and wetlands that would have been disrupted by conventional cut-and-cover construction techniques. The use of HDD also would eliminate any concerns of the Virginia Marine Resources Commission, which has primacy over submerged bottom lands in the State.

Preliminary layouts were established for each of the crossings. The initial considerations included minimizing the length of the HDD portion of the project to keep costs as low as possible. Proximity to the existing crossings was also important. The length of the connecting piping was kept as short as possible while trying to provide ample separation to minimize electrical interference from the existing pipelines with the locating devices used during construction.

For the larger crossing there appeared to be an opportunity to locate one end of an HDD installation on the extended peninsula on the east side of the river below Abraham Point (see Figure 3). This location would shorten the HDD portion of the project by more than 300 m (about 1000 ft). This fact had obviously been noticed long before the consideration of this project, since the peninsula contained one of the two Civil War fortifications at the crossing site, as well as the landfall of the existing pipeline.

There was not enough additional space on the peninsula to assure adequate working space for the installation, so the proposed HDD portion was extended.

The initial field investigations into the historic resources returned mixed results. The entry sites on the east side of the river had either been disturbed or held very little promise of significant new information. The archaeologists recommended that no additional action was required for those sites, and the SHPO concurred with their recommendation. The story on the west side of the river was quite different. The individual sites near the larger crossing had been combined into one large site. Figure 4 shows the redesignated site.

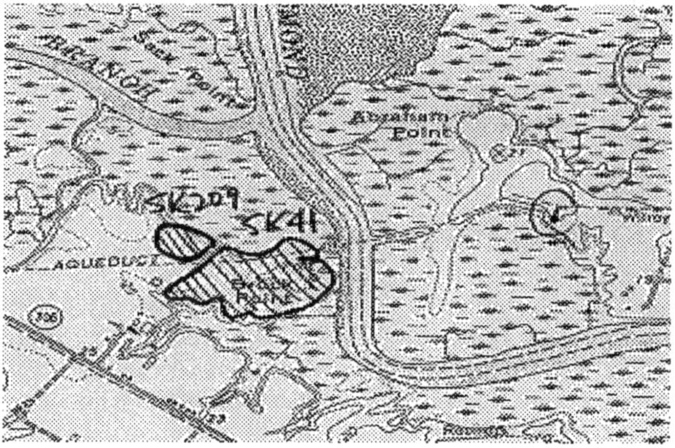

Figure 4. Redesignated Site 44SK41

The use of both sites on the west side of the river now involved minimizing the impact to larger sites rather than trying to miss smaller sites altogether. Site 44SK11, the site of the smaller of the crossings, covers nearly 23 ha (57 acres). Site 44SK41, the site of the larger crossing, is nearly 9.3 ha (23 acres). Both sites occupy farm fields that have been actively cultivated for many years.

The archaeologists characterized both sites as having major significance in the historical record of the area. Both sites were said to offer unique and rare opportunities to intensively study Early Contact period Native American village sites. Although no features were found during this phase, site 44SK41 was given a moderate to high probability of finding deposits relating to the Civil War battle for Hill's Point Battery in the spring of 1863. The archaeologists concluded that both sites were eligible for nomination to the National Register of Historic Places under criterion D (capable of yielding important information about the past), and that site 44SK41 was also eligible under criterion A (associated with significant events in the broad patterns of national history).

The choice of a construction technique was now fairly obvious. Using cut-and-cover would disrupt over 630 m (2050 ft) of designated wetlands. It would also disturb a trench at least 3 m (10 ft) wide across two designated historic sites from the point of exit from the wetlands to the connections to the existing pipelines. Using HDD would eliminate any impacts to the wetlands to be crossed and would reduce the impacts to the historic sites to the receiving mud pit and the connecting trenches to the existing pipelines.

A final design consideration now appeared. Both crossings would be equipped with corrosion control systems. An impressed current system had been anticipated from the beginning of the HDD design. The remaining question was one of power supply. Because of the remote nature of the sites next to the river, the closest available power supply for either site was more than 760 m (2500 ft) away. Although the power company would provide the service, making the installation was considered an unreasonable expense for the amount of power required (5 amps for one crossing and 7 amps for the other). Solar panels were incorporated into the design to eliminate the need to run power lines to provide the service.

Installation

The Corps of Engineers has issued permits for the construction of both crossings using the HDD technique. The issues concerning wetlands were avoided completely. Historic resources have been identified, and with the concurrence of the Department of Historic Resources of the State of Virginia, construction of the mud pit and the connecting pipelines withing the historic sites has been permitted. Still remaining is the Phase III recovery of artifacts from the designated construction areas. This activity is expected to be completed in the early spring of 1997.

Because of the relative accuracy of the HDD technique, the Phase III recovery has been limited to the areas expected to be excavated during the installation. For both sites this is limited to a 6-m (20-ft) square mud pit and the trenches required to connect the crossing to the existing pipelines. The total excavated area is expected to be less than 0.1 % of site 44SK11 and about 0.3 % of site 44SK41. Disturbance outside of the recovery area during construction will require additional archaeological recovery. To help safeguard those areas not excavated during the construction activities, the contractor will be required to place supporting mats in those areas where the soil could be disturbed below a depth of 200 mm (8 in).

Conclusion

When presented with the requirement of recrossing the Nansemond River, the City of Norfolk considered alternative methods of installation. Although the existing pipelines had been constructed using conventional cut-and-cover techniques years ago, the likelihood of having a similar project permitted today is very small. The Corps of Engineer's permitting process mandates that they eliminate potential impacts to wetlands wherever possible, and minimize those impacts that cannot be eliminated (in many cases accompanied by an appropriate level of mitigation). A similar approach is used when considering the potential impacts to historical sites.

By understanding the purpose and direction of the permitting agencies, the City was able to quickly focus on the key issues in the final design of the project. The potential use of cut-and-cover was quickly abandoned, and the design concentrated on optimizing the HDD design. With HDD selected the wetlands issues essentially disappeared. The location of the entry and exit points became the key elements in determining the nature of the project.

The concept of relocating the crossings to areas of little historic value was not an available option. The smaller crossing is paralleling the existing pipeline and tieing into the upland pipelines on each side of the river. The larger crossing is part of a pipeline replacement, which is using the existing routing and easements for construction.

By working with the permitting agencies during the early stages of the design process, the City of Norfolk was able quickly settle on an easily permitable construction technique. The balance of the final design process was limited to those issues that could minimize the historic impacts or reduce the overall cost of the project.

References

Hogan, J. L. & Fuller, R. G. Western Branch Pump Station #2 and Raw Water Intake #2, Final Report, Volume 1 (June 1992).

Layman, B. C.; Higgins, K. A. ; & Vergeres, J. D. Summary of Geotechnical Exploration, 42-inch and 48-inch Nansemond River Crossings, Suffolk, Virginia (January 1996).

McDonald, B. M.; Laird, M. R.; & Johnson, P. S. Nansemond River Archaeology: Phase II Archaeological Significance Evaluation of Sites 44SK11, 44SK41, 44SK50, and 44SK210 and a Phase I Archaeological Survey of 2,000 feet Along the Nansemond River 42" and 48" Utility Line Replacement, City of Suffolk, Virginia (July 1996).

Sprinkle, J. H., Jr. & Reinhart, T. R. Phase I Archaeological Investigations Hillpoint Farms, Suffolk, Virginia. Department of Anthropology, College of William and Mary, Williamsburg, Virginia (June 1988).

Virginia Department of Historic Resources. PROJECT REVIEW IN VIRGINIA. A Guide for Historic Property Owners, Government Agencies and Private Citizens. (June 1995).

Case History:
Horizontal Directionally Drilled Pipeline Crossing of the Red River at Shreveport/Bossier City, Louisiana

Charles W. Hair, III[1]
Member, ASCE

Abstract

Evolving technology relative to the planning, design, and installation of horizontal directionally drilled (HDD) pipeline installations is examined via this paper's discussion of a state-of-the-art river crossing recently completed in northwestern Louisiana. Although a comparatively small, 20 mm (8 in.) nominal diameter liquids-product pipe was involved; this HDD placement - including a subsurface horizontal curve - extended for more than 1.67 km (5,400 ft.) beneath a railroad, two highways, a residential neighborhood, a mainline flood protection levee, and an environmentally sensitive surface area in addition to the navigable river channel. Because of the technologically challenging length/geometry and non-standard surface features subtended; salient features crucial to the crossing's success were the site characterization - design procedures inherent to project inception together with the cooperative effort between the owner, various regulatory bodies, and the constructor throughout HDD execution.

Assuming an understanding of trenchless technology on the part of the reader, the paper concentrates on the site history and river/subsurface conditions leading to selection of HDD as the crossing implementation methodology. Crafting of the installation's overall geometry together with the decision to subdrill the non-alluvial, manmade surface features - especially the flood protection levee - are discussed. The right-of-way/permit acquisition processes at federal, state, and local levels plus the actual construction - with concentration on solutions to encountered problems - are then described. Concluding the paper are an analysis of the effects this installation has had/is having on HDD design and construction - especially in light of ongoing research - plus several recommendations/suggestions for improving the HDD engineering art form.

Project Description

Location. Site positioning in northwestern Louisiana is within the municipal confines of Shreveport and Bossier City. There; Texas Eastern Products Pipeline Company's (TEPPCO's) almost seven decades old Ark-La-Tex Pipeline transits the navigable, section/flow-managed, and leveed Red River.

---

[1]President, Louis J. Capozzoli & Associates, Inc., 10555 Airline Highway, Baton Rouge, Louisiana 70816.

New Crossing. TEPPCO's objective was to place a 20 mm (8 in.) nominal diameter products pipeline across the Red River and attendant natural/artificial bank surface features. Such alignment - in a right-of-way offset from that of the existing crossing - was to be skewed across the river's course and to feature a mid-length horizontal curve. Overall geometry - i.e. the line's horizontal extent and vertical penetration - was based on the facility's service life not being disrupted by forthcoming natural alluvial activity (meandering/scouring) nor humankind's future inchannel/onbank construction. Also of crucial importance were that:

- the mechanical stability (landslide resistance) and hydraulic integrity (water tightness) of a riverside flood protection levee be maintained;
- the crossing's construction be performed in site-specific materials and conditions amenable to the HDD placement methodology; and
- the pipe installation's adverse affects on the right-of-way's natural and artificial surface features be negated/minimized.

Overall intent of this essentially "non-standard" crossing's design and execution was an economical, river channel/site surface impact reduced, and environmentally compatible placement not subject to alluvial activity nor manmade disturbance for at least 50 years subsequent to construction.

Existing Crossings. Numerous other facilities - i.e. submarine telephone cables, aerial/underwater pipelines, highway/railroad bridges, etc. - presently span the Red River in the vicinity of TEPPCO's site. Most significant of these to the project at hand were TEPPCO's long-standing pipeline crossing plus another such installation belonging to others positioned slightly more than 11 km (about 7 mi.) downstream. While both facilities were placed via the conventional (cut-and-cover dredging) technique, TEPPCO's crossing apparently has not been affected by the river. Conversely, the downstream alignment was exposed by flood-induced channelbed scouring/channelbank caving roughly seven years ago. This occurrence resulted in an emergency line replacement via HDD.

Site Evaluation. Development of an environmentally compatible, long-lived pipeline installation - especially if by HDD - entails definition of both the in situ earth materials as well as the site-specific obstacle. The overall process is graphically presented by figure 1.

In this particular case, the obstacle was posed by the Red River as well as the onbank works of man - the levee, highways/parkways, a residential street, a railroad, etc. - and an adjoining ecologically fragile undeveloped area. Consequently, geotechnical site characterization had to define:

- the potential for river channel and bank surface configuration changes, both natural as well as manmade;
- the HDD construction procedure's applicability and conduct particulars; plus
- the means for countering any adverse effects of the crossing's construction on the site's natural and artificial features.

Incorporation of study findings into the crossing's planning, permitting, and implementation once again demonstrated the efficacy of designing/executing pipeline placements based on understandings of both the site's physical make-up as well as the obstacle's dynamic and environmentally sensitive characteristics.

Site Conditions

General. The Red River - a major component of the Mississippi River System - is site specifically entrenched in the southwestern edge of it's alluvial valley floor: an 8 to 11 km (5 to 7 mi.) wide, generally northwest-to-southeast

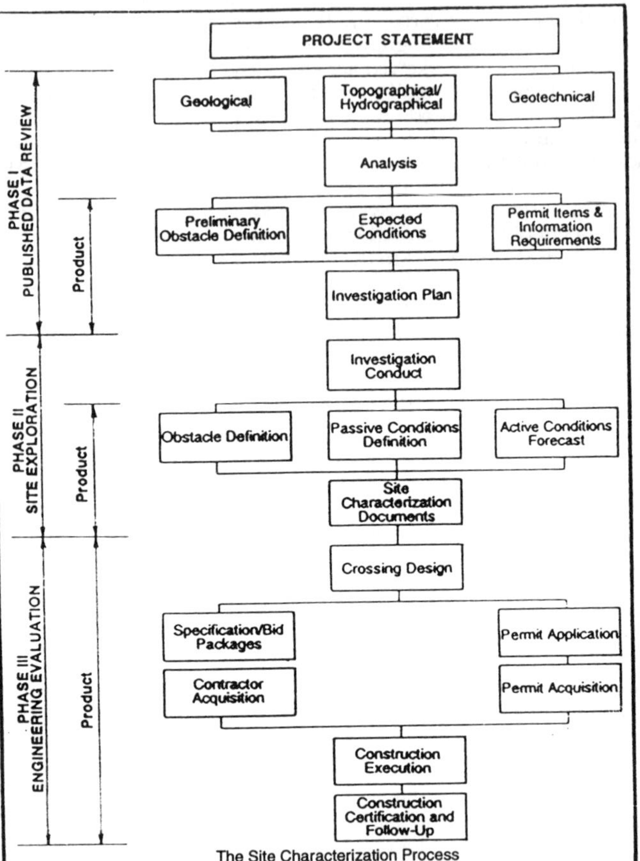

The Site Characterization Process
After *Hair -1996, ASCE*
Figure 1

oriented, floodplain. Formation of the valley was by alluvial scouring of Marine (saltwater sea) deposited, highly overconsolidated, very stiff to hard strength clay and very dense sand. Followon alluvial sedimentation - in thicknesses ranging between 30 and 45 m (100 and 150 ft.) - now partially fills the valley: elevation of the present day floor near TEPPCO's crossing site varies between 50 and 52 m (165 and 170 ft.), National Geodetic Vertical Datum (NGVD). The valley's edge bluffs - composed of Marine materials - extend upward 15 to 23 m (50 to 75 ft.) higher, i.e. to between elevations 67 and 73 m (220 and 240 ft.), NGVD. Although somewhat masked by onbank construction; numerous former channels, cutoff old courses, and meander scrolls (curvilinear ridges/swales remaining in the "wake" of a laterally migrating channel) dominate the topography near TEPPCO's site. These natural features - ranging from dry, sediment filled depressions to water bearing sloughs, oxbow lakes, and active streams - indicate the Red's past propensity for spontaneous horizontal course relocations across its valley floor floodplain. Of greater significance, though, are manmade channel cut-offs: closest of these to the site formed *Wright Island* on the Red's southwestern side in the mid 1800s. Such artificial activity - by establishing the river in apparently competent, erosion resistance earth material - has seemingly had a stabilizing effect on course meandering.

Channel. The new pipeline crossing was to transit the upstream third of a northwest-to-southeast orientated, approximately three kilometer (two mile) long, slightly bowed river reach. Flow direction in the nominally 300 m (1,000 ft.) wide by 6 to 8 m (20 to 25 ft.) deep (at normal stage) *active channel* - usually water filled component of the nearly 1,600 foot wide *gross channel* or *meander belt* - is southeastward. Roughly $2^1/_2$ km ($1^1/_2$ mi.) upstream of the intended alignment, the channel describes a comparatively acute curve from the west. Nearly the same distance downstream, the river executes a rather gentle bendway to the south.

At the right-of-way, the active channel cross-section features a thalweg (lowest point) probably established near the center of its positionally stable, "flattened U" shaped section. However, slowly developing sloughing of the left descending - i.e. northeastern - bankslope reveals a propensity for *attack* (ablation) of that side coincident with nominal *building* (accretion) of the course's opposite edge. These tendencies are confirmed by the sloughing/landsliding and scallop-plan bank failures embankment set-back visible on the river's leveed northeastern edge coupled with natural levee formation on the other side. The existing TEPPCO alignments' apparent security confirms that progression of such activity is relatively slow.

Banks. Both sides of the active channel steeply rise above the normal water surface. The essentially batture-less northeastern side's 30 foot high slope is surmounted by a 2.4 m (8 ft.) high artificial levee. For 760 m (2,500 ft.) landward of this; the level ground surface is occupied by a residential neighborhood, a highway/railroad, and then TEPPCO's pump station. Of interest is that a multi-lane elevated highway - the *Teague Parkway* - will soon be extended just landward of the levee. By contrast, only a natural levee is atop the southwestern side's crest. A nearly 275 m (900 ft.) wide expanse of treed meander scrolling compose the adjacent gross channel's bank surface. The valley floor edge is defined by a public boat launch ramp access road and adjacent 4-lane *Fant Parkway*. Landward of the parkway, the ground surface is covered with scrub brush interspersed with small to medium size trees.

Subsurface. Although undetected anomalies - undecayed logs/stumps, gravel pockets, etc. - may exist, generalized conditions are defined in table 1 and graphically displayed on figure 2. Such natural layering typifies Red alluvial valley stratification at this latitude. Marine constituents are members of the *Wilcox Group*. Of prime interest is that extensive gravel deposits and/or undecayed wood concentrations were not encountered by the subsurface exploration. Additionally, conduct of the soil sample borings necessitated only normal materials/procedures: vertical penetration of subsurface conditions did not require extensive casing nor excessive amounts of drilling mud for bore hole maintenance.

Groundwater. Immediately adjacent to the active channel, the water table is denoted by the Red River's stage. By about 30 m (100 ft.) landward, the phreatic surface rises to within 1 m (3 ft.) of the ground surfaces. During periods of rain and/or flooding, however, the water table can rise to, or even above, the bank surfaces on either side of the flood protection levee.

Stratum Nomenclature	Southwestern Bank	Channel	Northeastern Bank	Stratum Description
	Inclusive Elevations - Meters, NGVD			
Flood Protection Levee and Highways/Railroad. Manmade earthen embankments.	Crest (53±) to 52.4	N/A	Crest (56±) to 52.4/51.8	Relatively steep-sloped ridges typically consisting of compacted silty clay borrowed locally.
Topstratum. Recent Alluvium deposited from/in the Red's present day gross channel.	From The Above to 48.5/47.8	Surface/Mudline (51.2/39.6) to 37+	From The Above to 46.3/43.3	Natural levee of medium strength clay/firm silt overlying soft to medium strength clay/silty clay and loose to firm sandy silt.
Substratum. Ancient Alluvium deposited from/in remote (previous) Red River gross channels.		From The Above To 35.1/28.0		Firm to dense sand containing looser/denser pockets plus a gravel constituency at depth.
Marine. Deposition from Paleocene age saltwater seas.		From The above to The Exploration's Extent		Mainly hard strength clay and very silty clay. The clay's strength characteristics are the result of past desiccation which probably produced slickensides: drying crack "fractures" which have subsequently "healed".

Table 1. Geotechnical Profile

Potamological/Geotechnical Analyses

The site's pertinent natural/artificial characteristics - plus geological history/geotechnical conditions affecting crossing installation - are:

- the Red's minimal potential for reconfiguration/relocation in the vicinity of the crossing
- HDD's site specific applicability vis-a-vis the in situ surface/subsurface conditions and features.

The following paragraphs summarize key details.

Potamology. Humankind's activities regarding the Red River have minimally altered the stream's natural alluviation tendencies. Site specific consequence of:

- bolstering the channel's entrenchment conditions (bed/bank materials)
- manipulating the course's formerly unconstrained, natural geometry
- managing the river's flow

is that a comparatively high degree of alluvial equilibrium is possible. However, the Red's penchant for attacking its banks at specific points has caused/can generate localized instabilities. Overall result is that the Red at TEPPCO's crossing site is a **relatively stable alluvially mature** - i.e. **river** presently bent on slowly filling its' previously scoured valley.

Within the coming 50 years, Red River activity relative to this location should stem from minor *gross channel relocation* due to active channel cross-section expansion. Preservation/enlargement of humankind's "improvements" should continue such regimen and thus preclude the course from undergoing major configuration changes. In view of this, the Red's active channel at TEPPCO's alignment was projected to:

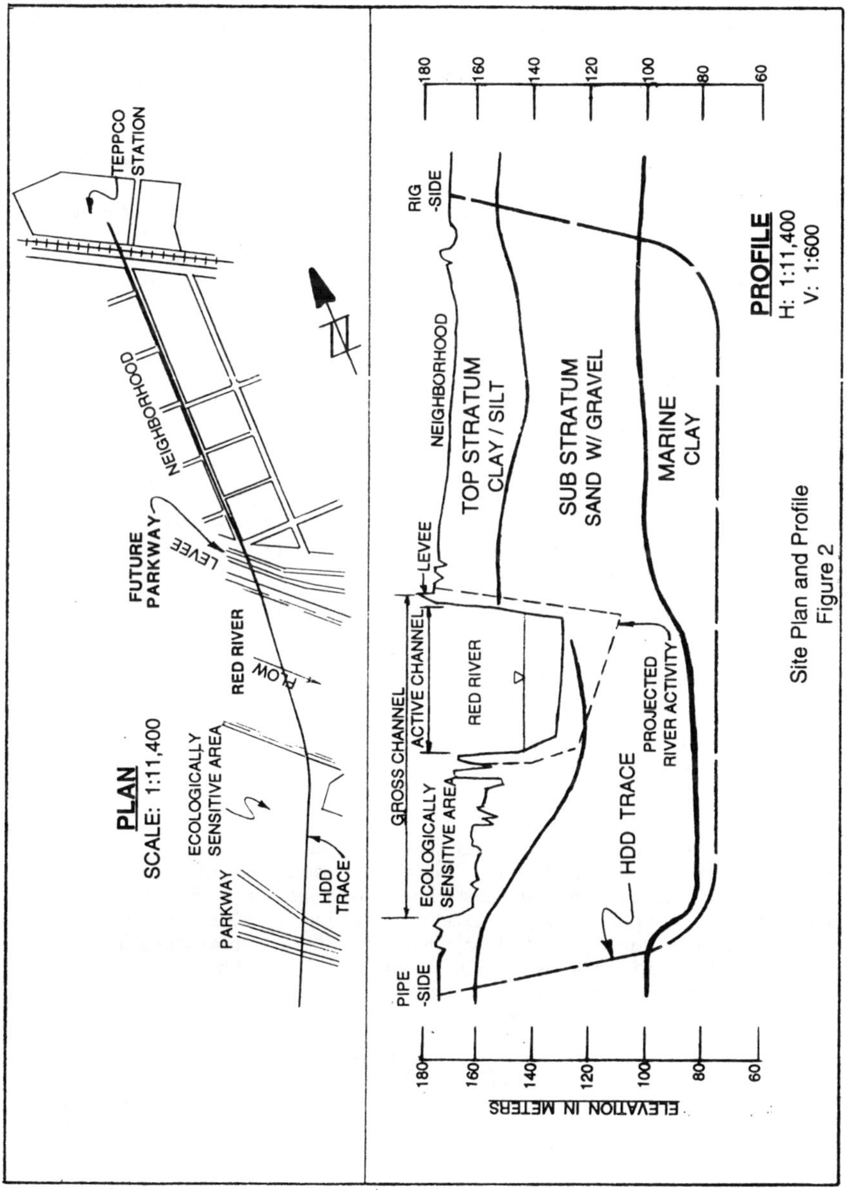

Site Plan and Profile
Figure 2

- horizontally displace between 15 and 30 m (50 and 100 ft.) in either direction
- vertically scour, in times of extreme flooding, to about elevation -33.5 m (-110 ft.), NGVD, i.e. slightly into the *Substratum* layer

during the next half century. Occurrence rate of this alluviation will depend on: the future extent of "artificial" channelization; humankind's tenacity in containing/repairing the effects of "natural" activity; plus the severity and frequency of rainfall/flooding.

Geotechnology. In terms of crossing construction: practically all subsoil types and conditions encountered in situ are highly amenable to HDD. *Topstratum/Substratum Alluvium* and *Marine* clays and sands - since only minimal gravel is apparent - pose few problems to such technique's efficient conduct. Furthermore, the deeper *Marine* clay, besides being an excellent medium for drilling the line's horizontally curved run, potentially offers absolute security from river activity together with almost complete protection from the generation of effects on the artificial/natural surface features. Based on these factors; negotiating anomalous impediments (buried logs, gravel pockets, etc.), avoiding channel stabilization measures/future onbank construction (embankments, bridge piling, etc.), and preserving the flood protection levee are this installation's major design/construction challenges.

Crossing Design

Since suitable site conditions were demonstratably prevalent, and because protection from alluvial activity/environmental disturbance would result, HDD was the method chosen for installing TEPPCO's new pipeline crossing. Figure 2 portrays the resulting design.

Crossing Geometry. Avoidance of the site's river obstacle plus sensitive onland features was achieved by establishing ground surface penetration points respectively 520 and 855 m (1,700 and 2,800 ft.) from the present-day channel's southwestern and northeastern edges.

Slightly steeper than normal penetration slopes - ranging between 12 and 16 degrees from the horizontal - were employed. Such steepened angles were necessary to achieve maximum pipe depth as rapidly as possible. The line's small diameter coupled with good steerability afforded by the subsurface soils were relied upon to offset the adverse effects of lengthening the sagbend curves.

The line's horizontal run was slated for the *Marine* clay at, or below, elevation 24.4 m (80 ft.), NGVD. Besides providing security from natural alluviation and channel maintenance construction, at depth placement in this material also offered good to excellent steering precision together with circumvention of future parkway bridge support piling. Minimum radius for the horizontal curve and sagbends was nominally 24.4 m (800 ft.) based on the *100 feet of bend radius per inch of carrier pipe diameter* rule-of-thumb.

In sum, the horizontally curved HDD pipeline crossing established according to the foregoing geometry was intended to extend for approximately 1,650 m (5,400 ft.), horizontally, and penetrate, to between 24 and 27 m (80 and 90 ft.), vertically, below the bank surfaces. Based on available hydrographic data, roughly 15 m (50 ft.) of undisturbed "cover" over the emplaced line would result beneath the active channel thalweg.

Crossing Permitting. The other salient features of this project involved right-of-way selection and levee subdrilling. The former aspect was based on "traditional" construction and therefore generated the subsurface horizontal curve beneath the river. In the future, lack of surface disruption by HDD may allow "straight-shot" installations and thereby dispense with the need for a "fixed" right-of-way at the surface.

Originally regarded as a stumbling block, drilling beneath the levee proved to be neither a physical nor an administrative concern. Set-back from the levee - both horizontally as well as vertically - was such that the embankment was not in any wise affected. Also, TEPPCO's openness with the Bossier Levee District throughout the project laid to rest regulatory concerns about compromise of the flood protection embankment's integrity and/or stability.

Crossing Conduct

Execution. Figures 3 and 4 picture details. Salient considerations were:

- congestion on the northeastern bank made the need to control the penetration point location there highly important
- more extensive clear space was available on the southwestern bank for layout/makeup of the carrier pipe prior to pullin.

Consequently, HDD rig set-up was on the northeastern bank within the confines of TEPPCO's pump station. Pilot hole drilling was southward/southwestward with reaming/carrier pipe pullin in the opposite directions. Of importance at this particular site was that subsoil conditions along the geotechnically recommended HDD trace were such that the pullin operation was not unduly jeopardized by the carrier pipe's make-up in two sections.

In terms of execution experience, the contractor - Land and Marine, Inc. of Houston, Texas - found the *Marine clay* to contain more cemented/partially-cemented sand than encountered by the geotechnical exploration. A down-hole mud motor - rather than a jet bit - was therefore needed to negotiate such formation. While directional drilling was slowed by this factor, the overall crossing - from pilot hole initiation to carrier pipe pull-in completion - required only about three weeks of effort. Additionally, subsurface steering was fairly precise owing to both the earth material conditions as well as to the presence of electronic position trackers placed on the rather broad expanse of subdrilled bank surfaces.

Drilling Mud. Especially important to this construction's success was the contractor's correct "handling" of the drilling mud: although drilling in clay "spontaneously" creates a slurry, "manufactured" drilling mud is usually necessary when penetrating sand. At this project site, subsurface conditions were such that occurrence of mud seeps onto the ground surface or riverbed was essentially impossible. Furthermore, drilling slurry returns were maintained for most of the crossing's length. Finally, grout injected at either end of the bore subsequent to carrier pipe pullin virtually insured site integrity restoration.

Summary

The HDD placement of this 20 mm (8 in.) nominal diameter pipeline crossing of the Red River and adjacent bank surface features - horizontally extending for about 1,650 m (5,400 ft.), and vertically penetrating to roughly 15 m (50 ft.) below the active channel bottom - did not affect/will not be disturbed by alluvial activity nor man-made construction for at least the next 50 years. Success of this state-of-the-art installation stemmed from a thorough site characterization, astute crossing design/permitting efforts, and pragmatic application of construction expertise.

Upper View. Looking southeastward. The HDD trace extends from near the tanks in the photograph's lower left corner to just beyond the parkway interchange in its' upper right corner. **Lower View.** Looking westward. Note layout of the site plus proximity of the HDD rig to the subdrilled railroad/highway.

Rig-side Conditions
Northeastern Bank
Figure 3

Upper View. Looking northward. HDD trace is positioned in the photograph's central third. Carrier pipe makeup area is the cleared space visible in the foreground. **Lower View.** Looking northwestward. Note that the carrier pipe has been made up in multiple segments.

Pipe-side Conditions
Southeastern Bank
Figure 4

Structural Rehabilitation of Rigid Pipes

George McAlpine[1], Affiliate, ASCE

Abstract

This paper discusses a method of structural rehabilitation of rigid pipes with PVC and cementitious grout liner materials. A design procedure for the structural rehabilitation of rigid pipes is proposed and discussed .

Introduction

The subject rehabilitation system consists of a spirally wound profiled PVC liner which is grouted in place by filling the annulus (0.5 - 3.0 inch) with a high strength cementitious grout. The grouting is accomplished through holes in the crown of the PVC liner every 5 - 10 feet and is performed in several lifts to minimize the stresses on the liner during the grouting operation. Once the grout is in place and cured, it integrates the PVC and host pipe into a composite structure (Ahmad and McAlpine 1994); this integration is accomplished through mechanical locking to the host pipe (via joints, cracks, and rough surfaces) and to the "T"s of the PVC profile. Use of a superplasticizer allows the use of a cement rich mix with low water content (i.e., high strength) while achieving excellent flow characteristics which insures complete fill of all voids in the annulus, including cracks and joints in the pipe. If structural demands require, the grout may be specially blended (e.g., polymeric mix) or steel wire mesh may be placed in the annulus before installing the PVC liner and grouting. In very large diameter pipes (≥ 84") the system can be installed in panels arched across the corroded top portion of the pipe only. This design flexibility allows the system to address a wide range of structural rehabilitation situations.

Most rehabilitation of gravity flow sewer and storm drains is in rigid pipes such as concrete or brick structures as opposed to flexible pipes such as ductile iron or steel pipes. This paper discusses the rehabilitation of rigid buried structures and more specifically, circular concrete pipes. Further, this paper addresses only the case of "fully deteriorated" rehabilitation design (see ASTM F 1698); that is, designing a rehabilitation system to restore the required load carrying capacity to the deteriorated host structure. Note that this approach is true rehabilitation of the host structure and not a "pipe within a pipe" (although this approach is also possible with this system of rehabilitation). Also note that, strictly speaking, the host pipe cannot be structurally "fully deteriorated" as it is successfully carrying its load at the time of rehabilitation;

[1]President, Danby of North America, Inc. P.O. Box 5127, Cary, NC 27512

otherwise, the structure would have collapsed. The operative portion of the ASTM F 1698 definition is that the host pipe "...is expected to reach this condition over the design life of the rehabilitated pipe.", or if left unrehabilitated it will eventually become, truly, "fully deteriorated". Standard industry practice is to use the terms "partially deteriorated" and "fully deteriorated" to describe host pipe conditions and this then dictates appropriate rehabilitation design. These terms were adopted for use principally by flexible plastic liners but are now pipe rehabilitation industry standard terms.

This paper describes a rehabilitation method that results in a rigid composite of PVC/grout/existing pipe which, according to its ASTM installation standard practice (F 1698), "should have strength at least equal to that required to sustain the loads (with safety factor) specified by the applicable project specifications.". As each project is unique in the way the pipe has deteriorated and the extent of deterioration, it is not possible to generalize about the specific rehabilitation design; e.g., whether or not to include new steel reinforcing in the rehabilitation process. Thus, the specific design for each project should be customized in cooperation with the owner's engineer.

Concrete Pipe Deterioration

Concrete pipes in need of structural rehabilitation normally have experienced severe corrosion of the concrete, usually in the crown above the daily high flow level, may be cracked and slightly oval (≤ 10%). In severe situations in reinforced concrete pipe (RCP) the corrosion will extend to or beyond the depth to the inner reinforcement. Once the steel reinforcement in RCP is exposed, its corrosion is rapid and its moment carrying capacity is unreliable at best. If the pipe is carrying moment generating loads, which must be assumed, the pipe will eventually fail. The rigid pipe fails in flexure at the points of maximum flexural stress; that is, at the crown on the inner surface and at the springline on the outer surface of the pipe. RCP will continue to carry its loads after flexural cracking until the load reaches the ultimate strength of the existing pipe wall, which may be considerably less than its specified magnitude. This assumes that the corrosion is limited to the upper half of the pipe and has not yet caused the reinforcement to become ineffective either by removing the concrete anchoring the steel or by eating through the steel itself.

Once the rigid pipe fails (load exceeds its ultimate strength) the pipe will begin to deflect, acting as a hinged circular-sectioned structure with the crown moving down and the springlines moving out (Serpente 1994). This deflection is resisted only by the surrounding soil. Clearly, the job of rehabilitation is to prevent flexural failure (and the resulting deflection) of the rigid pipe. The method to accomplish this rehabilitation under discussion here is to replace the missing concrete (and steel if required and possible) and then protect from corrosion with a plastic barrier to the corrosive environment.

The rehabilitation system being discussed results in a composite pipe wall structure consisting of concrete, steel (for RCP or where it is installed as part of the rehabilitation process in man-entry sized structures), cementitious grout, and plastic (PVC). The plastic is mechanically anchored in the grout and the grout is (adequately) bonded/anchored to the host pipe wall. Insuring adequate bonding/anchoring of the grout to the pipe wall requires that the pipe surface be cleaned to remove all corrosion products and expose competent concrete. This can be accomplished by high pressure (5,000 - 15,000 psi) water jetting; in larger diameter pipes (≥ 60" ID) this may have to be done using manually directed hand-held jet nozzles. If needed, water jetting may be supplemented with mechanical scraping to handle particularly difficult situations.

The water jetting also flushes the exposed surface removing residual corrosive acid, increasing the pH toward the desired neutral value of 7.0. Due to small amounts of acid in the pores of the cement paste, and to a lesser degree in the aggregate, the surface pH may remain in the range of 6.0 - 7.0. Considering the small total volume of acid present and the high pH (≥12.5) of the large volume of cement rich grout to be added at this surface, this slight surface acidity should present no problem. This rehabilitation system and this cleaning process has been evaluated in controlled experiments and in actual field demonstration with periodic inspections and found to be successful (Redner, Hsi, and Esfandi 1994). No evidence has been found of deleterious effects of residual acid. The field demonstration in 600 feet of 78" RCP with a surface pH = 2 has been inspected periodically since its installation in January 1991. This system was first installed in Australia in 1984 and in the USA in 1989.

This surface cleaning normally leaves a very rough surface consisting of exposed large aggregate which is still well anchored in the pipe wall. The fluid grout easily flows into this rough surface producing a mechanical bond/lock, insuring that any movement of the pipe surface is directly transmitted to the hardened grout; thus future strains in the pipe wall will be resisted by the grout and the entire structure will act as a composite material. In those cases in which the host pipe's rebar has been damaged by acid corrosion, the installation of steel welded wire fabric with mechanical anchors into the pipe wall will provide additional mechanical anchoring of the grout to the pipe.

Composite Material Analysis

The analysis of this structure requires the use of the transformed-section method of engineering mechanics as is depicted in Figure 1. The basic operation of the transformed-section method is to convert all materials into one material by transforming the actual areas of each material into an equivalent area with a common modulus of elasticity. For example, if we assume the modulus of elasticity for concrete is $E_c = 4x10^6$, steel $E_s = 30x10^6$, grout $E_g = 3x10^6$, and plastic $E_p = 3x10^5$, and we transform all materials to concrete, the area of concrete is unchanged, steel area is 7.5 times the physical area, grout area is 3/4 the physical area and the area of the plastic is reduced by a factor of 3/40. This is accomplished by changing the width of each material so as to maintain their centroids unchanged in the geometry of the problem. The example of Figure 1 assumes that the corrosion has removed 1.0 inch from the inner wall in the top half of the pipe and has exposed the rebar; the rehabilitation replaces the missing concrete and adds another 1.0 inch of 5000 psi grout, plus the plastic of 0.36" moment of inertia equivalent thickness of the profile. The addition of the rehabilitation materials has two primary effects on the stress levels: 1) the neutral axis (NA) is shifted toward the inner surface, and 2) the composite moment of inertia I_{comp} is increased. The following equations formalize the analysis (using 48" ASTM C 76 Class II, Wall B RCP: 5" wall, 1" cover over both steel cages, inner cage area = 0.18 in^2/ft, outer cage area = 0.11 in^2/ft, assume spacing of 4" for rebar separation, f'c = 4000 psi).

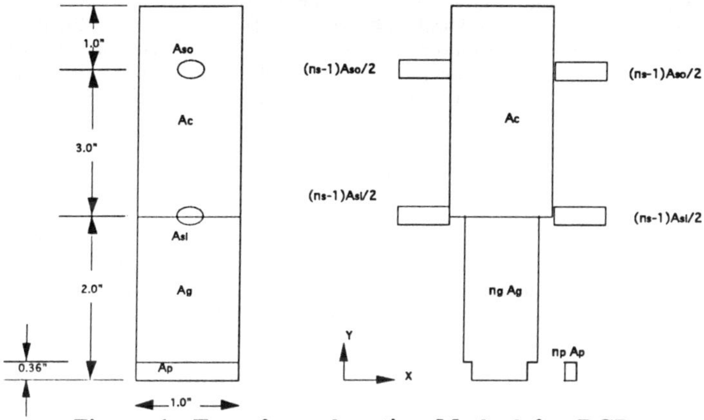

Figure 1 Transformed-section Method for RCP

$$Y_{NA} = (\Sigma\, Y\, A) / \Sigma A = 3.114" \tag{1}$$

where Y_{NA} is the distance to the neutral axis from the X axis of the coordinate system (inner surface of PVC) and Y is the distance to each centroid of each transformed area $A=n_xA_x$, $n_x=E_x / E_c$.

$$I_{comp} = \Sigma\, I'_x = 16.415 \tag{2}$$

where

I'_x = moment of inertia of transformed material = $n_xI_x + (Y_{NA} - Y_c)^2\, n_xA_x$,
Y_c = distance from X axis to centroid of area.

It is assumed that the cross section of the wall remains a plane while bending, and the materials obey Hook's law; the strain varies linearly from the neutral axis and the flexural stress at a distance y from the neutral axis due to moment M is given by

$$\sigma_x = n_x\, M\, y / I_{comp} \tag{3}$$

Repeating the transformed-section method for the deteriorated host pipe (no grout or PVC) yields Y_{NA} = 1.928 and I_{comp} = 5.722.

Table 1 compares the responses of the two structures to moments by tabulating values of σ_x/M at maximum values of y_x.

Material	Maximum y_x Deteriorated; Rehabed	Deteriorated σ_x/M	Rehabed σ_x/M	Reduction Ratio
Pipe Concrete	2.072; 2.886	0.3622	0.1758	0.485
Outer Cage	0.964; 1.886	1.2640	0.8619	0.682
Inner Cage	-1.790; -0.976	-2.3455	-0.5090	0.217
Grout	- ; -3.114	-	-0.1258	-
PVC	- ; -3.114	-	-0.01423	-

Table 1 Structural Responses to Moments at Pipe Crown

The benefit of the rehabilitation is evident from the reduction in the stress response to future additional flexural loads (moments) at the crown of the pipe. The stress response at the springline can be determined by applying a -M; i.e., pipe concrete and outer cage reinforcing are in tension (- stress), inner cage and grout/PVC are in compression (+ stress). The above analysis ignores the thrust forces which add to the compressive stress and subtract from the tensile stress; further this analysis is only valid if the pipe concrete has not cracked under tensile stress due to flexural loading. After cracking, the analysis of stress assumes all tensile forces are carried by the tension reinforcement and the compressive forces are carried by part of the compression portion of the section.

Once the magnitude of the load induced moment M has been determined then the state of stress of the deteriorated host structure can be determined. For example, assume M = +1000 in-lbF/in at pipe crown and -1000 at springline: the critical value is at the springline where the tension stress in the outer concrete is -362 psi, which is approaching the modulus of rupture of the 4000 psi concrete ($7.5\sqrt{4000} = 474$). An additional 112 psi of tensile stress in the rehabilitated structure, bringing the total tensile stress to the 474 psi modulus of rupture, would be produced by 112/0.1758 = 639 in-lbF/in additional moment. Thus the -1000 in-lbF/in moment at the springline is being carried with a safety factor of 1.639; without rehabilitation the safety factor would be 474/362 = 1.309. This analysis is based on the failure criteria being flexure cracking of the concrete; this is not true failure in terms of collapse and most undeteriorated reinforced concrete pipe has ultimate load capacity of about 50% more than the load producing concrete cracking. However, an elementary but tedious algebraic exercise to determine the shift in the neutral axis and the stresses after concrete cracking at the springline (M = -1309) shows that the tensile stress in the outer cage reinforcements is about 81% of the assumed yield strength of the steel (65,000 psi) or a safety factor over ultimate load of 1.23. After rehabilitation as described above, the safety factor is increased to 1.42. Increasing the grout thickness and compressive strength will give a small increase in the safety factor. Current design practice (ASCE 15-93, 1993) for factory made RCP uses a load factor of 1.3 and strength reduction factors of 0.95 for reinforcement tensile yield and 0.9 for concrete radial tension; 1.3/0.9 = 1.44 = safety factor for 0.01" crack and 1.3/0.95 = 1.37 = safety factor for ultimate strength.

Of course, the calculated safety factors for the deteriorated pipe are illusory because the deterioration is on-going and the pipe will eventually fail if no action is taken to stop the deterioration. Further, the effectiveness of the rehabilitation in securing a satisfactory safety factor depends on effecting the rehabilitation while there is still load capacity in the host structure, i.e., adequate safety factor. Clearly, the first contribution of rehabilitation is the halting of the deterioration which preserves the structural capacity then existing; enhancing that capacity is an added bonus.

Rehabilitation Design

In general, the rehabilitation situation (pipes installed 25 -150 years ago) is highly uncertain in the knowledge of burial and soil conditions and pressure distributions and does not support or justify detailed assumptions and loading analysis. In fact, it is rehabilitation industry practice to calculate loads on buried flexible pipes using prism loads in Eq. A.20 of AWWA C950-88, which may not be appropriate for rigid pipes (or rigidly encased flexible pipe either). It is clear, however, that the rehabilitation must preserve and enhance moment handling capacity (i.e., flexural strength) of the deteriorated host concrete pipe if it is to remain a rigid structure.

Making conservative simplifying assumptions appropriate to the rehabilitation situation, the recommended design method for rehabilitation is as follows:

° Compute vertical loads on the structure (using prism loads (WRc 1990, page III/40) unless better information is available).

° Apply the calculated loads to the structure assuming uniform horizontal and vertical pressure distribution over full diameter of pipe with K = horizontal pressure/vertical pressure; usually K = 0.4 (WRc 1990, page III/45). Again, if better information is available, use appropriate model (moments may vary by as much as a factor of 4 to 1.0, depending on assumptions about K and bedding of rigid pipe).

° Complete a structural analysis using transformed-section method, both before and after rehabilitation.

° Design wall thickness and grout compressive strength to achieve required safety factors by iterating the analysis for different combinations of grout thickness and strength.

This is a reasonable design method for the following reasons:

° The assumed pressure distribution allows reasonable modeling of soil side support and moment generating conditions.

° Using the modulus of rupture of the grout (alone) as the flexural strength to resist bending stress on the inner surface at the crown is 50 - 100% less than the flexural load carrying capability of the PVC/grout composite with reasonably small strains; see Figure 2.

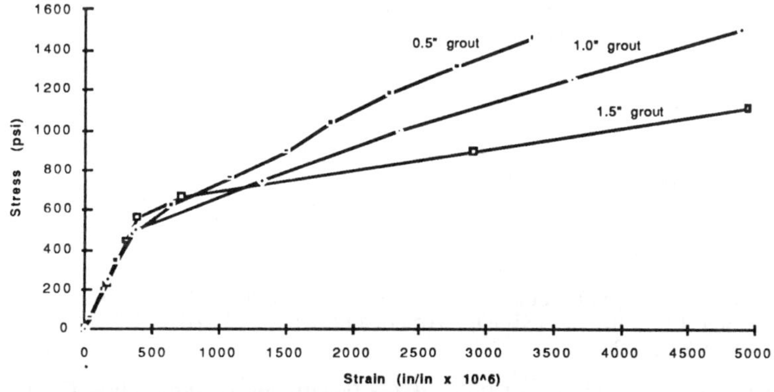

Figure 2 Flexural Stress/Strain for PVC/Grout Composite Beams

The data for Figure 2 are from laboratory tests following ASTM D 790-92 for three point loading and measuring beam deflection versus load. The composite beams consisted of 6" wide, 24" long PVC profile 0.5" high ("T" height) with 5000 psi (nominal) grout filling the space between the "T"s (0.5" grout case of Figure 2) and solid grout above the top of the "T"s. The "1.0" grout" and the "1.5" grout" cases of Figure 2 had 0.5" and 1.0" of grout, respectively, above the top of the "T"s. The composition of the cross sectional area was 75% grout and 25% PVC for the 0.5" grout case. The load was applied to the grout face of the beam (Ahmad and

McAlpine 1994) which would be consistent with flexural loads on the liner system at the crown of a rehabilitated pipe.

° Compressive strength and flexural strength (modulus of rupture) used are based on 28-day values. The *Standard Handbook for Civil Engineers* (Merritt 1983, page 8-3) states "Concrete may increase significantly in strength after 28 days, particularly when cement is mixed with fly ash. Therefore, specification of strengths at 56 or 90 days is appropriate in design." This is true for both the grout and the pipe concrete.

The assumption is made in this design method that the grout does replace missing concrete in the host pipe wall and fills cracks that may have been acting as hinged joints, thus making the structure act as a homogeneous rigid pipe. The reasonableness of this assumption has been demonstrated in D-Load and Large Soil Cell tests (Ahmad and McAlpine 1994).

Conclusions

A design method has been presented for the structural rehabilitation of rigid pipes using high strength cementitious grout in conjunction with a plastic (PVC) liner which serves as formwork for the grout and a chemical barrier to prevent corrosion. The design variables are the thickness and strength of the grout; when needed, steel wire mesh can also be used in man-entry size pipes. Although this paper discusses only concrete pipes, the method should be applicable to other rigid structures such as circular brick pipe; non-circular structures may require modifying the details of the method or using an appropriately modeled finite element analysis.

Appendix-References

Ahmad, S. and McAlpine, G., 1994, "Structural Characteristics of the Danby Pipe Renovation System", *Proceedings, Third Materials Engineering Conference, Materials Engineering Division/ASCE, INFRASTRUCTURE: New Materials and Methods of Repair*, November 13-16, 1994, San Diego, Calif., American Society of Civil Engineers (ASCE), New York, NY.

ASCE 15-93, "Standard Practice for Direct Design of Buried Precast Concrete Pipe Using Standard Installations (SIDD)", American Society of Civil Engineers (ASCE), New York, NY, 1993.

ASTM F 1698 - 96, "Standard Practice for Installation of Poly(Vinyl Chloride) (PVC) Profile Strip Liner and Cementitious Grout for Rehabilitation of Existing Man-Entry Sewers and Conduits", American Society for Testing and Materials, 100 Barr Harbor Dr., West Conshohocken, PA 19428, 1996.

ASTM D 790 - 92, "Standard Test Method for Flexural Properties of Unreinforced and Reinforced Plastics and Electrical Insulating Materials", American Society for Testing and Materials, 100 Barr Harbor Dr., West Conshohocken, PA 19428, 1992.

AWWA C950-88, "AWWA Standard for Fiberglass Pressure Pipe", American Water Works Association, 6666 West Quincy Avenue, Denver, Colorado 80235, 1989.

Merritt, Frederick S.(Editor), 1983, *Standard Handbook for Civil Engineers*, Third Edition, McGraw-Hill, New York, NY.

Redner, Hsi, and Esfandi, 1994, "Evaluation of Protective Coatings for Concrete", County Sanitation Districts of Los Angeles County, Whittier, CA, February 1994. (Contact Mr. John Redner, 920 South Alameda Street, Compton, CA 90221.)

Serpente, Robert F., P.E., 1994, "Understanding the Modes of Failure for Sewers", pp. 86-100, *URBAN DRAINAGE REHABILITATION PROGRAMS AND TECHNIQUES*, Edited by William A. Macaitis, American Society of Civil Engineers (ASCE), New York, NY.

Water Research Centre (WRc), 1990, *Sewerage Rehabilitation Manual.*, WRc, Swindon, England (UK).

Sewers and Manholes with Polymer Concrete

Thomas D. Bloomfield, Dipl.-Ing.[1]

Abstract

This paper describes recent developments in the trenchless construction industry using polymer concrete pipe. These developments include the first microtunneling installations in the United States, Canada and Kuwait; two curved microtunneling projects in Germany; and the development of a polymer concrete liner pipe to slipline deteriorated circular and semi-elliptical pipes.

Background

Polymer concrete pipe was developed in Germany in the 1960's. This pipe was developed to meet the growing needs of industrial and municipal clients for a strong corrosion resistant pipe. Several pipe manufacturing companies in Europe and the United States also began developing corrosion resistant pipe products during this period. One of these pipe products is a centrifugally cast fiberglass reinforced plastic mortar pipe. This pipe called Hobas was first manufactured in Switzerland in 1960. It was first introduced in the United States in 1984. This pipe currently has widespread use in the trenchless market where it is used in microtunneling and large diameter sliplining projects.

A reinforced plastic mortar pipe was developed in the United States in the 1960's and 1970's. This pipe made of sand, polyester resin and fiberglass was designed primarily for pressure applications such as water supply and irrigation systems. This pipe was used extensively in the United States in the 1960's and 1970's. Problems developed on several major projects in the 1970's and 1980's. These problems were traced to inadequate design, poor quality control and other deficiencies.

In Germany, non-reinforced polymer concrete pipe was designed for use in industrial plants. It was resistant to chemical attack from both acidic and caustic effluents inside the pipe and from chemical attack on the outside of the pipe. It also had high axial and compressive strengths without adding steel or fiberglass. Polymer concrete pipe was used infrequently in Germany because of the lack of sophisticated

[1]Export Manager, Meyer Rohr & Schacht GmbH Otto-Brenner-Strabe 5, D-212327 Luneburg

manufacturing production facilities and its high cost of production. In 1986 Meyer Betonwerk GmbH, who had continued to develop polymer concrete pipe production, began manufacturing polymer concrete pipe for microtunneling and direct burial applications. Polymer concrete pipe was greatly influenced by the developments that were taking place in the construction industry at this time. In 1983, microtunneling reached Europe. This trenchless installation market exploded in the 1980's and is still growing in 1997. This installation method required high strength corrosion resistant pipes, such as polymer concrete.

Microtunneling Developments in North America

During the summer of 1996, Nada Pacific Corporation of Caruthers, California became the first contractor in the United States to install polymer concrete microtunneling pipe. The Peralta Boulevard Sewer Replacement Project in Fremont, California for Union Sanitary District called for the installation of 872 meters (2,860 feet) of 457 mm (18 inch) inside diameter pipe by microtunneling. Ground conditions consisted of medium stiff sandy clays and clayey sands, loose to medium dense silty and sand soils and medium dense to dense gravels. Groundwater was not encountered during subsurface exploration. The longest planned drive was 140 meters (460 feet). Fourteen reaches were originally planned to be microtunneled. Contract documents specified only clay or centrifugally cast fiberglass reinforced plastic mortar pipe. Clay was selected by Nada Pacific, the low bidder. Subsequent to the bid, Nada, who had been closely following the development of polymer concrete pipe, requested a substitution. The immediate reaction to Nada's request was a firm "No" by the consultant, Montgomery Watson and by Union Sanitary District. The reason given was unfamiliarity with Polymer concrete microtunneling pipe and that it had not been installed in the United States. Nada Pacific was persistent and persuasive. They requested calculations and project information from Europe. Included in this information was Montgomery Watson's European office evaluation and approval of polymer concrete for microtunneling projects in Europe and the Middle East. At a partnering meeting subsequent to receipt of the pipe calculations, Nada Pacific again requested the substitution of polymer concrete pipe. This time their request was granted. Nada Pacific had ordered 2.44 meter (8.0 foot) long segments of 625 mm (24.6 inch) outside diameter polymer concrete pipe to match the size of the cutting head on their 660 mm (26 inch) Akkerman microtunneling machine and their 2.44 meter (8.0 foot) long slurry pipes. Most of the standard pipe lengths in Europe come in one, two or three meter lengths.

Nada Pacific had expressed a concern in the partnering meeting that cobbles or boulders nestled in the soils directly adjacent to the pipe or pushed away from the cutting head and rolling along side, might impart point loads on the pipe, causing it to crack or break. They were also concerned that jacking loads on the longer drives might exceed the allowable 173 metric tons (190 tons) of the clay microtunneling pipe on five or more drives exceeding 110 meters (360 feet). Because of the larger axial capacity, increased ductility and general durability it was felt that polymer concrete pipe would provide a more reliable installation, especially for the longer drives.

The manufacture of the mold in Germany took approximately five weeks. Pipe production and testing of the first shipment of 136 meters (448) feet, took two more weeks. Finally pipe shipment in 12.2 meter (40 foot) containers from Bremerhaven to Oakland took four weeks. Before the arrival of the first polymer pipe from Germany, Nada Pacific started the project using clay microtunneling pipe. The first drive was unsuccessful and the machine had to be dug up. The second drive was also unsuccessful. The project team decided to open cut four reaches which were determined to have significant amounts of cobbles. On July 26th the polymer concrete pipe arrived and the fourth drive began. Drive 3 had been completed with steel pipe. Drive 4, a 110 meter (360 foot) long drive was completed in two days without any problems. A total of five drives, including one of 140 meters (460 feet) were completed. Jacking forces increased to between 182 and 282 metric tons (200 and 310 short tons), which were within the capacity of the polymer pipe. The fiberglass sleeves and rubber gaskets enabled the pipe to successfully pass the pressure tests after installation. Union Sanitary District has recently bid another microtunneling project in which polymer concrete pipe was specified.

In January 1996 the City of Vancouver, British Columbia, Canada advertised for bids for the Clark Drive Trunk Sewer Project. The project called for the installation of 500 meters (1640 feet) of 800 mm (31.5 inch) internal diameter microtunneling pipe in four drives in a heavily congested area of Vancouver. The project specifications called for centrifugally cast fiberglass reinforced plastic mortar pipe or a contractor's alternate. The contractor's alternate was added to the specifications because polymer concrete pipe had been proposed by the manufacturer's agent. However, before polymer concrete pipe could be accepted a significant amount of technical information on the pipe had to be submitted by the contractor and approved by the City of Vancouver. One item alone, a drive under six sets of railroad tracks, required the stamp of a structural engineer in British Columbia. The requirement was for a minimum equivalent to ASTM-C76-Class V reinforced concrete pipe. After review of the polymer pipe calculations, whose design was based on the previous approval of the German Railways. The pipe's design compressive strength was two times that of C76-Class V RCP. Approval for use of the pipe was given in April 1996.

The first drive was 81 meters (266 feet) long under the six sets of railroad tracks. The ground conditions consisted of glacial till, large boulders, conglomerate, rock, sandstone and construction debris. The bedding under the railroad tracks often includes boulders, cobbles, railroad spikes and tree trunks, all of which were encountered. Microtunneling was selected because of the large number of utilities in this location. On the third drive the ground was so hard that P. Baratta Construction had to blast the reception shaft open with dynamite. Three drives exceeding 183 meters (600 feet) including the last one of 200 meters (656 feet) were installed without any problems caused by the pipe. The last drive had just started when a sudden rainstorm flooded the pushing pit with 8 meters (26 feet) of water. nevertheless, the project was completed successfully and the pipe passed the pressure test.

Kuwait Experience

Microtunneling began in Kuwait in 1992 using an Iseki Unclemole. In all 2 kilometers (1.24 miles) were microtunneled, using pipes of 250 milimeters to 600 milimeters (10 inches to 24 inches) diameter, at depths of 5 meters (16 feet) and drive lengths of up to 50 meters (164 feet). In 1994 a total of 5 kilometers (3.1 miles) were microtunneled, using pipes with the same range of diameters and at depths of up to 11 meters (36 feet) and drive lengths of up to 80 meters (262 feet).

Projects in design or under construction, total approximately 190 kilometers (118 miles). The projects to be microtunneled total approximately 17 kilometers (10.6 miles). Pipe sizes range up to 1200 milimeters (47.2 inches) diameter, at depths of up to 18 meters (59 feet) and drive lengths of over 120 meters (394 feet).

A number of factors in Kuwait point to the high use of no-dig techniques: flat topography which requires deep sewers to reduce the number of pumping stations, unstable ground which requires microtunneling or costly deep trench support, high water table and high permeability which requires frequent well point dewatering for open cut construction. In addition there are numerous road crossings, congested underground utilities, and frequent surface obstructions.

The very corrosive sewage and ground water requires high quality, corrosion resistant materials. The majority of the no-dig replacement work in Kuwait has been carried out with microtunneling machines with slurry spoil removal and vitrified clay jacking pipes. Pipe sleeves have been specified with high grade stainless steel collars for corrosion resistance.

However problems with varied ground and eccentric loads have been encountered. In order to overcome the above mentioned problems the contractor started to use steel pipes for the pipe jacking and tried to replace the steel pipes with vitrified clay pipes after completing the drive. Usually the pipes are checked by closed circuit television cameras (CCTV) immediately after the drive and before removing the microtunneling machine. This enables immediate action in case damaged pipes are found. The damaged pipes will then be pushed out and replaced by new pipes. On one drive the vitrified clay pipe was not strong enough to restart the jacking and push the steel pipes out. The line was finally abandoned and the steel dummy pipes were left in the ground.

In 1995 a new polymer concrete composite jacking pipe was designed and manufactured to comply with the Ministry of Public Works specifications. The inner liner of this polymer concrete pipe consists of a glass reinforced plastic (GRP) pipe of 7 mm (.28 inches) thickness made of vinyl ester resin and using ECR glass to guarantee long term corrosion resistance under the severe Kuwaiti soil conditions. The GRP sleeve with the corresponding double lip elastomeric seals insure a leak free joint good for up to 2 bar (28.6 psi) of outside pressure.

A DN 900 mm (35.4 inches) internal diameter polymer concrete microtunneling pipe was used for the first time in April 1996. The polymer concrete composite jacking pipe had a wall thickness of 116 mm (4.57 inches) and an allowable jacking load of 516 metric tons (568 short tons).

The first polymer concrete composite jacking pipe were used only as dummies being pushed out after the drive and replaced with vitrified clay. The same pipes were used for three drives of 120 meters (394 feet) each without any problems. When trying to push out the polymer concrete composite jacking pipes with vitrified clay pipes on one drive, several vitrified clay pipes broke and had to be replaced. On the third drive it was decided to leave the polymer concrete composite jacking pipes in the ground.

In view of the good performance of polymer concrete composite jacking pipe and the compliance with the specification, the Ministry of Public Works approved the polymer concrete composite jacking pipe for the construction of sewers in Kuwait on August 7, 1996.

By October 31, 1996 a total of approximately 1000 meters (3,280 feet) of the Sulaibiya Trunk Sewer had been completed with polymer concrete jacking pipes at depths between 16 meters (52 feet) and 18 meters (59 feet), and drive lengths up to 126 meters (413 feet). No problems occurred during the jacking of the pipe and the camera inspection of the completed drives showed no problems.

On a second site in Area 4 of the Sulaibikhat Sewer Renovation Project, seven drives totaling 510 meters (1,673 feet) of 600 mm (24 inch) polymer concrete pipe had been microtunneled by October 31, 1996. A Soltau microtunneling machine installed this pipe at depths ranging from 8 meters (26 feet) to 12 meters (39 feet).

This project included one drive where the previously mentioned steel dummy pipes were pushed out. The maximum jacking load recorded on this drive was over 215 metric tons (237 short tons). The wall thickness of the 600 mm (24 inch) polymer concrete composite jacking pipe used was 80 mm (3.15 inches) and the allowable jacking load was 183 metric tons (201 short tons).

Curved Microtunneling Drives in Germany

In 1994 a 212 meter (695 feet) long curved drive with a radius of 115 meters (377 feet) had been completed successfully around Morsbroich Castle in Leverkusen Germany using 1 meter long (39.4 inch) polymer concrete pipes designed to withstand axial forces of up to 600 metric tons (660 short tons), having a wall thickness of 145 mm (5.71 inches).

The good experience with polymer concrete made the City of Leverkusen decide to use the same pipe material for the 162 meter long (531 feet), 1400 mm (55.1 inch) diameter curved drive with a radius of 300 meters (984 feet), starting straight and driving into the curve until crossing a double railway track at a cover depth of 10 meters (32.8 feet).

The pipe was designed to the requirements allowing axial jacking forces of 866 metric tons (953 short tons), having a wall thickness of 160 mm (6.30 inches). The length of the pipe was 2.5 meters (8.2 feet) to achieve a maximum angular deflection of 0.477°. Special care had been taken to protect the fiber glass reinforced sleeve by a 10 mm (0.39 inch) recess, i.e. with an OD of the spigot 20 mm (0.78 inch) smaller than the OD of the pipe.

In the previous Morsbroich Castle Project segmented jacking force distribution rings were fitted in each pipe joint. The knotless wood ring segments on the outer radius of the curve were thicker than the segments on the inner side of the curve. In this project wood rings with the same thickness over the whole circumference were used with thickness of 18 mm (0.71 inch) to enable the wood being compressed on one side to achieve the angular deflection. It is necessary to provide high quality knotless wood to guaranteed the even distribution of jacking forces.

Special lifting lugs facilitates the handling of the pipes on the site. Bentonite injection nipples were installed in every pipe at the position of 9:00, 12:00 and 3:00 o'clock. The jacking forces remained low due to the lubrication and smooth surface of the pipe material. The project was started January 7, 1997. As of February 5, 1997, about 120 meters (394 feet) were pushed with advance rates of 10 meters (32.8 feet) per day without any problems in the ground. The curve negated the use of laser alignment for directional control and required the pipe position to be monitored up to four times per day using electromagnetic detection.This was the first project in which a 1400 mm (55.1 inch) internal diameter polymer concrete pipe is being jacked under a railway track.

Polymer Concrete Liner Pipe Developments

Over the last year polymer concrete liner pipe to slipline deteriorated circular and semi-elliptical pipe has been designed. Circular liner pipe has been manufactured and tested. Semi-elliptical liner pipe has been designed to meet the full ground loadings required to renovate a fully deteriorated reinforced concrete or brick sewer. Field tests for this semi-elliptical liner pipe have been proposed and are expected to take place this year.

KEY WORDS

polymer - synthetic compound of unusually high molecular weight.

concrete- a construction material consisting of conglomerate gravel, aggregate in a mortar or cement matrix.

spalling - a chipping, fragmenting or flaking of a pipe.

GRP - glass reinforced plastic.

ECR glass - type of corrosion resistant fiberglass.

polyester - synthetic resins which are light, strong and weather resistant; examples= boat hulls, swimming pools and molded parts.

-

quartz - a hard crystaline, vitreious mineral found worldwide as a component of sandstone and granet.

elastomer - any of various polymers having the elastic properties of natural rubber.

betonite - clay mixture used to lubricate jacking pipe during installation.

microtunneling - installation of new pipelines accurately without manned entry using remotely controlled, steerable tunneling machines in all types of ground conditions.

Direct Design of Concrete Pipe Installed by Jacking

Frank J. Heger, Fellow, ASCE[1]

Abstract

Structural design criteria for precast concrete pipe installed by jacking are provided in a new ASCE standard practice that has recently been approved for Society ballot by the ASCE Direct Design of Concrete Pipe Standards Committee. These design criteria are presented in a summary form in this paper.

Introduction

The ASCE Direct Design of Concrete Pipe Standards Committee has recently completed its work on a new ASCE Standard Practice for Direct Design of Concrete Pipe Installed by Jacking in Trenchless Construction. Design procedures are called "direct" because they are for concrete pipe in installed conditions, rather than the more traditional "indirect" procedures that are based on defining equivalent three edge bearing test strengths. This paper provides an overview of the more significant provisions in the proposed standard practice for direct design of jacked concrete pipe.

Design Responsibilities

The standard practice begins by defining the responsibilities of the three principal parties that will be involved in developing the pipe-soil installation design: Owner, Pipe Manufacturer and Jacking Contractor. The Owner must define the overall requirements for the installation: use of pipeline, inside diameter, plan and profile, alignment and tolerances, earth cover dimensions, characteristics and unit weight of in-situ soil around pipe, water table elevations and fluctuations, provisions for permanent grouting of annular space outside the pipe, pipe joint performance, live and surcharge loads at surface, and any other special requirements. The Owner may also provide a complete design for the pipe; however, the more typical procedure is for the Pipe Manufacturer to design the pipe to meet the Owners' performance requirements based on the design procedures specified in the standard practice. In this case, the Pipe Manufacturer determines the pipe wall thickness, concrete design strength, reinforcement amount, type and arrangement,

---

[1] Senior Principal, Simpson Gumpertz & Heger Inc., Consulting Engineers, 297 Broadway, Arlington, MA 02174

laying length, joint details, and limiting longitudinal strength for application of jacking forces. The design for longitudinal strength requires coordination with the installation design by the jacking contractor.

The Jacking Contractor determines the jacking installation requirements, including maximum jacking load and maximum angular change in alignment, and the entire construction system used in the jacking process. It is essential that the Jacking Contractor work with the Pipe Manufacturer to achieve an overall installation design that is based on pipe requirements that are within the Manufacturer's equipment and materials capability and on jacking requirements that can be realistically met at the installation site. Both the Pipe Manufacturer and the Jacking Contractor must submit the final results of their collaborative effort to the Owner for his review and acceptance before the start of their manufacturing and construction work.

Limit States Design

The pipe design procedure given in the proposed standard practice is a "limit states" procedure that specifies minimum requirements for both strength and serviceability. Minimum strength requirements are specified using "load and resistance factors." Serviceability requirements are defined by a "crack control factor" that when taken as 1.0 represents reinforcement stress levels that resulted in maximum crack widths averaging about 0.01 in. in tests on representative concrete pipe in many three edge bearing tests. This is the same criteria that has been used as a serviceability limit in the "Standard Practice for Direct Design of Buried Precast Concrete Pipe Using Standard Installations (SIDD) - ASCE 15-93."

The load factors specified for circumferential design are the same as those given in ASCE 15-93. In addition, the following load factors are given for longitudinal effects produced by the maximum jacking forces:

Concentric application of jacking force: $LFJ_1 = 1.5$
Eccentrics application of jacking force: $LFJ_2 = 1.2$

The Owner may specify, or the pipe designer may use, more conservative load factors if a greater margin of safety is deemed desirable because of uncertainties associated with a particular installation.

The resistance factors (strength reduction) specified for design are also the same as those given in ASCE 15-93, with the addition of a resistance factor, $\phi_j = 0.90$ for longitudinal compression produced by jacking.

Transverse Loads

Earth Load is usually the principal transverse load on a pipe that is installed by jacking since such installations seldom have sufficiently shallow cover to be subject to significant effects from surface live load. Exceptions are some installations that are jacked under railroad tracks where the jacked concrete pipes may experience significant stresses from the heavy railroad live loads.

The earth load on a jacked pipe is specified to be determined by an analysis that accounts for soil-structure interaction. However, in contrast to recent availability of soil-structure interaction research results for pipe in embankment installations, no similar modern research using Finite element models for practical determination of earth loads on jacked pipe was made known to the Committee. In view of this, the traditional analysis of Marston/Spangler (ACPA, 1987) is recommended in the Proposed Standard Practice as an acceptable procedure based on soil-structure interaction concepts as developed for a narrow trench having the width of the pipe outside diameter plus the amount of overbore on each sides of the pipe from the cutting edge of the excavation equipment used to advance the pipeline. Based on this procedure, the earth load, W_e, is determined by multiplying the weight of the column of earth above the bore diameter, B_{Th}, by a coefficient termed the vertical arching factor, VAF, as given by:

$$W_e = [VAF]\, w\, \frac{B_{Th}\, H}{U_n} \tag{1}$$

$$[VAF] = \frac{1 - e^{-\alpha}}{\alpha} \tag{2}$$

$$\alpha = 2.0\, K\, \mu'\, \frac{U_n\, H}{B_{Th}} \tag{3}$$

U_n is a coefficient to obtain consistent units between variables and has the values given in the notation section for the usual metric and english units.

Table 1 gives values for typical soil unit weights and friction unit forces, Ku', that have been used with the soil types listed in the Table.

A procedure for determining vertical earth load suggested by Terzaghi is used in Europe and is identical to the Marston/Spangler approach used in the Practice, except that instead of the vertical sliding planes (trench walls) assumed by Marston, it assumes more conservatively a sliding plane inclined at 30 degrees from the vertical. With this procedure, Eqs. (1) and (2) above still apply but Eqs. (3) is modified to:

$$\alpha = 1.155\, K\, \mu'\, \frac{U_n\, H}{B_{Th}} \tag{4}$$

The standard descriptions and recommended unit weight for in-situ soil types and friction coefficients used in European practice are also given in Table 1. See (Hornug, 1989) for a more detailed explanation of European practice for determining the soil weight supported by a jacked pipe.

The earth load given by Eqs. (1), (2), (3) and (4) above may be reduced if the soil above the pipe is cohesive and if reliable values for the long term soil cohesion, c, can be determined. Otherwise the earth load should not be reduced from that given by Eq. (1). If a reliable estimate of c can be made, the reduced vertical arching factor, VAF_R, is:

$$VAF_R = \left(1 - \frac{2.0\, U_n\, c}{w\, B_{Th}}\right) [VAF] \tag{5}$$

Table 1
Design Values for Soil Unit Weight and Ku' for Standard Descriptive In-situ Soil Types

Design System (1)	Soil Type (1)	Unit Wt. of Soils				Ku'
		Above Water		Below Water		
		kN/m³	lb/ft³	kN/m³	lb/ft³	
M/S	Granular	19	120	-	-	.192
E	Non-cohesive	20	127.6	11	70.2	.158
M/S	Sand & Gravel	19	120	-	-	.165
E	Slightly Cohesive	20	127.6	11	70.2	.134
M/S	Saturated Top Soil	19	120	-	-	.150
E	Cohesive - Mixed Soil & silt	20	127.6	10	63.2	.111
M/S	Ordinary Clay	19	120	-	-	.130
E	Cohesive - Except Silt	20	127.6	10	63.2	.089
M/S	Saturated Clay	21	130	-	-	.110

Note: 1. M/S is Marston/Spangler System (ACPA 1987)
E is European System (Hornug 1989)

If the European assumptions are used, Eq. (5) is modified as follows:

$$VAF_R = \left(1 - \frac{1.155\ U_n\ c}{w\ B_{Th}}\right) [VAF] \tag{6}$$

Values of long term cohesion that have been used for design of jacked pipe are given in Table 2 (ACPA, 1987).

The horizontal earth load on a pipe in a jacked installation is taken as:

$$P_{eh} = [HAF]\ w\ \frac{B_{Th}\ H}{U_n} \tag{7}$$

$$[HAF] = K_2\ [VAF] \text{ without cohesion} \tag{8}$$

$$[HAF_R] = K_2\ [VAF_R] \text{ with cohesion} \tag{9}$$

K_2 varies with the type of installation and is given in the standard and in Table 3 for installations defined for the construction state and the long term design state, as given in the Table. Somewhat less conservative assumptions for K_2 are used in European practice (Hornug, 1989). These are also given in Table 3.

Table 2
Design Values of Cohesion for Descriptive In-Situ Soil

Soil Type		kN/m²	lb/ft²
Clay	Soft	2.	40
	Medium	12.	250
	Hard	48	1,000
Sand	Loose Dry	0	0
	Silty	5	100
	Dense	14	300

Table 3
Ratio of Lateral Force to Vertical Load and Bottom Bedding Angle for Common Jacked Pipe Installation Conditions

Installation		System (1)	Ratio of Lateral Force to Vertical Load K_2	Bedding (2) Angle, β, degrees
State	Condition			
During construction without, or with, annular space filled with Bentonite or other lubricant	Without	M/S E	0.25 (0.3)	45 (2)
	With	M/S	0.40	90
After construction, without, or with, pressure-injected cementitious grout	Without	M/S E	0.25 (0.4)	45 (2)
	With	M/S E	0.50 (0.50)	120 (2)
After construction, without or with, Bentonite	Without	M/S	0.25	45
	With	M/S	0.40	90

Notes: 1. Use M/S with Marston Spangler earth loads.
Use E ( ) with European earth loads.
2. European uses 50% uniformly distributed with ϕ = 180° and 50% cosine distributed. See (Hornug, 1989)

The pipe loads produced by live load and surface surcharge load or jacked pipe installations are determined using the procedures given in ASCE 15-93 for live loads on installations with placed soils around the pipe.

Pressure Distribution for Transverse Loads

Simplified assumptions are used for earth pressure distributions for loads applied to the pipe by the typical design loads: pipe weight, vertical earth weight, lateral earth pressure and surface live and surcharge loads. Generally, vertical loads are uniformly distributed over the pipe outside diameter above the pipe and lateral loads are uniformly distributed over the outside height and placed symmetrically on each side of the pipe. Support reactions are uniformly distributed over a bottom bedding angle that varies with each of the installation conditions as shown in Table 3. The specific simplified vertical and lateral load pressure distributions for load and support pressure distributions are given in the standard practice (ASCE, 1997). Pressure distributions used in European practice are presented in (Hornug, 1989).

Stress Analysis for Effects of Transverse Loads

Moments, thrusts and shears are determined for all load effects by analysis of the pipe wall as an elastic isotropic (uncracked) thin ring. Analysis is facilitated by use of the PC computer program PIPECAR (SGH, 1995). Alternatively, for circular pipe, coefficients for determining moments, thrusts and shears at governing locations at the invert, crown and springline of the pipe ring for flexure, and at a location where M/Vd equals 3.0 for shear (diagonal tension) for each type of specified load are given in the Standard Practice (ACPA 1997).

Design of Transverse (Circumferential) Reinforcement

The design process usually involves selecting one of the three standard precast concrete pipe wall thickness types, designing flexural reinforcement based on tensile strength and crack width control criteria, and checking for adequacy in radial tension and concrete compression at points of maximum positive and negative moment and for adequate shear (diagonal tension) strength at the location of critical shear where M/Vd equals 3.0. If the pipe wall is inadequate for any of the latter three strength criteria, stirrup reinforcement is designed to provide the necessary radial or diagonal tension strengths. The design equations for flexural reinforcement and shear, radial tension and flexural compression checks are the same as the reinforcement design given in ASCE 15-93 and are described in more detail in (ACPA, 1994).

Non-circular reinforcing cages are not permitted in jacked pipe because during the jacking process, pipe sections can rotate circumferentially about their longitudinal axis. Pipe with inside diameters greater than 36 in. are required to have two circular cages with inner and outer cages located at nominal concrete cover from the inner and outer surfaces, respectively.

Design for Axial Forces from Jacking

Adequate longitudinal compressive strength of the pipe wall and joint region to resist the compressive effects of the applied jacking forces is a major design requirement for jacking pipe. Two jacking conditions are addressed in this part of the Standard Practice. The first is concentric application of the maximum jacking forces. Under this condition, the maximum uniform allowable (unfactored) compressive stress on the contact surfaces at the pipe joints through which the applied jack forces are distributed is limited to:

$$f_{p1m} = \frac{0.85\,\phi_j\, f_c'}{LFJ_1} \tag{10}$$

When the specified capacity reduction factor for jacking, ϕ_j = 0.9, and the specified minimum load factor for jacking with concentric load, LFJ_1 = 1.5 are used, the maximum compressive stress on the joint contact area when concentrically applied jacking forces are assumed is limited to:

$$f_{p1m} = 0.51\, f_c' \tag{11}$$

Since the jacking ring is assumed to distribute the individual jack loads uniformly around the circumference of the contact surface, the ring must be sufficiently stiff to perform this function.

The second jacking condition addressed in the Standard Practice is eccentric application of the maximum jacking force. Under this condition, the maximum allowable (unfactored) compressive stress at the point of highest stress on the contact surface, based on an assumed linear variation of compressive stress and strain across the contact surface, is limited to

$$f_{p2m} = 0.85\phi_j\, f_c' \tag{12}$$

When the specified capacity reduction factor for jacking, ϕ_j = 0.9, and the specified minimum load factor for jacking with eccentric load, LFJ_2 = 1.2 are used, the maximum compressive stress on the joint contact area when eccentrically applied jacking forces are assumed is limited to:

$$f_{p2m} = 0.65\, f_c' \tag{13}$$

This maximum allowable compressive stress is increased for Condition 2 over that for Condition 1 because with varying compressive strains across an eccentrically loaded cross section, redistribution of high peak stress occurs due to plastic behavior of the highly compressed concrete. Either Condition 1 or Condition 2 may govern the actual maximum jacking force that can be applied to the pipe. The amount of eccentricity assumed to determine the maximum allowable jacking force that can be applied to the pipe depends on the Jacking Contractor's capability to maintain the alignment and to control accidental eccentric application of the jacking forces. Also, eccentric application of jacking compression often is needed to facilitate angular changes in alignment at planned bends, or to correct misalignments.

Some engineers and jacking contractors use safety factors (Load Factor/capacity reduction factor) that are larger than the above minimum values specified in the proposed ASCE jacking practice.

Manufacturing Tolerances

Certain manufacturing tolerances are required to minimize dimensional variations that could make it more difficult to maintain the alignment during jacking, or that might increase the required jacking force, or that might increase the compressive stress under the design maximum jacking force. These are the end squareness of each pipe section, the uniformity of outside diameter of each pipe section, and the variations in thickness of pipe wall and joint contact surfaces. These tolerances are all given in a manufacturing appendix to the Practice (ASCE, 1997).

Pipe Joints

Adjacent pipe sections must be separated by a cushioning material to facilitate a uniform transfer of stress between contact surfaces and to minimize radial splitting pressures associated with high bearing pressures. Typically, plywood is used with a minimum thickness of 3/4 in. for pipe inside diameters larger than 36 in. and with a minimum thickness of 1/2 in. for pipe inside diameters of 36 in. and smaller. The cushion is usually held back about 1/8 in. from the joint edges to minimize spalling adjacent to the edge.

Joint details are not specified in the Standard Practice (ASCE 1997). Joint configurations that have been used in the past are shown in the Commentary to the Practice.

Conclusions

The new ASCE Standard Practice for Direct Design of Precast Concrete Pipe for Jacking in Trenchless Construction will provide owners, consulting engineers, pipe manufacturers and jacking contractors a common basis for specifying and designing concrete pipe for jacking installations that reflects the current state-of-the-art in knowledge about the forces and their structural effects that determine the behavior of the pipe in the installed state. It is intended to contribute to the development of more rational, safer and more cost effective practices for the design of concrete pipe for trenchless construction. The use of this type of pipe installation is becoming more prevalent in urban areas where the excavation of open trenches for new pipe installation is disruptive and expensive.

Notation

c = cohesion, kN/m^2 (lb/ft^2)
B_{Th} = outside diameter of pipe plus overbore each side, mm (in.)
f_c' = design compressive strength of concrete, MPa (lbs/in.^{2})
f_{p1m} = maximum allowable compressive stress on joint contact area when concentric application of jacking forced is assumed, MPa (lbs/in.^{2})
f_{p2m} = maximum allowable compressive stress on joint contact area when eccentric application of jacking force is assume MPa (lbs/in.^{2})
H = design height of earth above top of pipe, m (ft)
HAF = horizontal arching factor, ratio of total horizontal force from lateral soil pressure acting on pipe to vertical earth load, W_e acting on pipe
K = earth pressure ratio on sliding plane above pipe
K_2 = ratio of lateral force to vertical load on jacked pipe

LFJ_1	=	minimum load factor when concentric application of jacking force is assumed
LFJ_2	=	minimum load factor when eccentric application of jacking force is assumed
M/Vd	=	ratio of moment to shear times depth of concrete from compression face of pipe wall to centerline of tensile reinforcing
U_n	=	coefficient to obtain consistent units between variables (= 1000 for SI units and 12 for English units)
VAF	=	vertical arching factor, ratio of earth load on pipe, W_e, to weight of column of earth over the excavated diameter
VAF_R	=	reduced vertical arching factor to account for cohesion
w	=	unit weight of soil, kN/m^3 (lb/ft^3)
W_e	=	total weight of earth acting on pipe, N/m, (lb/ft)
α	=	coefficient calculated from Eq. (2)
ϕ_j	=	capacity reduction factor for compression produced by jacking thrust

References

ACPA, 1994, American Concrete Pipe Association, *Concrete Pipe Technology Handbook*, Second Printing, Vienna, VA.

ACPA, 1987, American Concrete Pipe Association, *Concrete Pipe Design Manual*, Seventh Printing, Vienna, VA.

ASCE, 1997, American Society of Civil Engineers, "Standard Practice for Direct Design of Precast Concrete Pipe in Trenchless Construction," Draft for Society Ballot, New York, NY.

ASCE, 1993, American Society of Civil Engineers, "Standard Practice for Direct Design of Precast Concrete Pipe in Standard Installations," New York, NY.

Hornug, K., and Kittel, D., 1989, *Structural Analysis of Buried Pipes*, Baurerlag GMBN, Wiesbaden und Berlin (in German and English).

SGH, 1995, Simpson Gumpertz & Heger Inc., "PIPECAR PC Computer Program," McTrans Center for Microcomputers in Transportation, University of Florida Transportation Research Center, Gainesville, FL.

SUBJECT INDEX

Page number refers to the first page of paper

AUTHOR INDEX

Page number refers to the first page of paper